DATE DUE

BRODART, CO.   Cat. No. 23-221-003

Bioluminescence
Chemical Principles and Methods

Bioluminescence

Chemical Principles and Methods

Osamu Shimomura

Formerly Senior Scientist at the Marine Biological Laboratory, Woods Hole, Massachusetts

World Scientific

NEW JERSEY · LONDON · SINGAPORE · BEIJING · SHANGHAI · HONG KONG · TAIPEI · CHENNAI

Published by

World Scientific Publishing Co. Pte. Ltd.

5 Toh Tuck Link, Singapore 596224

USA office: 27 Warren Street, Suite 401-402, Hackensack, NJ 07601

UK office: 57 Shelton Street, Covent Garden, London WC2H 9HE

Library of Congress Cataloging-in-Publication Data
Shimomura, Osamu 1928–
    Bioluminescence : chemical principles and methods / Osamu Shimomura
        p. cm.
    Includes bibliographical references (p.).
    ISBN 981-256-801-8
        1. Bioluminescence. 2. Chemiluminescence. I. Title.

QH641.S52 2006
572'.4358--dc22

                                                                    2006049843

British Library Cataloguing-in-Publication Data
A catalogue record for this book is available from the British Library.

Typeset by Stallion Press
Email: enquiries@stallionpress.com

Printed by Mainland Press Pte Ltd

Preface

In studying the chemical aspects of bioluminescence, comprehensive reviews of practical use were scarce in the past except on luminous bacteria. This is a considerable inconvenience and disadvantage to researchers. In fact, I have been frequently frustrated myself by the need to search for old articles published 30–40 years ago to find data on some basic properties of bioluminescent substances, such as the absorption spectra and luminescence activities of luciferins. In the absence of any compendium of the substances and reactions involved in bioluminescence, researchers will have to spend their precious time delving through the literature in order to find the needed information. Such a situation may discourage new investigators who are interested in the chemical study of bioluminescence, and might hamper them from actually taking up a project. Upon consideration of these matters, I decided to write this book.

The present book describes all the significant studies and findings on the chemistry of the more than 30 different bioluminescent systems presently known, accompanied by over 1000 selected references. It includes descriptions of the purification and properties of bioluminescent compounds, such as luciferins, luciferases and photoproteins, and the mechanisms of luminescence reactions. To make the book more useful than a mere review volume and to save researchers time in looking into original references, I have included a considerable amount of original experimental methods, data and graphs. In addition, I have included some new data and experimental methods unavailable elsewhere. I hope this volume will be useful to researchers and students, and it will be my greatest pleasure if this book contributes to the finding of new luciferin structures and new luminescence mechanisms.

I am grateful to J. Woodland Hastings, Satoshi Inouye and Yoshihiro Ohmiya who kindly read a draft version of this book and provided me with valuable suggestions and advice. I also would like to express my sincere thanks to Steven Haddock, John Brinegar and Sachi Shimomura for their help in correcting my English, and Sook Cheng Lim for editing this book.

I have been extremely fortunate to be able to continue my research on bioluminescence for 50 years without interruption. It was made possible with the help of many people and the continued support from the National Science Foundation; I am particularly indebted to Toshio Goto, Yoshito Kishi, Benjamin Kaminer and J. Woodland Hastings for their kind help. All my work has stemmed, however, from the initiatives taken by my three mentors: the late Professors Shungo Yasunaga (Nagasaki University), Yoshimasa Hirata (Nagoya University) and Frank H. Johnson (Princeton University). Yasunaga, in 1955, advised me to shift my specialty from pharmacy to chemistry and he arranged for me to work at Hirata's organic chemistry lab; Hirata gave me the very difficult problem of crystallizing *Cypridina* luciferin, which eventually rewarded me with the experience and knowledge necessary as a researcher; and Johnson, in 1961, gave me the subject of *Aequorea* and helped me for 20 years in solving the problems of aequorin and other bioluminescent substances.

With my great respect, I dedicate this book to the memory of my three mentors.

Osamu Shimomura

Contents

Abbreviations, Symbols and Definitions

ε	Molar absorption coefficient (absorbance of a 1 M solution in a 1 cm-path cell)
λ	Wavelength
μm Hg	1/760,000 of one atmospheric pressure
A	Absorbance [$\log(I_0/I)$]; optical density (OD); for 1 cm light path if not specified
ADA	N-(2-Acetamido)-2-iminodiacetic acid
AMP	Adenosine 5′-monophosphate
ATP	Adenosine 5′-triphosphate
Bis-tris	bis(2-Hydroxyethyl)aminotris(hydroxymethyl)methane
BSA	Bovine serum albumin
CAPS	3-(Cyclohexylamino)-1-propanesulfonic acid
CHES	2-(Cyclohexylamino)ethanesulfonic acid
CIEEL	Chemically initiated electron-exchange luminescence
CTAB	Hexadecyltrimethylammonium bromide
Da	Dalton (1/12 of the mass of one atom of ^{12}C)
DEAE	Diethylaminoethyl
Diglyme	2-Methoxyethyl ether
DMF	Dimethylformamide
DMSO	Dimethylsulfoxide
DNA	Deoxyribonucleic acid
DTT	Dithiothreitol
E. coli	*Escherichia coli*
EDTA	Ethylenediaminetetraacetic acid
EGTA	Ethyleneglycol-bis(2-aminoethylether)-N,N,N′,N′-tetraacetic acid
FAB	Fast atom bombardment ionization

FAD	Flavin adenine dinucleotide
FMN	Riboflavin 5'-monophosphate
$FMNH_2$	Riboflavin 5'-monophosphate, reduced form
FPLC	Fast protein liquid chromatography (Pharmacia)
FWHM	Full width at half maximum
GFP	Green fluorescent protein
HEPES	N-(2-Hydroxyl)piperazine-N'-(2-ethanesulfonic acid)
HPLC	High performance liquid chromatography
I	Intensity
K_d	Dissociation constant
K_m	Michaelis constant
LBP	Luciferin binding protein
LCC	Lauroylcholine chloride
LU	Light unit
m/z	Mass to charge ratio
MES	2-(N-Morpholino)ethanesulfonic acid
MOPS	3-(N-Morpholino)propanesulfonic acid
M_r	Relative molecular weight (dimensionless number)
NAD^+	Nicotine amide adenine dinucleotide, oxidized form
NADH	Nicotine amide adenine dinucleotide, reduced form
NMR	Nuclear magnetic resonance
NTA	Nitrilotriacetic acid
PAGE	Polyacrylamide gel electrophoresis
pCa	Minus log of molar Ca^{2+} concentration, $-\log[Ca^{2+}]$
pI	Isoelectric point
pK_a	Minus log of acidic dissociation constant
quantum yield	The number of photons emitted divided by the number of molecules reacted
SB3-12	3-(Dodecyldimethylammonio)propanesulfonate
SDS	Sodium dodecylsulfate
SOD	Superoxide dismutase
TAPS	N-[Tris(hydroxymethyl)methyl]-3-aminopropanesulfonic acid
TLC	Thin-layer chromatography
Tris	Tris(hydroxymethyl)aminomethane
UV	Ultraviolet
YC	Yellow compound

Introduction

The emission of light from animals and plants has inspired the curiosity and interest of mankind ever since the ancient times of Aristotle (384–322 B.C.) and Pliny (A.D. 23–79). It has been the target of investigations by a number of great naturalists, physicists and physiologists (Harvey, 1957). Using a vacuum pump he built, Robert Boyle (1627–1691) showed that the luminescence of meat and fungi requires air. Benjamin Franklin (1706–1790), who found that lightning is caused by electricity, hypothesized that the phosphorescence of the sea is an electrical phenomenon, but he did not hesitate to change this opinion when he discovered that the light in seawater could be filtered off with a cloth. Paolo Panceri (1833–1877) is noted for his publications on the anatomy and histology of various types of luminous organisms, and Raphaël Dubois (1849–1929) discovered luciferin and luciferase. The secrets of the chemistry of bioluminescence, however, began to be uncovered only in the 20th century. Out of necessity, Eilhardt Wiedemann (1888) had created the term "luminescence," meaning the emission of cold light, and Harvey (1916) used the term "bioluminescence," luminescence from living organisms, possibly for the first time.

Today, bioluminescence reactions are used as indispensable analytical tools in various fields of science and technology. For example, the firefly bioluminescence system is universally used as a method of measuring ATP (adenosine triphosphate), a vital substance in living cells; Ca^{2+}-sensitive photoproteins, such as aequorin from a jellyfish, are widely utilized in monitoring the intracellular Ca^{2+} that regulates various important biological processes; and certain analogues of *Cypridina* luciferin are employed as probes for measuring superoxide anion, an important but elusive substance in biological systems. In addition, the green fluorescent protein (GFP), which was discovered together with aequorin, is utilized as a highly useful marker protein in the field of biomedical research. All of these applications of bioluminescent systems have been derived from the chemical studies of bioluminescence, in efforts to understand this phenomenon.

Some apparent peculiarities have been long known in the phenomenon of bioluminescence. One of them is the distribution of luminous organisms on the phylogenetic tree. There is no obvious rule or reason in the distribution of luminous species among microbes, protists, plants, and animals. Harvey (1940, 1952) expressed it in this way: "It is as if the various groups had been written on a blackboard and a handful of damp sand cast over the names. Where each grain of sand strikes, a luminous species appears. It is an extraordinary fact that one species in a genus may be luminous and another closely allied species contain no trace of luminosity." The Cnidaria and Ctenophora have received the most sand; thus, many members of the former phylum and nearly all of the latter are luminous. On the other hand, certain phyla contain no luminous organisms. There are also many instances of closely related genera in the same family wherein one genus is luminous but the other is nonluminous. Another well-known peculiarity is that vast numbers of known bioluminescent organisms are mostly marine dwellers, in contrast to a few that are terrestrial or freshwater inhabitants. The non-marine organisms can be readily named here due to their small number: the fireflies and beetles, earthworms, millipede *Luminodesmus*, limpet *Latia*, snail *Quantula*, the glow worms *Arachnocampa* and *Orfelia*, and luminous mushrooms.

Bioluminescence of organisms occurs in diverse forms of morphology, with various different mechanisms of light emission. Some animals have highly complicated light organs resembling eyes, in which light emission is under the control of the nerve systems and luminescence is emitted as needed or by stimulation (e.g., luminous fishes and squids). In some others, single cells contain all needed apparatus for light emission, and light is emitted continuously (luminous bacteria and fungi). There are also many that are intermediates of these two groups. The bioluminescence system of an organism may involve a series of interrelated chemical reactions, although the light is emitted only from the reaction that produces the singlet excited state of light-emitter, a reaction termed "light-emitting reaction." There are various different light-emitting reactions, but all involve the oxidation of a substrate (usually luciferin) that provides the energy for generating an excited state. So far as is known, every bioluminescence reaction is basically a chemiluminescence reaction. It is remarkable that animals and plants have developed their functional abilities of bioluminescence, integrating the mechanisms of various disciplines such as chemistry, physics, physiology and morphology.

The Beginning of the Chemical Study of Bioluminescence

It is generally considered that the modern study of bioluminescence began when Dubois demonstrated the first example of a luciferin-luciferase reaction in 1885. He made two aqueous extracts from the luminous West Indies beetle *Pyrophorus*. One of the extracts was prepared by crushing the light organs in cold water, resulting in a luminous suspension. The luminescence gradually decreased and finally disappeared. The other extract was prepared by initially treating the light organs with hot water, which immediately quenched the light, and then it was cooled. The two extracts gave a luminescence when mixed together. He found nearly the same phenomenon with the extracts of the clam *Pholas dactylus* (Dubois, 1887). Dubois concluded that the cold water extract contained a specific, heat labile enzyme necessary for the light-emitting reaction, and introduced a term "luciferase" for this enzyme. He also concluded that the hot water extract, in which this enzyme has been destroyed by heat, contained a specific, relatively heat stable substance, which he designated "luciférine" (presently spelled luciferin). Thus, the luciferin-luciferase reaction can be viewed as a substrate-enzyme reaction that results in the emission of light.

Following the discovery of luciferin and luciferase by Dubois, the person who made the greatest contribution to the knowledge of bioluminescence was E. Newton Harvey (1887–1959) of Princeton University (biographical memoir: Johnson, 1967). He was initially a physiologist, but was quickly captivated by the phenomenon of bioluminescence, and his interest in bioluminescence grew into his lifelong project. Harvey traveled widely and studied the bioluminescence of a great variety of luminous organisms, producing over 300 publications. He understood the underlying foundation of the chemistry of bioluminescence reactions, despite the fact that little was known about the actual chemical reactions at the time. His book *Bioluminescence* published in 1952 is considered the bible of bioluminescence.

It was a common belief that all phenomena of bioluminescence were caused by the luciferin-luciferase reaction until the bioluminescent protein aequorin was discovered in 1962 (Shimomura *et al.*, 1962). When the terms luciferin and luciferase were found to be unsuitable for categorizing the two bioluminescent proteins, aequorin and another from the tubeworm *Chaetopterus*, a new term "photoprotein" was introduced to supplement the term luciferin (Shimomura and Johnson, 1966). Further explanations for the terms luciferin and photoprotein are given below.

Luciferin

The term luciferin has never been strictly defined. Its precise meaning has changed from time to time, and may change in the future. Luciferin originally meant a relatively heat-stable, diffusible substance existing in the cooled, hot-water extract of luminous organs, as an essential ingredient needed for the emission of bioluminescence. Harvey discovered that the luciferin of the clam *Pholas* differs from that of the ostracod *Cypridina* (Harvey, 1920), and stated in his 1952 book "It is probable that the luciferin or luciferase from a species in one group may be quite different chemically from that in another." It became apparent by the end of the 1950s that the luciferins existing in *Cypridina*, the fireflies and luminous bacteria are chemically different from each other, generating a widely held view that the luciferins of all luminous species are different except in species biologically closely related. However, this view did not last long. Around 1960, a luciferin identical to the luciferin of *Cypridina* was discovered in the luminous fishes *Parapriacanthus* and *Apogon*. Moreover, it was followed by a series of discoveries during the period 1970–1980 that coelenterazine, a luciferin, exists in a wide range of luminous organisms that include various coelenterates, shrimps, squids and fishes.

Based on the presently available information, it seems appropriate to define "luciferin" as the general term of an organic compound that exists in a luminous organism and provides the energy for light emission by being oxidized, normally in the presence of a specific luciferase. The luciferase catalyzes the oxidative light-emitting reaction of the luciferin. It is an important criterion that a luciferin is capable of emitting photons in proportion to its amount under standardized conditions.

A bioluminescence reaction of a luciferin is basically a chemiluminescence reaction. The luciferin is an absolute requirement as the source of light energy, but the luciferase, an enzyme (protein), might not be needed if its role could be replaced by other substance(s). Recent studies at the author's laboratory suggest that a luciferase might not be involved in the light-emitting reaction of luminous fungi (see Chapter 9).

Photoprotein

In 1961, an unusual bioluminescent protein was discovered from the jellyfish *Aequorea*, and it was named aequorin (Shimomura *et al.*, 1962). Aequorin emits light in aqueous solutions merely by the addition of Ca^{2+}, regardless of the presence or absence of oxygen. The light is emitted by an intramolecular reaction of the protein, and the total light emission is proportional to the amount of the protein used. The properties of aequorin do not conform to the definition of luciferin or luciferase. Tentatively we thought aequorin could be an extraordinary exception occurring in nature. We discovered, however, another example of bioluminescent protein in 1966 in the parchment tubeworm *Chaetopterus* that emits light when a peroxide and Fe^{2+} are added; the total light emission was proportional to the amount of the protein used, like aequorin. Considering the possibility of finding many similar bioluminescent proteins from luminous organisms, we proposed a new term "photoprotein" to designate these unusual bioluminescent proteins (Shimomura and Johnson, 1966).

Thus, "photoprotein" is a general term of the bioluminescent proteins that occur in the light organ of luminous organisms and are capable of emitting light in proportion to the amount of the protein (Shimomura, 1985). A photoprotein could be an enzyme-substrate complex that is more stable than its dissociated components, enzyme and substrate. Due to its greater stability, a photoprotein occurs as the primary luminescent component in the light organs instead of its dissociated components. In the light organs of the jellyfish *Aequorea*, aequorin is highly stable as long as Ca^{2+} is absent, but its less stable components, coelenterazine and apoaequorin, are hardly detectable in the jellyfish.

Several different types of photoprotein are presently known, for example: the Ca^{2+}-sensitive types found in various coelenterates (aequorin, obelin, mnemiopsin) and protozoa (thalassicolin); the peroxide-activation types found in scaleworm (polynoidin) and the clam *Pholas* (pholasin); and the ATP-activation type found in a Sequoia millipede *Luminodesmus*.

Chemical Studies on Bioluminescence in the Last One Hundred Years

Bioluminescence is a complicated phenomenon. A complete understanding of the phenomenon will require studies in a wide range of disciplines, including morphology, cell biology, physiology, spectroscopy, biochemistry, organic chemistry, and genetics. In the past century, there have been very significant gains in the understanding of bioluminescence in all these disciplines.

Important findings, discoveries and breakthroughs in chemistry after the discovery of luciferin and luciferase by Dubois are chronologically listed in the table shown below. The chemical structures of luciferins and light-producing groups have been determined and the light-emitting reactions elucidated in considerable detail in the bioluminescence of eight different types of organisms, namely, the fireflies, the ostracod *Cypridina*, luminous bacteria, coelenterates, the limpet *Latia*, earthworms, krill and dinoflagellates. A new concept of

Major Progress in Research into the Chemistry of Bioluminescence

Year	Description
1885	Discovery of luciferin-luciferase reaction
1935	Benzoylation of *Cypridina* luciferin
1947	ATP requirement in firefly luminescence
1953	Requirement for long-chain aldehyde (luciferin) in bacterial luminescence
1954	FMNH$_2$ requirement in bacterial luminescence
1957	Crystallization of *Cypridina* luciferin
1957	Crystallization of firefly luciferin
1958	*Cypridina* luciferin in fishes; the first cross reaction discovered
1961–1963	Structure of firefly luciferin
1962	Discovery of aequorin and GFP (green fluorescent protein)
1966	Structure of *Cypridina* luciferin
1966	Concept of photoprotein
1968	Structure of *Latia* luciferin
1967–1968	Dioxetanone mechanism proposed in firefly and *Cypridina* luminescence
1968	Regulation of dinoflagellate luminescence by pH
1971	Dioxetanone mechanism confirmed in *Cypridina* luminescence
1974	Long-chain aldehydes identified in luminous bacteria
1975	Discovery of coelenterazine (a luciferin)
1975	Structure of light-emitting chromophore of aequorin
1975	Regeneration of aequorin
1976	Structure of earthworm luciferin
1977	*Renilla* luciferin identified to be coelenterazine
1977	Dioxetanone mechanism confirmed in firefly luminescence
1978	Coelenterazine-2-hydroperoxide in aequorin
1979	Structure of the chromophore of GFP
1981	Structure of the luminescence autoinducer of luminous bacteria
1984–1985	Firefly luciferase cloned
1985–1986	Bacterial luciferase cloned
1985–1986	Apoaequorin cloned
1988	Structure of krill luciferin
1988	Semisynthetic aequorins prepared
1989	Structure of dinoflagellate luciferin
1992	GFP cloned
1994	GFP expressed in living cells
1996	Crystal structure of bacterial luciferase
1996	Crystal structure of firefly luciferase
1996	Crystal structure of GFP
2000	Crystal structures of aequorin and obelin
2005	Crystal structure of dinoflagellate luciferase

"photoprotein" has been developed for the bioluminescent proteins, such as aequorin and obelin, that do not fit well with the definition of luciferin or luciferase, as already mentioned. It is surprising that many marine luminous organisms have been found to involve the identical luciferin, coelenterazine. In addition, it has been shown that the reaction mechanisms of the luminescence reactions of firefly luciferin, *Cypridina* luciferin and coelenterazine all involve the same type of intermediates, dioxetanes, possessing a four-member ring that consists of two carbon atoms and two oxygen atoms. Many luciferases have been purified and characterized. Helped by the advances in genetic technology, some luciferases and apophotoproteins have been cloned and their three-dimensional structures have been determined.

Despite the remarkable progress made, however, the trend shown in the table reveals a fact that cannot be interpreted favorably, at least to this author. In the third quarter of the 20th century, the structures of five different kinds of new luciferins have been determined, whereas, in the last quarter, only three structures, of which two are nearly identical, have been determined. None has been determined in the last decade of the century and thereafter, thus clearly indicating a declining trend, in contradiction to the steady advances in analytical techniques. The greatest cause for the decline seems to be the shift of research interest from chemistry and biochemistry into genetic biotechnology in the past 20 years.

Chemical Study of Bioluminescence in the Future

Bioluminescence still has many mysteries, which may yield many further insights into nature and science. In bioluminescence reactions, a luciferin generates the energy for light emission when oxidized. For that reason, luciferin is the most important element in bioluminescence; it can be considered as the heart of the bioluminescence reaction. Because of its importance, the author believes that the determination and identification of the structure of luciferins should be considered as one of the top targets in future research. The functional group of a photoprotein corresponds to a luciferin in its function, thus it is as important as a luciferin. Many luciferins and the functional groups of photoproteins remain to be determined, and at least two of them have been ready for structural work for many years: the luciferin of the Bermuda fireworm *Odontosyllis* and the functional chromophore of the photoprotein pholasin from the clam *Pholas*. Each of these two subjects has been briefly taken up from time to time at various laboratories during the past 30 years, but no structural information has been obtained for either of them.

Another important subject is related to the fashions of utilizing coelenterazine in bioluminescence reactions. Coelenterazine is widely distributed in marine organisms and plays a central role in many bioluminescent organisms. The compound is utilized at least in four different fashions: (1) unmodified form, as luciferin in many organisms; (2) disulfate form, as luciferin in the squid *Watasenia*; (3) peroxidized form, as the functional group of aequorin and obelin; and (4) dehydro-form, for regenerating the squid photoprotein symplectin. Many bioluminescent organisms contain substantial amounts of coelenterazine. However, some of them contain very weak luciferase activities and their utilization of coelenterazine do not match with any of the four cases given above. The examples of such organisms are: the shrimp *Sergestes*, the squids *Chiroteuthis*, and the deep-sea fish *Neoscopelus*. In these organisms, coelenterazine might be utilized in other, still unknown fashions. The study of this line would be important to fully understand the role of coelenterazine in bioluminescence.

In the luminescence systems that require a peroxide or an active oxygen species in addition to molecular oxygen (the scaleworm, the tube worm *Chaetopterus*, the clam *Pholas*, the squid *Symplectoteuthis*), their *in vitro* luminescence reactions reported are much slower and inefficient compared to their bright *in vivo* luminescence. The true, intrinsic activation factor in their *in vivo* luminescence should be determined, and the detailed mechanisms of oxidation should be elucidated.

Discovery of a new luciferin and a new mechanism will provide us with enormous benefit, as it was shown in the past. The work may not be easy; however, the author believes that it can be accomplished when the researcher has a firm determination to complete it. There is no established method or protocol for studying a new type of luciferin or photoprotein; thus, the method must be worked out. Some suggestions on experimental procedures are included in the Appendix.

The Contents of this Book

This book is devoted to the progress in the chemical understanding of bioluminescence, particularly the mechanisms involved in the light-emitting reactions. Though light emission from a luminous organism may involve a series of biochemical reactions integrated in a complex manner, the discussion in this book is focused on the light-emitting reaction, i.e. the chemical reaction that results in the emission of photons. The accessory reactions, such as the formation of luciferin from preluciferin, the biosynthesis of luciferin, and the reactions involved in nervous stimulation, are not discussed unless they are essential.

The methods of the isolation and purification of various luciferins, luciferases and photoproteins are described in detail as much as possible because of their importance. There have been considerable changes in the methods, techniques and materials used during the 50-year span covered by this book. However, the underlying principles of the purification methods have not changed significantly; the old methods and techniques are often very useful for the present research when the principles involved are understood.

The future is an extension of the past. The author believes that the process of the progress made in the past is as important as the findings and discoveries for the planning of future research. For this reason, a substantial weight is placed on describing historical accounts. Such information would also help researchers to get some idea of the effort that might be needed to isolate and identify a new luciferin.

The chapters in this book are arranged roughly in the chronological order of bioluminescence systems discovered, based on the date of the major breakthrough made in each bioluminescence system, such as the discovery of ATP in the firefly system (McElroy, 1947) and the identification of fatty aldehyde as the luciferin in luminous bacteria (Cormier and Strehler, 1953). This differs from Harvey's 1952 book, which is arranged in the order of taxonomic classification.

The description of luminous bacteria (Chapter 2) should be considered minimal because the works related to this subject are too enormous to be covered in this book; readers are referred to the excellent review articles by Hastings and Nealson (1977); Ziegler and Baldwin (1981); Hastings *et al.* (1985); Hastings (1986); Meighen and Dunlap (1993); Tu and Mager (1995); and Dunlap and Kita-Tsukamoto (2001). Bioluminescence due to the presence of symbiotic luminous bacteria is not covered in this book for the reason that the chemistry involved is identical to that of luminous bacteria. The reports of the data and information that are erroneous or irrelevant to the objectives of this book are not cited, except when considered essential, and reports of confirming nature are often omitted. For the general topics of bioluminescence, *Bioluminescence* by Harvey (1952) and *Bioluminescence in Action* edited by Herring (1978) are most useful even today, and are highly recommended. The references cited in the text and some additional relevant references are given in alphabetical order at the end of the book. For the convenience of researchers, some basic data and information that might be useful are included in the Appendix.

1

THE FIREFLIES AND LUMINOUS INSECTS

Since ancient times, the light emitted by fireflies and glow-worms has attracted the curiosity of people. Descriptions of the phenomena are frequently found in old poems, songs and folklores of many countries. Old scientific studies of these phenomena are also numerous, particularly after the 17th century. However, the chemical study was not begun until the early 20th century.

Although the class Insecta (in the phylum Arthropoda) contains bioluminescent organisms in four orders: Collembola, Hemiptera, Coleoptera and Diptera, biochemical studies have been carried out only with several types of organisms of the last two orders, Coleoptera and Diptera. In these orders, the adults have two pairs of wings: in Coleoptera, the front wings are modified as elytra (a heavy protective cover); and in Diptera, the hind wings are reduced to knobs. The order Coleoptera includes Lampyridae (fireflies), Phengodidae (railroad worms), and Elateroidae (click beetles such as *Pyrophorus*), and all the luminous species in this order utilize firefly luciferin in their light emission. The order Diptera contains the glow-worms *Arachnocampa* and *Orfelia*, and their bioluminescence systems are clearly different from that of Coleoptera.

According to Harvey (1952), there are considerable differences in morphology between males and females among the various genera of fireflies (lampyrids). In many of them, both males and females are winged and can fly, but usually the male has a larger light organ than the female. They use light signals to find each other. In some cases, females are luminous but males are not, and some species do not emit light at all. In *Lucidota* and *Pyropyga*, the larvae are luminous but the adult fireflies are not. The eggs of the fireflies are generally luminous, as was discovered as early as 1643 by Thomas Bartholin. The functions of the light emission of the fireflies and the behaviors of synchronous flashing have been reviewed by John B. Buck (Buck and Buck, 1976; Buck, 1978).

In northern and central Europe, the most common lampyrid is the glow-worm *Lampyris*. The female of this genus is wingless and has a bright light organ, which attracts the flying male that has a less bright light organ. In Italy and southern Europe, common fireflies are the genera *Luciola* and *Phausia*, of which both males and females are winged. In North America, the genera *Photinus* and *Photuris* are the most common. The females of *Photinus* has partially developed wings and does not fly. *Photuris* is carnivorous and eats fireflies of other species.

Lampyrids are abundant in Japan; Ohba (1997, 2004) lists nine genera of lampyrids containing more than 50 species. The most common of them are *Luciola cruciata* (Fig. 1.1) and *Luciola lateralis*. These two species live in water during their larval stage (one or two years), making a distinct difference from all other species of fireflies; the larvae are luminescent. These lampyrids belong to one of the two kinds of luminous organisms that inhabit freshwater; the other is the New Zealand freshwater limpet *Latia*.

The fireflies, railroad worms, and click beetles use the same luciferin in their luminescence reactions. Recent studies on the railroad worms and the click beetles have greatly contributed to the biochemical understanding of the firefly bioluminescence (see Section 1.2). Concerning luminous Diptera, significant progress has been made only recently.

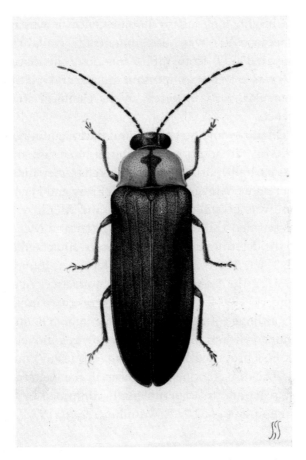

Fig. 1.1   The firefly *Luciola cruciata* (male) drawn by Sakyo Kanda (1874–1939), a pioneer of the study of bioluminescence in Japan, showing his extraordinary artistic talent (reproduced from Kanda, 1935).

1.1   The Fireflies

1.1.1   *An Overview of the Firefly Luminescence Reaction*

The luciferin-luciferase reaction of fireflies was first demonstrated by Harvey (1917), although the light observed was weak and short-lasting. Thirty years after Harvey's discovery, McElroy (1947) made a crucial breakthrough in the study of firefly bioluminescence. He found that the light-emitting reaction requires ATP as a cofactor. The addition of ATP to the mixtures of luciferin and luciferase resulted in a bright, long-lasting luminescence. It was not a simple experiment because ATP was not commercially available at the time. McElroy prepared ATP from rabbit muscles. The discovery of the cofactor ATP was extremely important indeed, and it cleared the way for the spectacular, rapid progress in the chemical study of firefly bioluminescence.

In 1949, McElroy and Strehler found that the luminescence reaction requires Mg^{2+} in addition to luciferin, luciferase and ATP. The luciferase was partially purified and various characteristics of the luminescence reaction were investigated by McElroy and Hastings (1955). The luciferase was crystallized by Green and McElroy (1956). The luciferin was purified and crystallized by Bitler and McElroy (1957), which led to the determination of its structure and chemical synthesis (White et al., 1961; 1963). The active luciferin was found to be in the D-form (Fig. 1.2); the L-form is practically inactive. According to a study by Lembert (1996), L-luciferin is a competitive inhibitor of firefly luciferase, although it can produce some light. The mechanism of the chemiluminescence of luciferin that involves a dioxetanone intermediate was first proposed by Hopkins et al. (1967) and McCapra et al. (1968). The dioxetanone mechanism in the luciferase-catalyzed luminescence reaction was experimentally confirmed by ^{18}O-labeling studies (Shimomura et al., 1977; Wannlund et al., 1978).

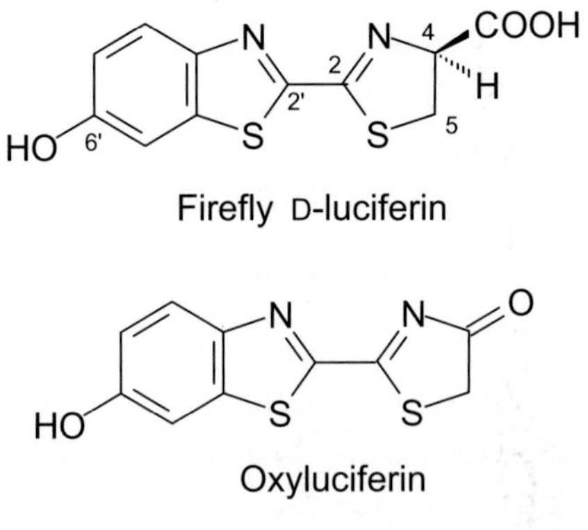

Firefly D-luciferin

Oxyluciferin

Fig. 1.2 Structures of firefly luciferin and oxyluciferin.

The following schemes represent the overall reaction of firefly bio-luminescence (McElroy and DeLuca, 1978), where E is luciferase; LH_2 is D-luciferin; PP is pyrophosphate; AMP is adenosine phosphate; LH_2-AMP is D-luciferyl adenylate (an anhydride formed between the carboxyl group of luciferin and the phosphate group of AMP); and L is oxyluciferin.

$$E + LH_2 + ATP + Mg^{2+} \longrightarrow E{\cdot}LH_2\text{-}AMP + PP + Mg^{2+} \quad (1)$$

$$E{\cdot}LH_2\text{-}AMP + O_2 \longrightarrow E{\cdot}L + CO_2 + AMP + Light \quad (2)$$

In the first step, luciferin is converted into luciferyl adenylate by ATP in the presence of Mg^{2+}. In the second step, luciferyl adenylate is oxidized by molecular oxygen resulting in the emission of yellow-green light, of which the mechanism is discussed in Sections 1.1.6 and 1.1.7. Both steps, (1) and (2), are catalyzed by luciferase. The reaction of the first step is slower than that of the second step, thus the first step is the rate-limiting step.

1.1.2 *Firefly Luciferin and Oxyluciferin*

Extraction and purification of luciferin. In the work of purifying and crystallizing firefly luciferin (Bitler and McElroy, 1957), McElroy used a unique method to gather the large quantity of fireflies needed for their research. In the now legendary story, they advertised for the purchase of fireflies at one cent per specimen. Children and youths in the neighborhood responded enthusiastically, collecting a huge number of the bugs for them. In this way, they easily obtained sufficient number of the firefly *Photinus pyralis* for their research.

The live fireflies are dried over calcium chloride in a vacuum desiccator, and then their lanterns are separated by hand. An acetone powder prepared from the dried lanterns is extracted with boiling water. The cooled aqueous extract is extracted with ethyl acetate at pH 3.0, and the ethyl acetate layer is concentrated under reduced pressure. The concentrated luciferin is adsorbed on a column of Celite-Fuller's earth mixture. The column is washed with water-saturated ethyl acetate, and eluted with alkaline water at pH 8.0–8.5. The aqueous eluate of luciferin is adjusted to pH 3.0 with HCl and luciferin is re-extracted with ethyl acetate. Luciferin is further purified by partition chromatography on a column of Celite using a mixture of butanol-chloroform-water (135:15:50) as the elution solvent, monitoring the absorbance values at 327 nm and 276 nm; the former value parallels the concentration of luciferin, whereas the latter indicates the protein concentration. The purity is judged by the absorbance ratio of 327/276. For crystallization, an aqueous solution of luciferin is extracted with ethyl acetate at pH 3.0. The ethyl acetate layer is washed with water, then evaporated to dryness. The residue of free (acidic) form of luciferin is dissolved in about 3.5 ml of acetone, and 1.5 ml of water is added. The acetone is slowly evaporated by bubbling a stream of nitrogen gas until crystals are formed. From 70 g of acetone powder (about 15,000 fireflies), 9 mg of crystalline luciferin were obtained.

In 1968, Kishi *et al.* reported another method to purify luciferin from the Japanese firefly *Luciola cruciata*. In this method, the first ethyl acetate extract of Bitler and McElroy was chromatographed on a column of cellulose powder using ethyl acetate-ethanol-water (5:2:3) as the eluting solvent, followed by chromatography on a column of DEAE-cellulose (elution with a gradient of NaCl concentration). They obtained 5.5 mg of crystalline luciferin from 233 g of the acetone powder of the abdomens (from 12,000 fireflies). The lower yield compared with that by Bitler and McElroy is probably due to the species used.

Properties of luciferin. The crystals are microscopic needles, which melt with decomposition at 205–210°C (Bitler and McElroy, 1957). It is a quite stable luciferin compared with some other luciferins, such as *Cypridina* luciferin and the luciferins of krill and dinoflagellates. It is not significantly affected by 10 mM H_2SO_4 and 10 mM NaOH at room temperature in air. The absorption spectral data of luciferin are shown in Fig. 1.3 (McElroy and Seliger, 1961). The molar absorption coefficient of the 328 nm peak in acidic solutions and that of the 384 nm peak in basic solutions are both 18,200 (Morton *et al.*, 1969). Luciferin is fluorescent, showing an emission maximum at 537 nm in both acidic and basic conditions, although the intensity of the fluorescence is lower in acidic solution than in basic solution (fluorescence quantum yields: 0.62 in basic condition, and 0.25 in acidic condition; Morton *et al.*, 1969). The chemical synthesis

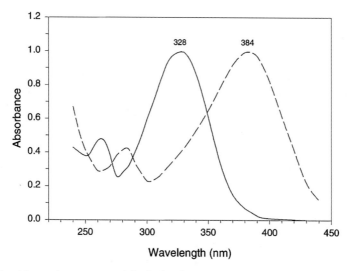

Fig. 1.3   Absorption spectra of firefly luciferin at pH 7.0 or below (solid line, λ_{max} 327–328 nm) and at pH higher than 9.0 (dashed line, λ_{max} 381–384 nm). Reproduced from McElroy and Seliger, 1961, with permission from the Johns Hopkins University Press.

of luciferin was accomplished by White *et al.* (1961; 1963); certain details and the improvements of the synthetic method are discussed by Bowie (1978) and Branchini (2000).

Oxyluciferin. Firefly oxyluciferin is an extremely unstable compound; it has never been isolated in a completely pure form (White and Roswell, 1991). A group in Nagoya synthesized the compound and its properties were investigated (Suzuki *et al.*, 1969; Suzuki and Goto, 1971). The fluorescence of oxyluciferin in DMSO in vacuum in the presence of potassium *t*-butoxide is yellow-green (λ_{max} 557 nm), the same emission maximum as the chemiluminescence of luciferin in DMSO in the presence of potassium *t*-butoxide, suggesting that oxyluciferin is the light emitter in the chemiluminescence of luciferin. In the bioluminescence reaction, the absorption peak of synthetic oxyluciferin at pH 7 (382 nm; Suzuki *et al.*, 1969) closely coincides with the absorption peak of the luciferase-oxyluciferin complex in the spent luminescence solution (Fig. 1.4; Gates and DeLuca, 1975), suggesting that oxyluciferin is the light-emitter as in the case of chemiluminescence. However, the fluorescence emission maximum of the spent

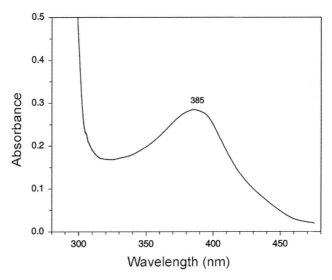

Fig. 1.4   Absorption spectrum of a spent luminescence solution of firefly luciferin containing luciferase-oxyluciferin after dialysis in 0.1 M potassium phosphate, pH 7.8. Replotted from the data of Gates and DeLuca, 1975, with permission from Elsevier.

luminescence solution is found at 523 nm, clearly different from the bioluminescence maximum at 562 nm (Fig. 1.5). Gates and DeLuca attributed this difference to the environmental change of the oxy-luciferin molecule that is caused by the conformational change of luciferase after the emission of light.

1.1.3 *Firefly Luciferase*

Firefly luciferase has been purified, crystallized and partially characterized by Green and McElroy (1956). The acetone powder prepared from the dried lanterns of *Photinus pyralis* is extracted with water containing 1 mM EDTA, at pH 7.8. The luciferase extracted is purified by calcium phosphate gel adsorption, ammonium sulfate fractionation, and then crystallized by dialysis against a low ionic strength buffer. Although the ammonium sulfate precipitated fractions are stable in the frozen state, the crystalline luciferase is inactivated by freezing and thawing. For the storage of luciferase, the crystals are dissolved and precipitated with 2.4 M $(NH_4)_2SO_4$ in the presence of 1 mM EDTA

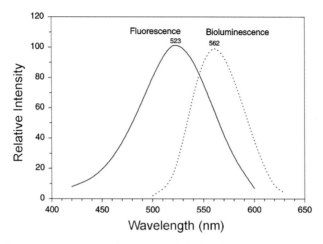

Fig. 1.5   Fluorescence emission spectrum of the luciferase-oxyluciferin complex in the same solution as in Fig. 1.4 (solid line), compared with the luminescence spectrum of firefly luciferin measured in glycylglycine buffer, pH 7.6 (dotted line). The former curve from Gates and DeLuca, 1975; the latter from Selinger and McElroy, 1960, both with permission from Elsevier.

at a pH between 7.5 and 8.0, then stored at 4°C. For use, the suspension is centrifuged, and the precipitate is dissolved in 1 mM EDTA at pH 7.9. This solution is stable for days at 4°C at a protein concentration of about 10 mg per ml. The molecular weight of luciferase was estimated at 100,000 (however, see next page), the isoelectric point is found at pH 6.2–6.3, and the absorbance of a 1 mg/ml solution at 278 nm is 0.75. A purification method of luciferase involving high-performance liquid chromatography (HPLC) was reported by Branchini and Rollins (1989).

Using the American firefly *Photinus pyralis*, the cloning of luciferase was achieved first by *in vitro* translation of RNA by Wood *et al.* (1984), followed by the expression of the cDNA in *Escherichia coli* by De Wet *et al.* (1985, 1986). The purification and cDNA cloning of firefly luciferase are also reported for several other species of fireflies: Japanese fireflies *Luciola cruciata* (Tatsumi *et al.*, 1989), *Luciola lateralis* (Tatsumi *et al.*, 1992) and *Pyrocoelia miyako* and *Hotaria parvula* (Ohmiya *et al.*, 1995); European firefly *Luciola mingrelica* (Devine *et al.*, 1993); and American firefly *Photuris pennsylvania* (Ye *et al.*, 1997; Leach *et al.*, 1997).

The role of the sulfhydryl groups of luciferase in the firefly bioluminescence reaction has been a target of intensive investigation since the early 1960s. DeLuca *et al.* (1964) found that luciferase contains 6–8 sulfhydryl groups per 100 kDa of protein. The catalytic activity of luciferase is completely lost by treatment with p-mercuribenzoate. However, in the presence of excess amounts of luciferin, Mg^{2+} and ATP for forming the luciferin-luciferase-ATP complex, two sulfhydryl groups are protected from p-mercuribenzoate and the catalytic activity is maintained, implying the necessity of these sulfhydryl groups for the activity. Despite these results, Alter and DeLuca (1986) concluded that the sulfhydryl groups of luciferase are not essential for the activity of luciferase. They also reported that the treatment of luciferase with methyl methanethiosulfonate produces an enzyme that causes the emission of red light, differing from the native luciferase that results in yellow-green light. It seems possible that the modified enzyme contains a distorted active site. Later, Ohmiya and Tsuji (1997) confirmed that the sulfhydryl groups are nonessential based on the results of the replacement of the cysteine residues by the site-directed mutagenesis.

The apparent molecular weights of both natural *P. pyralis* luciferase and an active luciferase obtained from *P. pyralis* by the *in vitro* RNA translation were 62,000 by SDS-PAGE (Wood *et al.*, 1984), in contrast to the value of 100,000 that had been widely referred to in the field for almost 30 years. Luciferases from other species of firefly probably have similar molecular weights. Presently, the molecular masses of firefly luciferases are considered to be 60–62 kDa.

Conti *et al.* (1996) solved the crystal structure of the *P. pyralis* luciferase at 2.0 Å resolution. The protein is folded into two compact domains, a large N-terminal portion and a small C-terminal portion. The former portion consists of a β-barrel and two β-sheets. The sheets are flanked by α-helices to form an $\alpha\beta\alpha\beta\alpha$ five-layered structure. The C-terminal portion of the molecule forms a distinct domain, which is separated from the N-terminal domain by a wide cleft. It is suggested that the two domains will close up in the course of the luminescence reaction.

1.1.4 *Assays of Luciferase Activity, ATP and Luciferin*

Various assay mixtures of different compositions have been used to measure the activity of luciferase and the amount of ATP (Leach, 1981). A typical mixture for luciferase assay contains 10–25 mM Tris-HCl or glycylglycine buffer, pH 7.5–7.8, 5 mM $MgCl_2$, 1–5 mM ATP, and 0.1 mM luciferin. Because luciferase at very low concentrations is rapidly inactivated, 0.5–1 mM EDTA and 0.1% BSA are included in some formulae to prevent the inactivation. Usually ATP is injected into the rest of the mixture to start luminescence, producing a sharp flash of light that diminishes rapidly (DeLuca and McElroy, 1978). The peak of the flash occurs about 0.3–0.5 second after the injection of ATP (Fig. 1.6), and the light intensity of the peak is proportional to the amount of luciferase in a wide range of luciferase concentration. If the measurement of flash height is difficult to carry out for some reasons (such as a slow response of recorder), the light intensity at 5 or 10 seconds after the ATP injection is measured instead of the flash height. Although the measured light intensity in this case

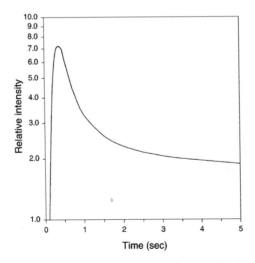

Fig. 1.6   The time course of luminescence reaction initiated by the injection of ATP. The light intensity first rises rapidly, reaching a maximum in 0.3–0.5 sec, followed by relatively rapid decrease for the first few seconds and then a much slower decay that lasts for several minutes or more. From McElroy and Seliger, 1961, with permission from the Johns Hopkins University Press.

is lower than the intensity of a flash, it is still proportional to the amount of luciferase as long as the same method and the same conditions are used. Luciferin and ATP can be assayed with appropriate modifications of the method.

1.1.5   General Characteristics of the Bioluminescence of Fireflies

The color of the luminescence of common fireflies varies slightly depending on the species, with their *in vivo* emission peaks in a range from 552 nm to 582 nm (Seliger and McElroy, 1964). The color of the *in vitro* luminescence using purified luciferin and *Photinus pyralis* luciferase (plus ATP and Mg^{2+}) under neutral or slightly alkaline conditions is yellow-green (λ_{max} 560 nm; Fig. 1.7), with a quantum yield of 0.88 ± 0.25 (Seliger and McElroy, 1959; 1960); in spite of the large error range, the quantum yield is clearly greater than those of *Cypridina* luciferin and coelenterazine (both about 0.3). The quantum yield and the color of luminescence are affected by the pH of the reaction medium (Fig. 1.8). Under acidic conditions, red luminescence (λ_{max} 615 nm) with a decreased light intensity is emitted. A similar red shift of luminescence is also observed by raising the reaction

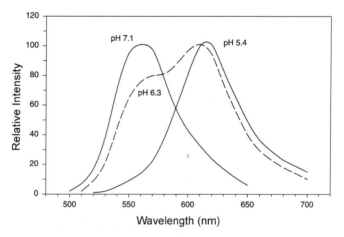

Fig. 1.7 Spectral change of the *in vitro* firefly bioluminescence by pH, with *Photinus pyralis* luciferase in glycylglycine buffer. The normally yellow-green luminescence (λ_{max} 560 nm) is changed into red (λ_{max} 615 nm) in acidic medium, accompanied by a reduction in the quantum yield. From McElroy and Seliger, 1961, with permission from Elsevier.

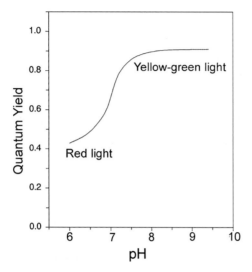

Fig. 1.8 Quantum yield of firefly bioluminescence as a function of pH. From McElroy and Seliger, 1961, with permission from Elsevier.

temperature, by carrying out the reaction in a glycylglycine buffer (pH 7.6) containing 0.2 M urea, and by adding a small concentration of Zn^{2+}, Cd^{2+}, or Hg^{2+} (Seliger and McElroy, 1964).

The optimum pH for the luminescence reaction is about 7.8, and the luminescence intensity is strongly affected by the buffer salt used

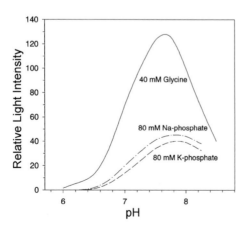

Fig. 1.9 Effects of pH and buffer on the activity of luciferase measured at the same luciferase concentration. The optimum pH with glycine buffer is approximately 7.8. From Green and McElroy, 1956, with permission from Elsevier.

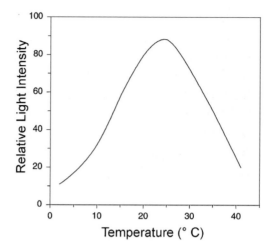

Fig. 1.10   Effect of temperature on the activity of luciferase. From McElroy and Seliger, 1961, with permission from Elsevier.

(Fig. 1.9; Green and McElroy, 1956). The optimum temperature for luminescence is 23–25°C (Fig. 1.10; McElroy and Strehler, 1949; Green and McElroy, 1956). In the presence of a low concentration of ATP, high concentrations of Mg^{2+} are inhibitory, while in the presence of a low concentration of Mg^{2+}, high concentrations of ATP are inhibitory (Fig. 1.11; Green and McElroy, 1956).

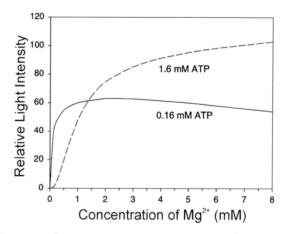

Fig. 1.11   Effect of Mg^{2+} concentration on the activity of luciferase in the presence of 0.16 mM and 1.6 mM ATP. From Green and McElroy, 1956, with permission from Elsevier.

The sharp flash in the firefly bioluminescence reaction (Fig. 1.6) is due to the formation of a strongly inhibitory byproduct in the reaction. The inhibitor formed is dehydroluciferyl adenylate, having the structure shown below at left. In the presence of coenzyme A (CoA), however, this inhibitory adenylate is converted into dehydroluciferyl-CoA, a compound only weakly inhibitory to luminescence. Thus, an addition of CoA in the reaction medium results in a long-lasting, high level of luminescence (Airth *et al.*, 1958; McElroy and Seliger, 1966; Ford *et al.*, 1995; Fontes *et al.*, 1997, 1998).

1.1.6   *Mechanisms of the Firefly Bioluminescence*

Seliger and McElroy (1962) discovered that the esters of firefly luciferin emit chemiluminescence. They reported that luciferyl adenylate (Rhodes and McElroy, 1958) emitted a red light (λ_{max} 625.5 nm) in dimethyl sulfoxide upon the addition of a base. The emission spectrum was dependent upon pH, producing a yellow-green light in the presence of a large excess of base. The observation of yellow-green light was also reported later by other authors (White *et al.*, 1969, 1971; however, see Section 1.1.7). The product of the luminescent oxidation of luciferin is oxyluciferin (the structure shown in Fig. 1.12), an extremely unstable compound. Hopkins *et al.* (1967) found that 5,5-dimethyloxylucferin, an oxyluciferin analogue having no H atoms at position 5, shows a red fluorescence in the presence of a base, coinciding with the red chemiluminescence spectrum of luciferyl adenylate. Considering these findings, the bioluminescence reaction of firefly was postulated as shown in Fig. 1.12 (Hopkins *et al.*, 1967; McCapra *et al.*, 1968; White *et al.*, 1969, 1971, 1975; Shimomura *et al.*, 1977; Koo *et al.*, 1978).

In the postulated bioluminescence mechanism, firefly luciferin is adenylated in the presence of luciferase, ATP and Mg^{2+}. Luciferyl adenylate in the active site of luciferase is quickly oxygenated at its tertiary carbon (position 4), forming a hydroperoxide intermediate (**A**).

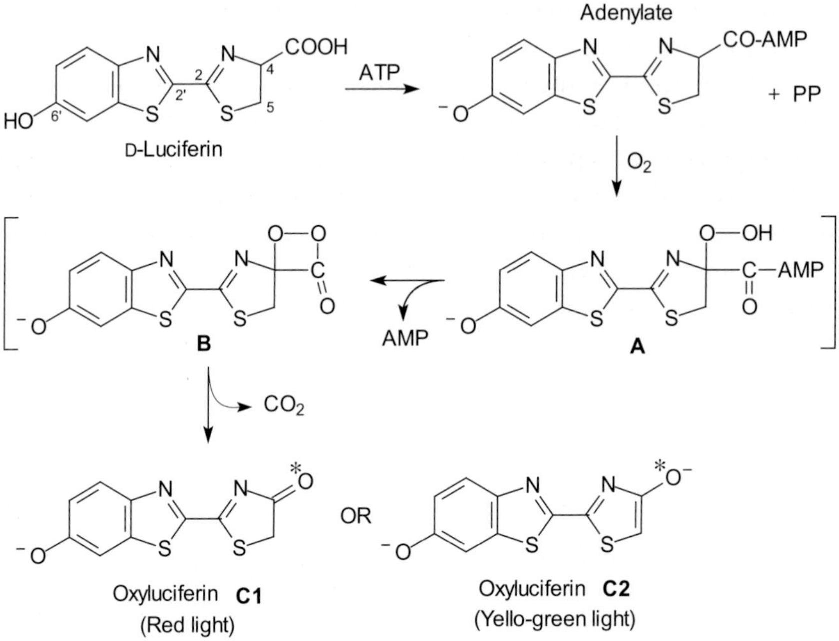

Fig. 1.12 Mechanism of the bioluminescence reaction of firefly luciferin catalyzed by firefly luciferase. Luciferin is probably in the dianion form when bound to luciferase. Luciferase-bound luciferin is converted into an adenylate in the presence of ATP and Mg^{2+}, splitting off pyrophosphate (PP). The adenylate is oxygenated in the presence of oxygen (air) forming a peroxide intermediate **A**, which forms a dioxetanone intermediate **B** by splitting off AMP. The decomposition of intermediate **B** produces the excited state of oxyluciferin monoanion (**C1**) or dianion (**C2**). When the energy levels of the excited states fall to the ground states, **C1** and **C2** emit red light (λ_{max} 615 nm) and yellow-green light (λ_{max} 560 nm), respectively.

The hydroperoxide forms a very unstable 4-membered dioxetanone ring (**B**), splitting off AMP. The dioxetanone decomposes by a concerted cleavage, yielding the keto-form oxyluciferin (**C1**) and CO_2, accompanied by emission of light. It is possible that the decomposition of dioxetanone results in light emission by the chemically initiated electron-exchange luminescence (CIEEL) mechanism (McCapra, 1977; Koo *et al.*, 1978). The formation of the dioxetanone intermediate was confirmed by ^{18}O-labeling experiments, by showing that one of the O atoms of the product CO_2 was derived from molecular oxygen (Shimomura *et al.*, 1977; also see Section 1.1.8).

Because luciferyl adenylate emitted a red chemiluminescence in the presence of base, coinciding with the red fluorescence of 5,5-dimethyloxylucferin, the keto-form monoanion **C1** in its excited state is considered to be the emitter of the red light. Thus, the emitter of the yellow-green light is probably the enol-form dianion **C2** in its excited state, provided that the enolization takes place within the life-time of the excited state. Although the evidence had not been conclusive, especially on the chemical structures of the light emitters that emit two different colors, the mechanism shown in Fig. 1.12 was widely believed and cited until about 1990.

1.1.7   Light Emitters in the Firefly Luminescence System

Bioluminescence of firefly luciferin can produce a wide range of colors when catalyzed by different luciferases obtained from various species of fireflies, with their emission maxima ranging from 535 nm (yellow-green) to 638 nm (red). Apparently, each spectrum is emitted from a single emitting species; they are not the composites of the yellow-green peak and the red peak (Seliger and McElroy, 1964).

White and Roswell (1991) investigated the fluorescence emission properties of the 5-methyl and O-methyl derivatives of oxyluciferin, and concluded that the structures **C3** and **C4** (shown below) should also be the candidates for the emitter of the yellow-green light, in addition to **C2**. Moreover, they have stated that the chemilumines-cence of the esters of firefly luciferin produces only red light and does not produce yellow-green light, even in the presence of a high con-centration of strong base. They wrote: "It is not clear how the earlier work produced contrary observation." Thus, it became difficult to understand why the chemiluminescence of luciferyl adenylate emits only red light, differing from the luciferase-catalyzed reaction that normally produces yellow-green light.

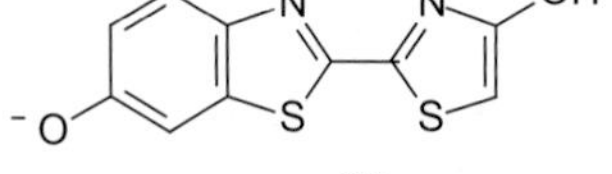

C3 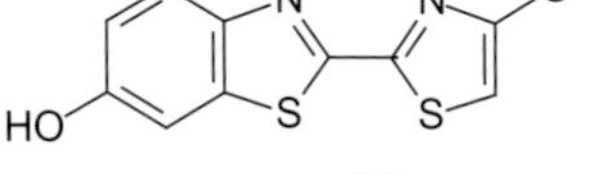   C4

McCapra *et al.* (1994) and McCapra (1997) suggested that the color variation could be caused by the conformational difference of the oxyluciferin molecule, when the plane of thiazolinone is rotated at various angles against the plane of benzothiazole on the axis of the 2-2′ bond; the red light would be emitted at 90° angle, reflecting its minimum structural energy.

However, Branchini *et al.* (2002) reported a surprising discovery that the adenylate of D-5,5-dimethylluciferin emits light in two different colors in the bioluminescence reaction catalyzed by two different luciferases, one from *Photinus pyralis* and the other from a green-emitting click beetle *Pyrophorus plagiophthalamus*. In the presence of Mg^{2+} and at pH 8.6, a yellow-green light (λ_{max} 560 nm) was produced with *P. pyralis* luciferase and a red light (λ_{max} 624 nm) was emitted with *P. plagiophthalamus* luciferase. In both cases, the reaction product was 5,5-dimethyloxyluciferin (shown below) that has no H atom on its C5; it cannot take the tautomeric enolized form, such as in C2, C3 or C4, that had been proposed to be the emitter of yellow-green light.

This finding by Branchini *et al.* (2002) clearly indicates that 5,5-dimethyloxyluciferin is able to emit the two different colors. This conclusion, however, does not rule out the involvement of the enolized oxyluciferin in the bioluminescence reaction of firefly.

Orlova *et al.* (2003) theoretically studied the mechanism of the firefly bioluminescence reaction on the basis of the hybrid density functional theory. According to their conclusion, changes in the color of light emission by rotating the two rings on the 2-2′ axis is unlikely, whereas the participation of the enol-forms of oxyluciferin in bioluminescence is plausible but not essential to explain the multicolor emission. They predicted that the color of the bioluminescence depends on the polarization of the oxyluciferin molecule (at its OH and O⁻ termini) in the microenvironment of the luciferase active site; the smaller the H–O polarization, the greater the blue shift of the absorption (and excitation). By this mechanism, the range of colors observed in the bioluminescence could be obtained with various forms of oxyluciferin. The most likely light emitter is keto-s-trans monoanion, but the enol-s-trans monoanion and keto-s-cis monoanion structures may also be involved. Their conclusion is in agreement with that of Branchini *et al.* (2002), in that the involvement of keto-enol tautomerism is not essential to explain the two different colors.

According to Branchini *et al.* (2004), luciferase modulates the emission color by controlling the resonance-based charge delocalization of the anionic keto-form of oxyluciferin in the excited state. They proposed the structure **C5** as the yellow-green light emitter, and the structure **C6** as the red light emitter.

C5   C6
(yellow-green)   (red)

It should be pointed out that the structure **C5** (yellow-green emitter) is identical to the structure **C1** that was previously assigned to the red light emitter.

1.1.8   A Note on the Dioxetanone Pathway and the ^{18}O-incorporation Experiment

In the luminescence reaction of firefly luciferin (Fig. 1.12), one oxygen atom of the product CO_2 is derived from the molecular oxygen while the other originates from the carboxyl group of luciferin. In the chemiluminescence reaction of an analogue of firefly luciferin in DMSO in the presence of a base, the analysis of the product CO_2 has supported the dioxetanone pathway (White *et al.*, 1975).

Contrary to the dioxetanone pathway, DeLuca and Dempsey (1970) proposed a mechanism of the bioluminescence reaction that involves a multiple linear bond cleavage of luciferin peroxide

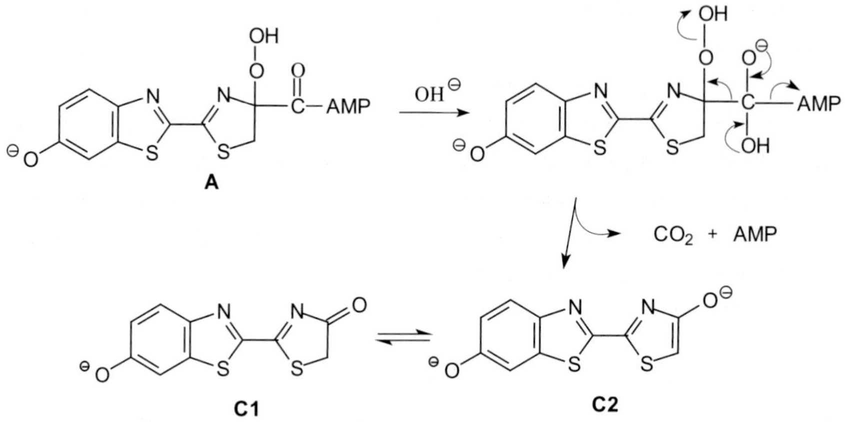

Fig. 1.13   A mechanism of the decomposition of luciferin-4-peroxide in the firefly bioluminescence reaction proposed by DeLuca and Dempsey (1970), which involves a multiple linear bond cleavage.

(Fig. 1.13). They carried out the luminescence reaction in $H_2{}^{18}O$ solvent under ${}^{16}O_2$ atmosphere, and also in $H_2{}^{16}O$ solvent under ${}^{18}O_2$ atmosphere, and analyzed the product CO_2 by mass spectrometry. Under the former condition, they found that one atom of ${}^{18}O$ was incorporated into the CO_2, whereas under the latter condition, no ${}^{18}O$ was found in the CO_2. Thus, they concluded that one O atom of the product CO_2 came from the solvent water, as the basis of their linear bond cleavage hypothesis. Although organic chemists are reluctant to accept this linear mechanism (e.g. White *et al.*, 1971), three follow-up papers involving four laboratories have confirmed and verified the Deluca and Dempsey 1970 report, supporting the linear bond cleavage hypothesis (DeLuca and Dempsey, 1973; DeLuca *et al.*, 1976; Tsuji *et al.*, 1977). Moreover, the same mechanism has also been applied to the bioluminescence and chemiluminescence of two kinds of imidazopyrazinone-type luciferins, *Renilla* luciferin (coelenterazine) and *Cypridina* luciferin (DeLuca *et al.*, 1971, 1976; Tsuji *et al.*, 1977), despite the fact that the bioluminescence reaction of *Cypridina* had previously been shown to involve the dioxetanone mechanism (Shimomura and Johnson, 1971).

However, the linear bond cleavage hypothesis of the firefly bio-luminescence was made invalid in 1977. It was clearly shown by Shimomura *et al.* (1977) that one O atom of the CO_2 produced is derived from molecular oxygen, not from the solvent water, using the same ^{18}O-labeling technique as used by DeLuca and Dempsey. The result was verified by Wannlund *et al.* (1978). Thus it was confirmed that the firefly bioluminescence reaction involves the dioxetanone pathway. Incidentally, there is currently no known bioluminescence system that involves a splitting of CO_2 by the linear bond cleavage mechanism.

It seems important to identify the factors that have led DeLuca, Dempsey, and others into a misjudgment. The following explanation is included here for future experimentalists (see also Section C6 in Appendix).

When gaseous CO_2 is equilibrated with aqueous buffer solution in a closed vessel, a large portion of the CO_2 is dissolved in the aqueous phase, mostly in the form of bicarbonate, maintaining the equilibrium of the following three phases:

$$CO_2 + H_2O \longleftrightarrow H_2CO_3 \longleftrightarrow HCO_3^- + H^+$$

Thus, the O atom of CO_2 is exchangeable with the O atom of H_2O. When the luminescence reaction is carried out in a $H_2^{18}O$ medium under an atmosphere of $^{16}O_2$, the $C^{16}O_2$ formed by the dioxetanone mechanism is spontaneously converted into $C^{16}O^{18}O$. If the reaction is carried out in a $H_2^{16}O$ medium under an atmosphere of $^{18}O_2$, the $C^{16}O^{18}O$ formed is spontaneously converted into $C^{16}O_2$. Thus, the result of ^{18}O-incorporation experiment can be obscurred by the exchange of O atom between CO_2 and H_2O. In addition to this exchange, the presence of contaminating CO_2 can also obscure the result. The occurrence of CO_2 is ubiquitous and clean air normally contains approximately 0.03% (v/v) of CO_2. In our experiments, care-fully prepared fresh buffer solutions contained 0.02–0.03 μmol/ml of CO_2 plus HCO_3^- even after vacuum degassing, and the amount was much greater when luciferase had been included (Shimomura *et al.*, 1977). Thus, the CO_2 produced from a small amount of luciferin (for example, 0.033 μmol in 3.5 ml: DeLuca and Dempsey, 1970) will be obscurred by the contaminating CO_2 even without considering the effect of the O atom exchange.

In the ^{18}O-incorporation experiment of *Cypridina* bioluminescence, the effects of the O atom exchange and contaminating CO_2 are clearly seen in the relationship between the amount of luciferin luminesced and the amount of ^{18}O atoms incorporated into the product CO_2 (Fig. 1.14; Shimomura and Johnson, 1973a). The experiments were done in glycylglycine buffer, pH 7.8, the same buffer as chosen by DeLuca and Dempsey (1970). The total volume of the reaction mixture was 4 ml, with 40 ml of gas phase (see the reaction vessel in Fig. A.5 in the Appendix). The data of the luminescence reaction with $^{18}O_2$ gas in the $H_2^{16}O$ medium indicates that at least 1 μmol of

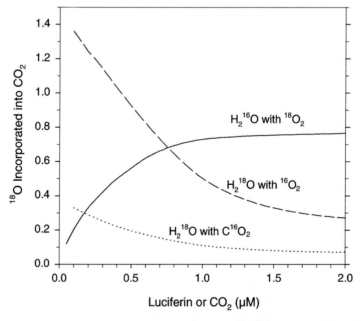

Fig. 1.14 Relationship between the incorporation of ^{18}O into product CO_2 and the amount of luciferin used, in the bioluminescence reaction of *Cypridina* luciferin catalyzed by *Cypridina* luciferase. The reactions were carried out in $H_2^{16}O$ medium with $^{18}O_2$ gas (solid line); in $H_2^{18}O$ medium with $^{16}O_2$ gas (dashed line); and the control experiment of the latter using $C^{16}O_2$ gas instead of luciferin and luciferase (dotted line), all in 20 mM glycylglycine buffer, pH 7.8, containing 40 mM NaCl. From Shimomura and Johnson, 1973a. Reproduced with permission from Elsevier.

luciferin has to be used to obtain a highly reliable conclusion (the theoretical maximum number of ^{18}O atoms to be incorporated is 0.85–0.90; see Section 3.1.8). In the luminescence reaction with $^{16}O_2$ gas in a $H_2^{18}O$ medium, the effect of oxygen atom exchange is even greater; the number of ^{18}O atoms incorporated is about 0.5 when 1 µmol of luciferin is used, and the value reaches 1.0 when 0.4 µmol of luciferin is used (theoretically, there should be no incorporation of ^{18}O in the dioxetanone mechanism). The ^{18}O-incorporation was much less in the control experiment that contained $C^{16}O_2$ instead of luciferin. These results would be reasonable when the following matters are considered: (1) the O atom exchange occurs only in the solution, not in the gas phase; and (2) in the following equilibrium reaction, the rate going to the left is twice that going to the right (Cohn and Urey, 1938).

$$H_2^{16}O + C^{16}O^{18}O \longleftrightarrow H_2^{18}O + C^{16}O^{16}O$$

Based on the data in Fig. 1.14, it is apparent that the O atom exchange and contaminating CO_2 can easily give misleading result and erroneous conclusion when a small amount of luciferin is used. The level of oxygen atom exchange is also affected by the pH and type of buffer used (Shimomura and Johnson, 1975a). The glycylglycine buffer is less suitable than Tris or phosphate buffer, due to the rapid increase in the amount of dissolved CO_2 plus HCO_3^- (Shimomura et al., 1977). See further details in Appendix C6.

1.2   Phengodidae and Elateroidae

The bioluminescence systems of Phengodidae (railroad worms) and Elateroidae (click beetles) are basically identical to that of Lampyridae (fireflies), requiring firefly luciferin, ATP, Mg^{2+} and a luciferase for light emission. However, there seem to be some differences. Viviani and Bechara (1995) reported that the spectra of the luminescence reactions measured with the luciferases of Brazilian fireflies (6 species) shift from the yellow-green range to the red range with lowering of the pH of the medium, like in the case of the *Photinus pyralis* luciferase (see Section 1.1.5), whereas the spectra measured with the luciferases of Elateroidae (5 species) and Phengodidae (3 species) showed no change with lowering of the pH of the medium.

1.2.1 *Phengodidae*

The railroad worm *Phrixothrix* is well known for displaying two different colors of luminescence from a single organism. This genus is widely distributed in Central and South America. The larva of *Phrixothrix* (and also the adult female) emits a greenish yellow light (λ_{max} 535–565 nm) from 11 pairs of luminous organs on the posterior lateral margins of the second to the ninth segment, and a red light (λ_{max} 600–620 nm) from the luminous area on the head (Viviani and Bechara, 1993). The adult male is a typical beetle and does not show a noticeable luminescence.

Viviani and Bechara (1993, 1997) investigated various railroad worm species of 8 genera of phengodidae collected in southeastern and west central Brazil, near the Parque Nacional das Emas. The bioluminescence systems of these phengodids were essentially the same as that of the fireflies, involving the same luciferin (firefly luciferin), ATP and Mg^{2+}. Their emission maxima of luminescence from the lateral and head organs are in the ranges of 535–592 nm and 562–638 nm, respectively. The color differences are probably due to the presence of luciferase isoenzymes (M_r about 60,000) according to the authors.

Gruber *et al.* (1997) reported the purification and cloning of the luciferases of a *Phengodes* species. Viviani *et al.* (1999) cloned the luciferases from the lateral light organs of *Phrixothrix vivianii* (emission λ_{max} 542 nm) and the head light organs of *Phrixothrix hirtus* (emission λ_{max} 628 nm).

1.2.2 *Elateridae*

The elaterid *Pyrophorus* is of special importance in the history of bioluminescence, because it was used by Dubois in his first demonstration of the luciferin-luciferase reaction in 1885. The Jamaican click beetle (*Pyrophorus noctilucus*) is commonly found in the West Indies. The beetle possesses two kinds of luminous organs. A pair of oval-shaped light organs is located on the head. These organs look like eyes, and emit very strong greenish luminescence. The second kind of light organ is ventrally located on the first abdominal segment, and it emits orange light only when flying or when the elytra are expanded. The bioluminescence system is chemically the same as those of fireflies and railroad worms.

Colepicolo-Neto *et al.* (1986) investigated the adults of 12 species of elaterids collected in Brazil. They found that the luminescence emission maxima of the abdominal light organs are shifted toward red (20–40 nm) relative to those of the head organs (λ_{max} 525–560 nm). Thin-layer chromatography under various conditions revealed that the luciferins extracted from both types of light organs are identical to firefly luciferin.

Wood *et al.* (1989) generated 11 cDNA clones from mRNA isolated from the abdominal light organs of a Jamaican click beetle *Pyrophorus plagiophthalamus*. When expressed in *E. coli*, these clones produced four types of luciferase distinguishable by the colors of bioluminescence they produce: green (λ_{max} 546 nm), yellow-green (λ_{max} 560 nm), yellow (λ_{max} 578 nm), and orange (λ_{max} 593 nm). Molecular cloning of the luciferase of a Brazilian click beetle *Pyrearinus termitilluminans* was also reported (Viviani *et al.*, 1999a).

1.3   Diptera

The order Diptera (flies) contains the glow-worms *Arachnocampa* and *Orfelia*. The bioluminescence systems of dipterans do not utilize firefly luciferin in their light-emitting reactions, differing from the bioluminescence systems of coleopterans. In dipterans, it is extremely intriguing that the bioluminescence system of *Arachnocampa* appears different from that of *Orfelia*: the former luminescence is activated by ATP, whereas the latter luminescence is stimulated by DTT but not by ATP.

1.3.1   *The Glow-worm Arachnocampa*

The glow-worm *Arachnocampa* is distributed in New Zealand and Australia. The larvae emit blue light continuously from their light organs located at the posterior extremity. They are found most often on the roofs of caves. The glow-worms usually stay on the horizontal network of mucous tubes suspended from rocks, and they hang down long sticky threads of "fishing lines" from the tubes to catch small insects. The spectacular view of glow-worms at the Waitomo Cave in New Zealand attracts hundreds of tourists everyday.

Earlier studies indicated that the bioluminescence emission maximum of the New Zealand glow-worm A. *luminosa* is 487–488 nm, and that the bioluminescence reaction probably requires ATP as a cofactor (Shimomura *et al.*, 1966), similar to the firefly luminescence reaction. According to Lee (1976), the luminescence emission spectrum of the Australian glow-worm A. *richardsae* (λ_{max} 488 nm) was not significantly influenced by pH in a wide range, i.e. 5.9–8.5, and firefly luciferin does not cross-react with the spent luminescence mixture, indicating differences from the firefly luminescence system. However, the luminescence was quenched by EDTA, but not by EGTA that does not chelate Mg^{2+}, suggesting that Mg^{2+} is probably required for the luminescence reaction, like the firefly luminescence.

The luciferin-luciferase reaction of *Arachnocampa* was first demonstrated by Wood (1993), by mixing a cold-water extract and a cooled hot-water extract. The cold-water extract was prepared with 27 mM Tricine, pH 7.4, containing 7 mM $MgSO_4$, 0.2 mM EDTA, 10% glycerol and 1% Triton X-100, and incubated with 1 mM ATP on ice for 18 hr. The hot-water extract was prepared by heating the cold water extract before the addition of ATP at 98°C for 5 min. The luminescence reaction was performed in the presence of 1 mM ATP.

Extraction and purification of luciferin and luciferase (Viviani *et al.*, 2002a) To isolate luciferin, the lanterns of the Australian A. *flava* were homogenized in hot 0.1 M citrate buffer, pH 5, and the mixture was heated to 95°C for 5 min. The mixture was acidified to pH 2.5–3.0 with HCl, and luciferin was extracted with ethyl acetate. Upon thin-layer chromatography (ethanol-ethyl acetate-water, 5:3:2 or 3:5:2), the active fraction of luciferin was fluorescent in purple (emission λ_{max} 415 nm when excited at 290 nm). To isolate the luciferase, the cold-water extract prepared according to Wood (1993; see above) was chromatographed on a column of Sephacryl S-300. On the same column, the molecular mass of luciferase was estimated at approximately 36 kDa.

Luminescence reaction (Viviani *et al.*, 2002a) The luciferin-luciferase luminescence reaction was carried out in 0.1 M Tris-HCl, pH 8.0, containing 2 mM ATP and 4 mM Mg^{2+}. Mixing luciferase with luciferin and ATP resulted in an emission of light with rapid onset and a kinetically complex decay. Further additions of fresh luciferase, after the luminescence has decayed to about 10% of its maximum value, resulted in additional luminescence responses similar to the initial one (Fig. 1.15). According to the authors, the repetitive light emission occurred in consequence of the inhibition of luciferase by a reaction product, as seen in the case of the firefly system (McElroy *et al.*, 1953). The luminescence spectrum showed a peak at 487 nm (Fig. 1.16).

1.3.2   *The American Glow-worm Orfelia*

The American glow-worm *O. fultoni* is found in the Appalachian Mountains. The larvae of *Orfelia* live on damp stream banks, and they

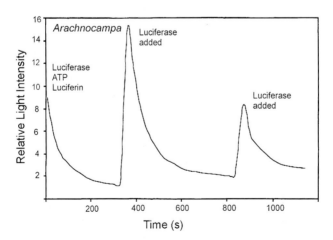

Fig. 1.15   Effect of successive additions of 10 µl of *Arachnocampa* luciferase (cold-water extract) to the assay mixture (90 µl) containing 2 mM ATP, 4 mM MgSO$_4$ and 5 µl of luciferin solution (hot-water extract made with 10 mM DTT). From Viviani *et al.*, 2002a, with permission from the American Society for Photobiology.

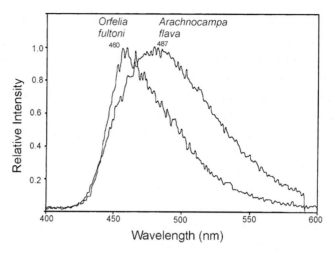

Fig. 1.16 The *in vitro* bioluminescence spectra of O. *fultoni* and A. *flava*. From Viviani *et al.*, 2002a, with permission from the American Society for Photobiology.

are luminous just like *Arachnocampa*. Viviani *et al.* (2002a) demonstrated a luciferin-luciferase reaction with a cold-water extract and a hot-water extract. The cold-water extract was prepared with 0.1 M phosphate, pH 7.0, containing 1 mM EDTA and 1% Triton X-100.

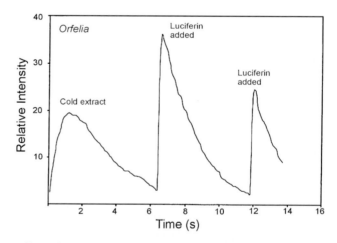

Fig. 1.17 Effect of successive additions of 10 μl of *Orfelia* luciferin solution (hot-water extract made with 10 mM DTT) to the assay mixture (90 μl) containing cold-water extract and 10 mM DTT after the luminescence has decayed. From Viviani *et al.*, 2002a, with permission from the American Society for Photobiology.

The hot-water extract was prepared by heating the cold-water extract at 95°C for 5–10 min in the presence of 10 mM DTT under argon gas. The luminescence reaction was performed in 0.1 M Tris-HCl, pH 8.0. The reaction was strongly stimulated by DTT and ascorbic acid, but not by ATP, indicating that the *Orfelia* luminescence system is different from the luminescence systems of the fireflies and *Arachnocampa* that require ATP for light emission. After the luminescence of a cold-water extract in the pH 8.0 buffer containing DTT had decayed to about 10% of the peak intensity, an addition of a hot-water extract caused an immediate increase in light emission (Fig. 1.17), suggesting that the decay of luminescence is caused by the depletion of luciferin. The molecular mass of the luciferase was estimated at about 140 kDa by gel filtration. The luminescence of *O. fultoni* is the bluest of all luminous insects (λ_{max}460 nm; Fig. 1.16).

CHAPTER

2

LUMINOUS BACTERIA

Although the luminescence of dead fish has been known since ancient times, the phenomenon became a target of scientific research only after the 16th century (Harvey, 1952, 1957). Robert Boyle (1668) found the dependence of luminescence on air, and Benjamin Martin (1761) discovered the necessity of salt for marine luminous bacteria although he thought he was studying the phosphorescence of the sea. The fact that light is produced by living bacteria became clear in the mid-nineteenth century (Harvey, 1952). Modern study of the bacterial luminescence reaction began with the discoveries of fatty aldehydes and $FMNH_2$ as the essential factors of the light emission (Cormier and Strehler, 1953; Strehler and Cormier, 1954). These discoveries eventually developed into the present detailed understanding of bacterial luminescence, in which J. Woodland Hastings and his associates played a major role.

Distribution of luminous bacteria. Luminous bacteria are widely distributed in the marine environment, and have been isolated from various sources, including seawater, the light organs and various other parts of marine luminous organisms, sometimes even from nonmarine sources as well. There are several major groups of luminous bacteria

(Hastings and Nealson, 1977; Nealson and Hastings, 1979; Ziegler and Baldwin, 1981; Dunlap and Kita-Tsukamoto, 2001).

Generic names. Although there has been a great deal of confusion in the generic names used in the literature, a bacterium can usually be correctly identified from the species part (the second part) of its scientific name. The marine forms of luminous bacteria are generally in the genera *Photobacterium*, *Beneckea* and *Vibrio*, while nonmarine forms are in the genera *Vibrio* and *Xenorhabdus*. All are gram-negative. The genus *Photobacterium* is predominantly symbiotic in light organs, making a distinction from *Beneckea*, although both genera are found as symbionts in the gut.

2.1 Factors Required for Bioluminescence

Until the middle of the 20th century, the classical luciferin-luciferase reaction applied to luminous bacteria was consistently negative (Harvey, 1952), and no knowledge was obtained on the chemistry of light emission. However, the biochemistry of bacterial luminescence began to be unveiled in the early 1950s by the efforts of two outstanding scientists, helped in part by the availability of newly developed, highly sensitive light detectors, i.e. photomultipliers. In 1953, Strehler discovered that NADH (DPNH) activates the luminescence of the extracts of luminous bacteria. This was quickly followed by the most important discoveries in the bacterial bioluminescence: the fact that the luminescence reaction requires $FMNH_2$ plus a long-chain, saturated aliphatic aldehyde (fatty aldehyde) in addition to the luciferase and molecular oxygen (Cormier and Strehler, 1953; Strehler and Cormier, 1953, 1954).

The way that the aldehyde requirement was discovered by Cormier and Strehler is extraordinary and noteworthy. First, they found that a boiled bacterial extract stimulated luminescence when it was added to a weakly luminescing NADH-activated bacterial extract. They thought that the stimulation was due to certain substances associated with the cell debris existing in the extract. Thus, they tested the extracts of various animal tissues in the hope of finding a substance that would give a strong stimulation effect. They found that the extracts of kidney cortex and liver from hog and beef strongly stimulated the luminescence. To identify the stimulation factor, they extracted 7 kg of dried hog kidney cortex with chloroform. The stimulation factor in the extract was purified by a complicated process that included repeated precipitation and extraction using various solvents, in addition to a seven-stage counter-current extraction method performed with separatory funnels, using hexane and methanol as the partition solvents. They finally obtained 1.5 g of purified stimulation factor, and called it "kidney cortex factor (KCF)." The elementary analysis of KCF and the properties of the 2,4-dinitrophenylhydrazone derivative indicated that KCF was hexadecanal. They did this work with an astonishing skill and precision in a very short period.

In addition to hexadecanal, Cormier and Strehler (1953) discovered that homologous aldehydes, such as decanal and dodecanal, were also active in stimulating bacterial luminescence. Thus, they showed that bacterial luminescence requires a saturated long-chain aldehyde, but the specific aldehyde that is actually involved in the *in vivo* luminescence remained unknown for the next 20 years.

In 1955, McElroy and Green conclusively showed that bacterial luciferase catalyzes a light emitting reaction in the presence of $FMNH_2$, a long-chain fatty aldehyde, and oxygen. The aldehyde is consumed in the luminescence reaction. Therefore, the total light emitted is directly dependent on the amount of the aldehyde spent. Thus, the aldehyde is considered to be a luciferin. The general schemes of the luminescence reaction presented by Cormier and Totter (1957) and Hastings and Gibson (1963) are shown below. The first scheme shows the *in situ* formation of extremely unstable $FMNH_2$ from FMN, by reduction with NADH in the presence of FMN-reductase, and the second scheme represents the light-emitting reaction.

$$FMN + NADH + H^+ \xrightarrow{\text{FMN-reductase}} FMNH_2 + NAD^+$$

$$FMNH_2 + RCHO + O_2 \xrightarrow{\text{Luciferase}} FMN + RCOOH + H_2O + Light$$

Since then, tremendous efforts have been made to elucidate the mechanism of this complex, multi-component luminescence reaction (reviews: Hastings and Nealson, 1977; Ziegler and Baldwin, 1981; Hastings *et al.*, 1985; Baldwin and Ziegler, 1992; Meighen and Dunlap, 1993; Tu and Mager, 1995). The publications relating to bacterial bioluminescence reaction are so numerous that it is not possible to go into detail in this book. Only a brief outline is given in the following sections.

2.2 Bacterial Luciferase

Cultivation of luminous bacteria. Nealson (1978) lists various culture media to culture luminous bacteria. Three examples from other sources are shown in Table 2.1. It is important to include 300–500 mM NaCl as a basic ingredient. For the growth of bacteria, liquid media must be adequately aerated by shaking or bubbling. Solid media containing agar are made in Petri dishes.

Table 2.1 Examples of Culture Media for Growing Luminous Bacteria

	Seawater Complete[a]	NaCl Complete[a]	Solid Medium[b]
Dist. water	200 ml	1,000 ml	1,000 ml
Sea water	800 ml		
NaCl		30 g	30 g
$Na_2HPO_4 \cdot 7H_2O$		7 g	
KH_2PO_4		1 g	
$(NH_4)_2HPO_4$		0.5 g	
$MgSO_4$		0.1 g	
$CaSO_4$			5 g
Glycerol	2 ml	2 ml	10 ml
Tryptone	5 g	5 g	
Yeast extract	3 g	3 g	
Agar	(10–20 g)	(10–20 g)	
Bacto-nutrient agar (Difco)			30 g
Temperature	30–34°C for *B. harveyi*	26°C for *P. fischeri* 18°C for *P. phosphoreum*	25°C

[a]From Hastings *et al.*, 1978; add 10–20 g agar to make solid medium.
[b]From Shimomura *et al.*, 1974a.

Extraction and purification. Bacterial luciferases have been extracted and purified from various species of luminous bacteria. The purification methods used include DEAE anion-exchange chromatography and ammonium sulfate fractionation for *B. harveyi*, *P. phosphoreum* and *P. fischeri* (Gunsalus-Miguel *et al.*, 1972; Hastings *et al.*, 1978); affinity chromatography for *B. harveyi* and *V. fischeri* (Holzman and Baldwin, 1982; Baldwin *et al.*, 1986); and high-performance liquid chromatography (O'Kane *et al.*, 1986).

Properties of luciferases. Bacterial luciferases isolated from all species have a heterodimer structure with a molecular weight of 76,000 ± 4,000 (Friedland and Hastings, 1967; Hastings *et al.*, 1969a). The heterodimer consists of two nonidentical subunits, α and β, which can be separated by chromatography in urea-containing buffers (Tu, 1978). The subunits α and β have molecular weights of 40,000–42,000 and 37,000–39,000, and are encoded by the *luxA* and *luxB* genes, respectively. The two genes from various species of luminous bacteria have been cloned and sequenced (Cohn *et al.*, 1985; Johnston *et al.*, 1986; Haygood and Cohn, 1986; Miyamoto *et al.*, 1986; Meighen, 1991; Tu and Mager, 1995). The individual subunits are inactive when separated, but luciferase activity is regained by the recombination of the subunits when urea is removed. Although the β subunit is required for luciferase activity, the catalytic center probably resides on the α subunit (Hastings, 1978). Bacterial luciferase from *V. harveyi* was crystallized (Swanson *et al.*, 1985) and its crystal structure was solved (Fisher *et al.*, 1996).

An approximate concentration of luciferase can be estimated from the absorbance at 280 nm using a specific absorption coefficient ($A_{0.1\%,1\,cm}$) of about 1.0 (Hastings *et al.*, 1978). The luciferases from *P. phosphoreum* and *P. fischeri* are inactivated at pH values below 6.0 and above 8.5 (above 9.5 for *Beneckea harveyi*), and at a temperature above 30–35°C (Nakamura and Matsuda, 1971; Hastings *et al.*, 1978). The chemical modification of the sulfhydryl group (cysteine) or imidazol group (histidine) causes inactivation of luciferase (Hastings and Nealson, 1977; Cousineau and Meighen, 1976, 1977). Multivalent anions (phosphate, sulfate, pyrophosphate, citrate, etc.) at concentrations of 0.1 M or higher significantly protect bacterial luciferase from inactivation by heat, urea, and proteases (Hastings *et al.*, 1978). For storage, luciferase should be precipitated with ammonium sulfate, and the precipitate kept in a freezer; otherwise, luciferase should be refrigerated in a phosphate buffer, pH 7.0, containing 0.5 mM DTT and 1 mM EDTA (Hastings *et al.*, 1978).

2.3   Long-chain Aldehyde

The requirement for a long-chain aldehyde in bacterial bioluminescence was discovered in 1953 by Cormier and Strehler, as already noted. Because of the dependency of the total light on the amount of hexadecanal and decanal used (Strehler and Cormier, 1954), it was postulated that the aldehyde is oxidized into the corresponding acid in the bacterial luminescence reaction (Cormier and Totter, 1957; Hastings and Gibson, 1963). However, the experimental proof of the oxidation reaction was difficult to obtain due to the very small amount of aldehyde existing in luminous bacteria. Fifteen years after the postulation by Cormier and Totter, the oxidation of an aldehyde into the corresponding acid in the luminescence reaction was experimentally demonstrated by means of mass spectrometry using tetradecanal (Shimomura *et al.*, 1972). The result was quickly confirmed by two independent groups using ^{3}H-labeled decanal and ^{14}C-labeled decanal (McCapra and Hysert, 1973; Dunn *et al.*, 1973).

To identify the specific aldehyde that is actually involved in the light-emitting reaction of living luminous bacteria, Shimomura *et al.* (1974a) extracted and purified the aldehyde from 40 g each of the bacterial cells of *P. phosphoreum*, *Achromobacter* (*Vibrio* or *Photobacterium*) *fischeri*, and an "aldehydeless" mutant of *A. fischeri*. The aldehyde fractions were purified, and then oxidized with Tollens' reagent (silver oxide dissolved in ammonia) to convert the CHO group into the COOH group. Then the acids obtained were analyzed by mass spectrometry. The results indicated that *P. phosphoreum* had contained a mixture of aldehydes: dodecanal (5%), tetradecanal (63%) and hexadecanal (30%), as shown in Table 2.2. Thus, tetradecanal was clearly predominant in

Table 2.2   Amounts of Saturated Fatty Aldehyde Extracted from 40 g of Luminous Bacteria (Shimomura *et al.*, 1974a)

	P. phosphoreum	A. fischeri	A. fischeri "Aldehydeless" Dark Mutant
Luminescence intensity of bacterial suspension, 1 mg/ml, at 20°C (quanta/s per ml)	6×10^{12}	1.1×10^{12}	1×10^8
Total amount of fatty aldehydes (nmol)	600	90	7
Fatty aldehyde (nmol) of			
10-carbon	< 1	< 1	< 1
11-carbon	< 1	< 1	< 1
12-carbon	30 (5%)	32 (36%)	1.5 (22%)
13-carbon	6 (1%)	2 (2%)	< 1
14-carbon	380 (63%)	29 (32%)	0.9 (13%)
15-carbon	6 (1%)	6 (7%)	0.5 (7%)
16-carbon	180 (30%)	18 (20%)	3.8 (54%)
17-carbon	< 1	2 (2%)	0.14 (2%)
18-carbon	< 1	< 1	0.07 (1%)

P. phosphoreum. In the case of *A. fischeri*, the total amount of aldehydes was only 15% of that from *P. phosphoreum*, which consisted of dodecanal (36%), tetradecanal (32%) and hexadecanal (20%). The contents of aldehydes having the carbon atoms of 10, 11, 13, 15, 17 and 18 were negligibly small in both bacterial species.

According to the data in Table 2.2, 1 g of *P. phosphoreum* cells continuously emit 6×10^{15} photons/s at a cell concentration of 1 g/liter at 20°C, whereas the amount of total aldehydes obtained from the same cells (1 g) is 15 nmol. This amount of aldehydes is capable of emitting a total light of 1.44×10^{15} photons, assuming the quantum yield of the aldehydes (see Section 2.6) to be 0.16. Considering that the light emission from luminous bacteria is continuous and in a steady-state, and that the overall yield of the aldehydes is probably 50–70% , these figures indicate that bacterial cells contain an amount of aldehydes that can sustain the luminescence for only about 0.3–0.5 s, suggesting that the long-chain aldehydes are continuously synthesized in the cells to compensate for the consumption of aldehydes by luminescence.

The "aldehydeless" dark mutants of luminous bacteria emit light when a long-chain aldehyde is added. In the dark mutants of one class of B. *harveyi*, luminescence is also stimulated by tetradecanoic acid with a high specificity (Ulitzur and Hastings, 1978, 1979a). In such mutants, the amount of light obtained by the addition of a small amount of either tetradecanal or tetradecanoic acid may be increased 60-fold by cyanide and other agents that block respiration. These data may suggest that the fatty acid product of the luminesce reaction is recycled and that tetradecanal is the natural aldehyde involved in the light emitting reaction of luminous bacteria. The biosynthetic mechanism involved in the reduction of tetradecanoic acid into tetradecanal has been investigated in the V. *fischeri* bacterial cells (Boylan et al., 1989).

2.4   Mechanism of Luminescence Reaction

The biochemical mechanism of bacterial luminescence has been studied in detail and reviewed by several authors (Hastings and Nealson, 1977; Ziegler and Baldwin, 1981; Lee et al., 1991; Baldwin and Ziegler, 1992; Tu and Mager, 1995). Bacterial luciferase catalyzes the oxidation of a long-chain aldehyde and $FMNH_2$ with molecular oxygen, thus the enzyme can be viewed as a mixed function oxidase. The main steps of the luciferase-catalyzed luminescence are shown in Fig. 2.1. Many details of this scheme have been experimentally confirmed.

$FMNH_2$ is produced in bacterial cells from FMN, by reduction with NADH and FMN-reductase (Duane and Hastings, 1975; Puget and Michelson, 1972; Watanabe et al., 1975). The free form of $FMNH_2$ is extremely unstable and instantly oxidized in solutions containing oxygen. In the presence of luciferase, however, $FMNH_2$ is deprotonated at its N1 and bound to luciferase, forming an $FMNH_2$-luciferase complex that is more stable than free $FMNH_2$ (Vervoort et al., 1986). The dissociation constant (K_d) of $FMNH_2$-luciferase complex is reported to be in the range of 0.1–2.0 μM (Meighen and Hastings, 1971; Watanabe et al., 1976; Meighen and Bartlet, 1980).

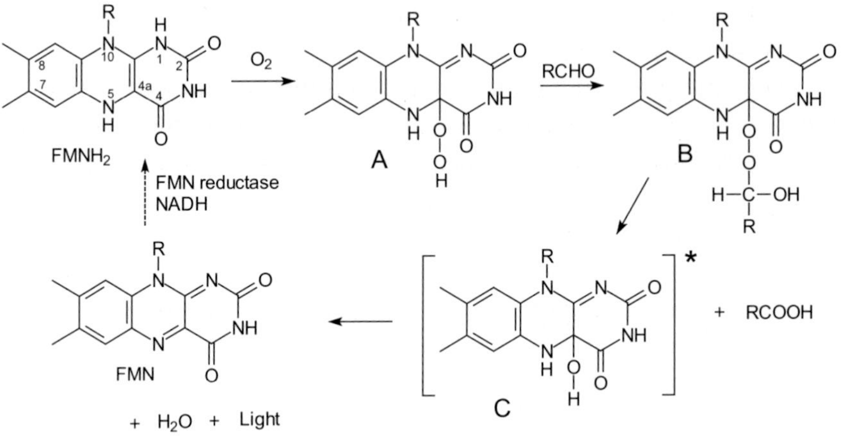

Fig. 2.1 Mechanism of the bacterial bioluminescence reaction. The molecule of FMNH$_2$ is deprotonated at N1 when bound to a luciferase molecule, which is then readily peroxidized at C4a to form Intermediate **A**. Intermediate **A** reacts with a fatty aldehyde (such as dodecanal and tetradecanal) to form Intermediate **B**. Intermediate **B** decomposes and yields the excited state of 4a-hydroxyflavin (Intermediate **C**) and a fatty acid. Light (λ_{max} 490 nm) is emitted when the excited state of **C** falls to the ground state. The ground state **C** decomposes into FMN plus H$_2$O. All the intermediates (**A**, **B**, and **C**) are luciferase-bound forms. The FMN formed can be reduced to FMNH$_2$ in the presence of FMN reductase and NADH.

The deprotonated flavin in the complex is readily attacked by molecular oxygen at C4a, giving 4a-hydroperoxide of the flavin-luciferase complex (intermediate A). This complex is an unusually stable intermediate, with a lifetime of tens of seconds at 20°C and hours at subzero temperatures, allowing its isolation and characterization (Hastings *et al.*, 1973; Tu, 1979; Balny and Hastings, 1975; Vervoort *et al.*, 1986; Kurfuerst *et al.*, 1987; Lee *et al.*, 1988).

In the presence of a long-chain fatty aldehyde, intermediate A (Fig. 2.1) is converted into intermediate B that contains a peroxyhemiacetal of flavin (Macheroux *et al.*, 1993). The apparent K_m value of the intermediate B produced with an aldehyde of 10–13 carbon atoms is approximately 200 μM when *P. phosphoreum* luciferase is used (Watanabe and Nakamura, 1972), and 1–10 μM when *P. fischeri* luciferase is used (Spudich and Hastings, 1963; Hastings and Nealson, 1977). The decomposition of intermediate B, through several steps, yields the excited state of 4a-hydroxyflavin-luciferase complex (intermediate C) and a fatty acid. The mechanism of the formation of the excited state possibly involves the chemically initiated electron exchange luminescence (CIEEL) mechanism (Schuster, 1979). The light emitter is considered to be luciferase-bound 4a-hydroxyflavin (intermediate C) (Ghisla *et al.*, 1977; Kurfurst *et al.*, 1987; Lei *et al.*, 2004).

2.5   Assay of Luciferase Activity (Hastings *et al.*, 1978; Baldwin *et al.*, 1986)

The activity of bacterial luciferase is measured by recording the light intensity when 1 ml of $FMNH_2$ solution is rapidly injected into 1 ml of 50 mM phosphate buffer, pH 7.0, containing 0.1–0.2% BSA, 0.1 mM long-chain aldehyde (decanal, dodecanal or tetradecanal) and a luciferase sample, at 20–25°C (Fig. 2.2). The peak light intensity is proportional to the amount of luciferase at a wide range of enzyme concentrations. $FMNH_2$ is extremely unstable under aerobic

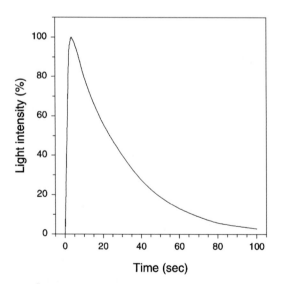

Fig. 2.2   Assay of luciferase by the injection of $FMNH_2$. The assay was initiated by the injection of 1 ml of 50 μM $FMNH_2$ solution into 1 ml of air equilibrated buffer, pH 7.0, containing 0.1% BSA, luciferase, and 20 μl of 0.01% sonicated suspension of dodecanal. From Baldwin *et al.*, 1986, with permission from Elsevier.

conditions; thus, any excess will be autoxidized within one second, preventing the enzymatic turnover of luciferase. The luminescence reaction initiated by the injection of $FMNH_2$ solution is terminated in 10 to 100 seconds depending on the chain length of aldehyde and the species of bacteria from which luciferase has been obtained (Hastings *et al.*, 1969a).

If the luciferase sample solution contains a flavin-reductase, luciferase activity can be measured by the addition of FMN and NADH, instead of $FMNH_2$. In this case, the turnover of luciferase takes place repeatedly using the $FMNH_2$ that is enzymatically generated; thus, the luminescence reaction continues until aldehyde or NADH is exhausted. A crude luciferase extracted from luminous bacteria usually contains a flavin-reductase.

$FMNH_2$ can be prepared by the following three methods (Hastings *et al.*, 1978), listed in order of preference:

(1) **Catalytic hydrogenation.** Aqueous solution of FMN ($50\,\mu M$) containing a small amount of platinized asbestos is babbled with hydrogen gas for 10–20 minutes, until the yellow color and green fluorescence of FMN completely disappear. A suitable setup for this purpose can easily be constructed using a small flask, a rubber septum and two syringe needles (used as the inlet and outlet of H_2 gas).

(2) **Photochemical reduction.** A deoxygenated aqueous solution of FMN ($50\,\mu M$) containing $2\,mM$ EDTA is irradiated with a long-wave UV light immediately before use (Nickerson and Strauss, 1960; Strauss and Nickerson, 1961). The reduction of FMN is accompanied by the formation of H_2O_2, which might be undesirable in some experiments.

(3) **Reduction by sodium dithionite.** A small amount of sodium dithionite, solid or in solution, is added to a luciferase solution made with $50\,mM$ phosphate, pH 7.0, containing $50\,\mu M$ FMN. The amount of dithionite used should be minimal but sufficient to remove oxygen in the solution and to fully reduce the flavin. The solution made is injected into an air-equilibrated buffer solution containing a long-chain aldehyde and luciferase to initiate the luminescence reaction. With this method, the reaction mixture will be contaminated by bisulfite and bisulfate ions derived from dithionite.

2.6   Quantum Yield of Long-chain Aldehydes

The quantum yields of the long-chain aldehydes reported in the *in vitro* bacterial luminescence vary widely (0.1–0.3; Hastings and Nealson, 1977). The values are apparently influenced by the calibration of the photomultipliers used. The reference light sources used for the calibration were: (I) luminol chemiluminescence reaction (Lee and Seliger, 1965, 1972; Lee *et al.*, 1966); (II) radioactive luminescence standard solution containing ^{14}C-labeled hexadecane and scintillators (Hastings and Weber, 1963; Hastings and Reynolds, 1966); and (III) the bioluminescence reaction of *Cypridina* luciferin (Johnson *et al.*, 1962). A comparison of the three standard light sources was done at the author's laboratory, using an identical luminescence sample and the same photomultiplier (Hamamatsu R136 and R928). The ratio of the quantum yields of the sample obtained with I, II and III was 1.0:2.8:1.06 (Shimomura and Johnson, 1967), showing a large deviation of the standard II. Assuming that the standards I and III are nearly correct, the quantum yield of firefly luciferin in the firefly bioluminescence reaction measured with the standard II would become more than 2, an unrealistic value. Thus, the standard II would have needed a recalibration. It should be pointed out, however, that the radioactive standard solution is highly convenient in monitoring the daily sensitivity fluctuations of light measurement apparatus.

The reported quantum yields of the long-chain aldehydes in the luminescence reaction catalyzed by *P. fischeri* luciferase are: 0.1 for dodecanal with the standard I (Lee, 1972); 0.13 for decanal with the standard I (McCapra and Hysert, 1973); and 0.15–0.16 for decanal, dodecanal and tetradecanal with the standard III (Shimomura *et al.*, 1972). Thus, the quantum yield of long-chain aldehydes in the bacterial bioluminescence reaction appears to be in the range of 0.10–0.16.

2.7   *In vivo* Luminescence of Luminous Bacteria

Overview. In the cultures of luminous bacteria, the bacterial cells are not luminous in their early stages of propagation. The formation of bioluminescence system is controlled by a substance called "autoinducer" that is produced by the cells of luminous bacteria.

The autoinducer is a low molecular weight compound that is easily leached from the cells into the culture medium. By the propagation of bacterial cells, the concentration of the autoinducer in the medium increases. When the concentration reaches a certain threshold, the biosynthesis of bioluminescence system begins, and the bacteria become luminescent. The process is also called quorum sensing (Fuqua *et al.*, 1994).

In the living cells of luminous bacteria, $FMNH_2$ is produced by the reduction of FMN with NADH catalyzed by FMN-reductase. This process is, in effect, the recycling of FMN. In the cells, a long-chain aldehyde is produced by the reduction of the corresponding long-chain acid, which is also a recycling process.

Another notable feature of the *in vivo* bacterial luminescence is seen in their emission spectra. Although the emission peak of *in vitro* bacterial luminescence is normally at about 490 nm, the *in vivo* emission peaks of various bacterial species and strains are significantly shifted from 490 nm, ranging from the shortest wavelength of 472 nm to over 500 nm. Some expanded notes concerning *in vivo* bacterial luminescence are given below.

Autoinduction (quorum sensing). Free-living luminous bacteria are commonly found in seawater near the shore, but their populations are very low, i.e. 1–5 bacteria per milliliter (Ruby and Nealson, 1978), and they are not luminous. As already mentioned, the synthesis of the components needed for light emission is controlled by a substance referred to as "autoinducer," a freely diffusible compound produced by luminous bacteria (Nealson *et al.*, 1970). In seawater, an autoinducer produced by the bacteria is diluted with a vast amount of seawater; thus, its concentration is infinitely low. When the bacteria are cultured in a culture medium, the autoinducer cannot leak out from the medium and will accumulate in the medium as the bacteria grow. The biosynthesis of the factors necessary for bioluminescence, including luciferase, begins when the concentration of the autoinducer in the medium reaches a certain threshold concentration (Eberhard, 1972). Long-chain acid reductase, which is needed for the recycling of long-chain aldehyde, is also co-induced by the autoinducer

(Ulitzer and Hastings, 1979; Riendeau and Meighen, 1980). The chemical structure of the autoinducer of *P. (Vibrio) fischeri* was determined by Eberhard *et al.* (1981) to be an acylhomoserine lactone (1), shown below. The autoinducer of *Vibrio harveyi* was found to be another acylhomoserine lactone (2) by Cao and Meighen (1989).

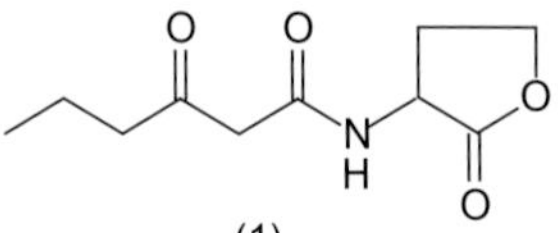

(1) 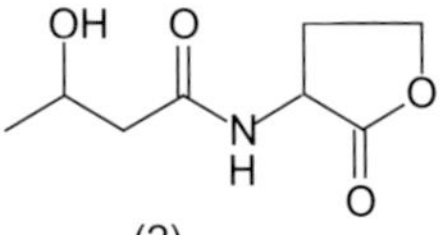   (2)

Formation of dark mutants. Luminous bacteria are easily transformed into dark mutants (or variants) that exhibit negligibly low levels of luminescence (Beijerinck, 1889, 1916; Keynan and Hastings, 1961; Nealson and Hastings, 1979). The transformation is commonly observed when bacteria are cultured at temperatures higher than the optimal. The mutants obtained in this manner contain very low levels of luciferase and, consequently, they emit very dim luminescence. Luminous bacteria can also be mutated by ultraviolet light radiation (Rogers and McElroy, 1955). The mutants obtained in this way are deficient in long-chain aldehydes, but without significant change in the luciferase content.

Energy transfer to fluorescent proteins. There are marked differences among the various bacterial species and strains in terms of the *in vivo* luminescence spectra. The emission maxima are spread mostly in a range from 472 to 505 nm (Seliger and Morton, 1968), but one of the strains, *P. fischeri* Y-1, shows a maximum at 545 nm (Ruby and Nealson, 1977), as shown in Fig. 2.3. However, the *in vitro* luminescence spectra measured with purified luciferases obtained from the various bacterial species and strains are all similar (λ_{max} about 490 nm). The variation in the *in vivo* luminescence spectra may be due to the occurrence of an intermolecular energy transfer that increases the efficiency of light emission.

Intermolecular energy transfer may take place from the excited state of 4a-hydroxyflavin-luciferase complex to a fluorescent protein,

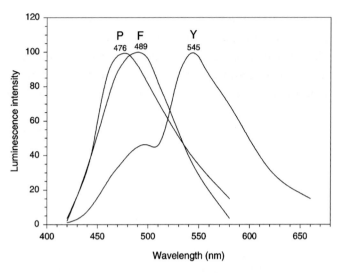

Fig. 2.3   Luminescence spectra of the living cells of luminous bacteria. F, *P. fischeri*; P, *P. phosphoreum* (Eley *et al.*, 1970); Y, *P. fischeri*, strain Y-1 (Ruby and Nealson, 1977). Reproduced with permission from the American Chemical Society and AAAS.

probably by the Förster type resonance energy transfer mechanism (Brand and Witholt, 1967; Lamola, 1969). This process requires a spectral overlap between the emission spectrum of the donor and the absorption spectrum of the acceptor (a fluorescent compound) as one of the conditions. The *in vivo* bacterial bioluminescence might always involve the process of energy transfer at least to some extent, even when the *in vivo* luminescence spectrum coincides with the *in vitro* luminescence spectrum emitted from the 4a-hydroxyflavin-luciferase complex (λ_{max} 490 nm); this is a highly likely possibility based on the information given below.

Two types of fluorescent proteins have been isolated from luminous bacteria and studied in detail. The first of them are the blue fluorescent lumazine proteins (LumPs) containing lumazine as their chromophores, which were isolated from *P. phosphoreum* and *P. fischeri* (Gast and Lee, 1978; Koda and Lee, 1979; O'Kane *et al.*, 1985). The second are the yellow fluorescent proteins (YFPs) containing a chromophore of FMN or riboflavin, isolated from *P. fischeri* strain Y-1 (Daubner *et al.*, 1987; Macheroux *et al.*, 1987;

Petushkov *et al.*, 1995). Both types of proteins have molecular weights of about 20,000 (Small *et al.*, 1980). When a lumazine protein is included in the bacterial *in vitro* bioluminescence reaction, the normal 490 nm luminescence maximum is shifted to 476 nm, accompanied by a significant increase in the light output (Gast and Lee, 1978; O'Kane and Lee, 1986). On the other hand, the blue-green bioluminescence emitted with the purified luciferase from *P. fischeri* strain Y-1 (λ_{max} 484 nm) is shifted to yellow (λ_{max} 534 nm) on addition of a YFP obtained from the same strain (Daubner *et al.*, 1987; Macheroux *et al.*, 1987). These spectral shifts of light emission clearly show the occurrence of energy transfer.

Lumazine   FMN   R = PO(OH)$_2$
                  Riboflavin   R = H

The absorption maxima of the recombinant LumP-type proteins from *P. leiognathi* that contain the prosthetic groups of lumazine, FMN, and riboflavin are at 420 nm, 458 nm and 463 nm, respectively. In these highly fluorescent proteins, the prosthetic groups are tightly bound to the apoprotein; their dissociation constants are: 0.26 nM with lumazine; 30 nM with FMN; and 0.53 nM with riboflavin (Petushkov *et al.*, 1995). In the case of YFP, the dissociation constant of the FMN-protein binding is reported to be 0.4 nM (Visser *et al.*, 1997). When these fluorescent proteins are bound with the luciferase complex containing the excited state of 4a-hydroxyflavin (intermediate C; emission maximum 490 nm), a resonance energy transfer takes place from the flavin to the fluorescent protein, resulting in the appearance of new luminescence emission peaks, at about

475 nm in the presence of LumP containing lumazineis and 540 nm in the presence of LumPs containing FMN or riboflavin.

The spectral shift in the latter case, from 490 nm to 540 nm, is readily understandable because of a good overlap between the emission spectrum of the intermediate C and the absorption spectrum of LumP-flavin (see Fig. 2.4). However, the former case of energy transfer, from 490 nm to 475 nm (Lee, 1993), has an apparent difficulty due to the energetics, because such a change requires an increase in the energy level. According to Tu and Mager (1995), however, there is a significant spectral overlap between the emission spectrum of C and the absorption spectrum of LumP for an occurrence of the resonance energy tansfer. The same authors also noted the possibility of an energy transfer from an intermediate luciferase complex, which precedes C and has an energy level higher than C, to the fluorescent protein.

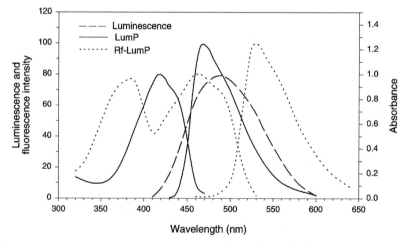

Fig. 2.4   The spectrum of bacterial luminescence measured with *B. harveyi* luciferase, FMN, tetradecanal and NADH, in 50 mM phosphate buffer, pH 7.0, at 0°C (dashed line; from Matheson *et al.*, 1981); and the absorption and fluorescence emission spectra of LumP (solid lines) and Rf-LumP (dotted lines) obtained from *P. leiognathi*, in 25 mM phosphate buffer, pH 7.0, containing 1 mM EDTA and 10 mM 2-mercaptoethanol, at room temperature (from Petushkov *et al.*, 2000, with permission from Elsevier). LumP is a lumazine protein, and Rf-LumP contains riboflavin instead of lumazine in the lumazine protein. Fluorescence emission curves are at the right side of the absorption curves.

3

THE OSTRACOD *CYPRIDINA (VARGULA)* AND OTHER LUMINOUS CRUSTACEANS

An overview of Crustacea. The class Crustacea belongs to the phylum Arthropoda together with the classes Insecta (such as the fireflies), Diplopoda (millipedes), and Chilopoda (centipedes). It is a very large class and embraces several important groups of bioluminescent organisms, such as ostracods, copepods, amphipods, euphausiids, and decapods (see Herring, 1985). The ostracods are identified by two hinged valves covering the small body, and include the genera *Cypridina* and *Pyrocypris*. These organisms eject luminous liquid into the seawater from the labral glands. Study on the bioluminescence of *Cypridina hilgendorfii* was started by E. N. Harvey about 90 years ago, and it is now one of the best studied fields in bioluminescence, as detailed in this chapter. The biochemistry of the *Pyrocypris* bioluminescence has been little studied, probably due to its resemblance to *Cypridina* bioluminescence, and also for the relative difficulty in obtaining a large quantity of specimens.

The subclass Copepoda contains at least 10 bioluminescent genera, such as *Metridia*, *Pleuromamma*, *Oncaea* and *Gaussia* (Harvey, 1952; Herring, 1978a,b). They are all very small planktonic forms and mostly free-swimming. Only a few studies have been made on the biochemistry of copepod bioluminescence.

The order Euphausiacea contains more than 80 species in 11 genera, and all but one are considered luminous (Harvey, 1952). The euphausiids are shrimp-like, but differ from the decapod shrimp in that the thoracic appendages are all biramose, i.e. their feet are split into two parts. They are pelagic and found throughout the oceans of the world, occurring in great swarms that come to the surface at night, and migrate into deep water in the daytime. The euphausiids are also known as krill; the word "krill" comes from the Norwegian meaning "young fish." The genera *Euphausia*, *Meganyctiphanes* and *Thysanoessa* are well known. They have complex photophores (light organs), and do not produce a secretory luminescence. In chemical aspects, euphausiids and dinoflagellates utilize tetrapyrrole compounds as luciferins, making a clear distinction from many other luminous crustaceans that utilize imidazopyrazinone compounds as luciferins.

The order Decapoda contains the shrimp, crayfish, lobsters and crabs, but bioluminescence is found only in pelagic shrimps. There are many genera and species of luminous decapods. Their first three pairs of thoracic appendages are modified as maxillipeds (feeding appendages) and the remaining five pairs are legs, from which the name Decapoda is derived. Some species possess photophores to emit light (such as *Sergestes*), some have organs to secrete a luminous liquid (such as *Heterocarpus*), and some possess both types of organs (such as *Oplophorus*).

Shrimps of the genus *Sergestes* normally emit light from their photophores. However, *S. lucens*, a species abundantly harvested in Japan as a delicacy "Sakura-ebi," has never been observed to emit light despite its specific name and the clear presence of photophores (Haneda, 1985). This species (body weight 0.4 g) undoubtedly contains some coelenterazine, but shows very little luciferase activity, which seems to be insufficient as evidence for the existence of a luciferase (Shimomura *et al.*, 1980; see the note at the end of Section 5.2 for the reason). Thus, the luminescence of *S. lucens* might involve a coelenterazine luminescence system different from an ordinary luciferin-luciferase system if this organism is really bioluminescent. According to Herring (1983), the luminescence of the *Sergestes* species can be stimulated by H_2O_2, an interesting observation in connection with the poor luciferase activity of *S. lucens*.

3.1   The Ostracod *Cypridina*

3.1.1   *Overview of Ostracoda*

Ostracods are tiny crustaceans characterized by a dorsally hinged bivalve carapace that covers the body. Ostracoda includes well-known bioluminescent genera such as *Cypridina* and *Pyrocypris* in the family Cypridinidae, and also the genus *Conchoecia* in the family Halocypridae. The ostracods of the former family use *Cypridina* luciferin in their light emission, whereas *Conchoecia* of the latter family emits light utilizing coelenterazine (Campbell and Herring, 1990; Oba *et al.*, 2004). The genus *Cypridina* contains 20–25 species (Harvey, 1952) which are widely distributed in various seas; for examples, *C. norvegica* (North Atlantic), *C. antarctica* (Antarctic), *C. spinosa* (South Korea to Japan), *C. hilgendorfii* (Japan and Southeast Asia), *C. bullae* (West Indies), and *C. (Vargula) tsujii* (southern California; Kornicker and Baker, 1977). The genus *Pyrocypris* is found only in the tropical or near tropical regions of the Indian and Pacific seas, and includes *P. acuminala*, *P. sinuosa*, *P. dentata* and *P. serrata*. The deep-sea ostracods of the genus *Gigantocypris* are large, up to 2 cm in length, but luminescence has not been reported for this genus. According to Harvey (1957), Godeheu de Riville (1760) was the first person who described the luminescence of an ostracod, possibly of the genus *Pyrocypris*, while Müller (1890) was the first to give a clear description of the light organ of ostracod. The best-known and most important luminous ostracod is *Cypridina hilgendorfii* Müller, named after Dr. Hilgendorff who collected the specimens in Japan and sent them to the Berlin Museum for identification. The ostracod *C. hilgendorfii* has been instrumental in the biochemical study of bioluminescence in the 20th century.

Generic names. There has been a shuffling in the generic names of ostracods, which caused considerable confusion in the field of bioluminescence. According to a detailed account of the classification of the ostracods by Poulsen (1962), Skogsberg (1920) divided Müller's genus *Cypridina* (Müller, 1912) into five subgenera, *Doloria*, *Vargula*, *Macrocypridina*, *Cypridina* and *Siphonostra*, placing virtually all the luminous species into the subgenus *Vargula*. At the same time he deleted Müller's genus *Pyrocypris* (Müller, 1890), replacing it with *Cypridina* (sub-genus), because he judged "*Cypridina Reynaudi* Milne Edwards 1840" as having priority over Müller's genus *Pyrocypris* (Müller, 1890), despite the fact that the descriptions given for *C. Reynaudi* were incomplete (Poulsen, 1962, p. 255). Nevertheless, Poulsen adopted most of Skogsberg's changes, and he changed the sub-genera into the genera. Therefore, according to Poulsen's classification, many bioluminescent species in Müller's *Cypridina* (such as *C. hilgendorfii* and *C. norvegica*) are in the genus *Vargula*, and Müller's genus *Pyrocypris* becomes genus *Cypridina*. Thus, the genus name *Cypridina* can indicates any of three different genera: *Cypridina* Müller, *Vargula* Skogsberg, and *Pyrocypris* Müller, depending on the classification chosen.

The Skogsberg–Poulsen classification was seldom adopted in the field of bioluminescence before 1980, and during that period a vast amount of important knowledge on bioluminescence was obtained from the studies on the ostracod *Cypridina hilgendorfii* Müller. However, the name *Vargula hilgendorfii* has become increasingly popular after a new ostracod discovered by F. I. Tsuji was named "*Vargula tsujii*" using the Poulsen classification (Kornicker and Baker, 1977).

In this book, the classification of ostracods by Müller is used; the genus names by the Poulsen classification are cited in parentheses only when appropriate. The author protests against the irresponsible shuffling of scientific names, and hopes for the revival of the Müller classification. In addition, the author believes that "*Cypridina* luciferin" is the proper name of this substance that is not necessarily bound to the genus name.

3.1.2   *Cypridina hilgendorfii Müller*

This species is abundant in shallow waters around Japan and Southeast Asia. The organism is egg-shaped and about 2 mm long (Fig. 3.1.1). An individual animal contains roughly 1 µg each of luciferin and luciferase. *C. hilgendorfii* lives in sandy bottom close to shore and comes out at night swimming around to feed on whatever is available. This ostracod is a voracious scavenger. If a large fish head tied to a string is thrown into the bottom of shallow water, the ostracods would nibble at the bait. After about an hour, when the bait is slowly pulled up to the surface, the tiny ostracods can be washed off from the bait. If the bait is left at the bottom too long, several hours or overnight, the ostracods would consume all the meat and other edible parts, leaving only the bone and hard skin. *Cypridina* has a large luminous gland consisting of gland cells of two kinds, one secreting luciferin and the other secreting luciferase. The animal squirts the luciferin and luciferase into the seawater in response to a predator or stimulation. Mixing of these two substances will give rise to a bright blue cloud of luminescence in the seawater, while the

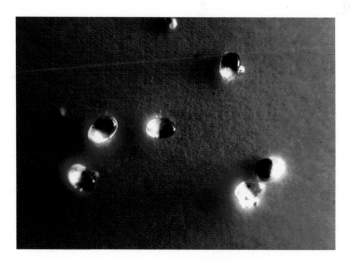

Fig. 3.1.1   The ostracod *Cypridina hilgendorfii* (photo by Dr. Toshio Goto).

ostracod swims away into the darkness of the surrounding area. The behavior suggests a defensive use of the luminescence.

History of dried *Cypridina*. Professor Newton Harvey of Princeton University observed the bioluminescence display of *C. hilgendorfii* during his first trip to Japan in 1916, on his honeymoon (Johnson, 1967). The luminescence from *Cypridina*'s tiny body was extremely brilliant. Harvey found out that dried *Cypridina* can emit a bright luminescence again whenever moistened. Because of its bright luminescence and its ease of preservation, he thought that this organism was ideally suited for the chemical study of bioluminescence. He made an arrangement to collect and dry the organisms. The dried material was shipped to Princeton. Thus, Harvey started his lifelong study of *Cypridina* bioluminescence with a large quantity of dried *Cypridina* sent from Japan. It should be noted that the dried *Cypridina* maintains its potency of luminescence almost permanently in the presence of a desiccant, although the drying process causes a large loss of luciferin due to oxidation when air-dried under the sun, as it was at the time. Thus, the quality of the dried material was indeed highly dependent on the weather at the time of drying.

Collection. During World War II, hundreds of gallons of the ostracod were collected and dried in Japan by the order of Japanese army for intended military use. The collection was done by volunteers and students in various parts of Japan, including Tateyama, 70 km south of Tokyo, where most of the ostracods for Harvey had been previously collected. A typical device used for the mass collection was a bowl containing fish bait which was enclosed by a cloth fastened to the edge of the bowl, with a small hole (~ 2 cm) at the center of the cloth. The bowl was sunk onto the sandy bottom for about one hour, and then pulled up to harvest the ostracods caught in the bowl. The techniques of collection and drying developed for Harvey were undoubtedly helpful in the military project. According to Dr. Yata Haneda, the military had planned to use the material as a source of low intensity light during the War in New Guinea and other places in the southern Pacific. One of the intended uses was to mark the backs of soldiers at night with the glowing substance, allowing soldiers to identify and follow one another silently through the dark jungles. It appears, however, that none of the material was actually utilized in the War, since some of the material was sunk with the transport ships by US submarines and the rest became quickly useless in the high humidity of the tropical climate, probably due to poor desiccation. Harvey died in 1959, leaving a large amount of dried *Cypridina* at his laboratory in Princeton. Some bottles of the *Cypridina* were transferred to my hands through Dr. Aurin Chase when I was in Princeton. The material is still at my laboratory and shows a good luminescence when moistened with water even after more than half a century of storage at room temperature.

Cultivation. A number of investigators in Japan cultivated *C. hilgendorfii* in their laboratories to use live animals in their research (e.g. Abe, 1994). The organism can be easily raised in a small glass aquarium for a period of several years that involves many generation changes. The aquarium should have a 2 cm layer of sand at the bottom and a simple aeration device; no temperature control is needed in a laboratory at room temperature. Both natural seawater and artificial seawater can be used; a water filtration device is desirable but not essential if the water is changed occasionally. The ostracod seems to feed on almost any animal tissues. Abe recommends commercial bait worms, such as lugworm, as well as any earthworms. He also notes that hog liver is the favorite of *Cypridina* although it tends to taint water. Dr. Y. Haneda chose food pellets for tropical fish to feed the ostracods. Whatever is chosen, it is important that the animals are not overfed.

3.1.3 Research on Cypridina Luminescence before 1955

During the period between 1916 and the early 1950s, Harvey, his associates and students made great efforts to study various aspects of the *Cypridina* bioluminescence, which was described by Harvey in his book *Bioluminescence* (1952). It should be pointed out here that: (1) the Princeton researchers had started from practically nothing, without any prior knowledge on any luminescence reaction, except the requirement of oxygen in the luminescence reaction; and (2) the dried organisms used in their experiments contained greatly reduced amounts of luciferin compared with the live organisms (10% at best, probably much less). Despite these disadvantages, the progress they made was remarkable. Harvey discovered a luciferin and a luciferase in *C. hilgendorfii*, which produced a strong luminescence when mixed together in aqueous solution, and he also found that these two substances are highly specific to each other (Harvey, 1922, 1926); the luciferin did not produce light by the addition of any substance other than the luciferases of *C. hilgendorfii*, *C. norvegica* and *Pyrocypris*, and the luciferase did not luminesce when substances other than the luciferin were added.

The purification of *Cypridina* luciferin was a formidable task in the Harvey era due to the extremely unstable nature of the luciferin; it is rapidly oxidized by air in a short period of time, in a matter of minutes if not seconds, especially when the material is impure. Despite the difficulty, Anderson (1935) made a remarkable breakthrough in the purification of luciferin. He devised a purification method that can remove both water-soluble impurities and fat-soluble impurities in two steps. The method was based on the fact that luciferin, a highly water-soluble substance, can be converted into a stable, fat-soluble form by treatment with benzoyl chloride. The fat-soluble form of luciferin is washed with water to remove all water-soluble impurities. Then the fat-soluble form is converted back into the original water-soluble luciferin by treatment with 0.5 N HCl. Now, fat-soluble impurities are removed from the aqueous layer by washing with ether, resulting in an acidic aqueous solution of significantly purified luciferin. The method is ingenious, and also completely rational on the basis of the structure of luciferin found later. It is surprising that such a method could be designed without the knowledge of the chemical structure of luciferin. Upon further purification, they obtained samples of purified luciferin that were suitable for measuring the ultraviolet absorption spectrum (Chase and Brigham, 1951; Tsuji, 1955; Tsuji *et al.*, 1955).

3.1.4   *Purification and Crystallization of Cypridina Luciferin*

Original procedure (Shimomura *et al.*, 1957). In the middle of the 1950s, Professor Yoshimasa Hirata of Nagoya University, Japan, embarked on the chemical study of *Cypridina* luciferin. For the study of the chemical structure of a luciferin, the luciferin must be isolated, purified and crystallized — the crystallization was a requirement to prove the chemical purity at the time. Prof. Hirata considered the subject to be too precarious for his students working for a degree, and assigned the task of purifying *Cypridina* luciferin to this author, a visiting researcher at his laboratory at the time. After some trial experiments, I made up a plan to extract 500 g of dried *C. hilgendorfii* (about 2 kg before dried) in the strict absence of air, hoping to obtain 2–3 mg of purified material for crystallization. It was 10 times the amount used at Princeton, and I thought the plan should work if the compound were really crystallizable. Prof. Hirata approved the plan and his glass blower made an oversized soxhlet apparatus for the extraction of luciferin. The whole set-up used for the extraction is shown in Fig. 3.1.2.

Thus, 500 g of dried, ground ostracods were defatted with benzene (1.2 l) under reduced pressure for two days, followed by extraction with methanol (1.2 l) in the same apparatus for two days under reduced pressure in a pure hydrogen atmosphere, at a temperature below 40°C. The hydrogen gas used had been passed through a red-heated quartz tube containing copper fragments to remove the contaminating oxygen; nitrogen gas and argon were unsuitable due to the difficulty in removing a trace of contaminating oxygen. The soxhlet was evacuated and the mantle heater was adjusted; when all stopcocks were closed, extraction with steadily boiling methanol continued for many hours without any further adjustment.

The extracted crude luciferin was purified by Anderson's benzoylation method mentioned above (Anderson, 1935), followed by partition chromatography on a cellulose powder column using a mixture of ethyl acetate-ethanol-water (5:2:3; Tsuji, 1955) as the solvent, again in a hydrogen atmosphere. The luciferin obtained (3 mg) was highly pure by paper chromatography, and appeared suitable for crystallization. I

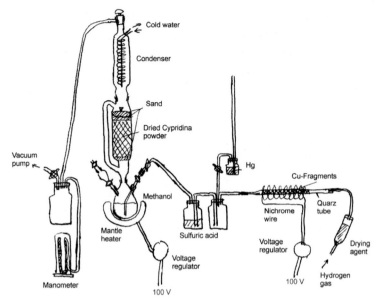

Fig. 3.1.2   The apparatus used in 1956 for the methanol extraction of *Cypridina* luciferin. The dried *Cypridina* (500 g) is extracted at a temperature lower than 40°C with refluxing methanol under reduced pressure for two days. The atmosphere inside the apparatus is completely replaced with hydrogen gas that was purified by its passing through a quartz tube containing red-heated copper fragments. The temperature of the mantle heater is adjusted, the system evacuated, and then all stopcocks are closed. The extraction with refluxing methanol continues for many hours without any further adjustment. From the author's 1957 notebook.

tried to crystallize luciferin with all possible combinations of solvents and salts that I could think of, but all my efforts ended up with the creation of amorphous precipitates, and any leftover luciferin became useless due to oxidation by the next day. The whole process of this experiment was day-and-night work for seven days, which was usually followed by 2–3 weeks of efforts to accumulate information on the chemical nature of luciferin using the oxidized materials. Then, I had to extract and purify a fresh batch of luciferin for my further efforts of crystallization. After I had used several batches of purified luciferin in unsuccessful crystallization, I finally saw the first crystals of *Cypridina* luciferin on one cold morning of February 1956; however, in an unexpected way.

On the previous night, I had some leftover purified luciferin after my crystallization attempts. I could not think of any further idea to improve crystallization, so I decided to use the material for amino acid analysis. Thus, I added an equal volume of concentrated hydrochloric acid to the luciferin solution. The color of the solution instantly changed from yellow to dark red. Because it was late at night, I went home without heating the sample. Next morning, I saw that the solution was discolored to light orange, an appearance that indicates the decomposition of luciferin by hydrolysis or oxidation. However, upon taking a closer look, I saw a small pinch of dark precipitate at the bottom of the test tube. Under a microscope, to my surprise, the precipitate was crystals — fine red needles. By paper chromatography, the crystals were confirmed to be active luciferin, not a decomposition product. That was 10 months after I began the crystallization experiments. The yield of crystalline luciferin was 2–3 mg from 500 g of dried ostracod (Shimomura *et al.*, 1957). The result indicated that the luciferin could be crystallized in a high concentration of hydrochloric acid, which I had not expected because luciferin had previously produced several amino acids upon hydrolysis, an indication of a peptide-like substance susceptible to acid hydrolysis.

Improved purification method of *Cypridina* luciferin. The purification of *Cypridina* luciferin became remarkably simple and easy after some of the properties of this substance were known. The following method was used to obtain the large quantity of luciferin needed for the study of its chemical structure. The method consists of three steps and takes less than eight hours to obtain crystallized luciferin.

(1) Extraction of luciferin with methanol from *Cypridina* frozen with dry ice. The chunks of frozen *Cypridina* are crushed into small pieces, and extracted with methanol with stirring. Small pieces of dry ice are added as needed to keep the temperature slightly below 0°C and also to prevent the oxidation of luciferin in a CO_2 atmosphere. Luciferin is easily extracted into methanol. The mixture is filtered on a Büchner funnel (with suction), protecting the luciferin in a CO_2 atmosphere by addition of small pieces of dry ice as needed.

(2) The filtrate is concentrated under reduced pressure with a rotary evaporator until most of the methanol is removed. Extract the aqueous residue with degassed butanol in a separatory funnel, of which the inside has been filled with argon in advance. Evaporate the butanol extract under reduced pressure until most of the water is removed.

(3) Alumina chromatography. A column of alumina is prepared using butanol, and the butanol solution of luciferin is added to the column. Luciferin (yellow) adsorbed at the top of the column is eluted stepwise with butanol-methanol mixtures containing increasing amounts of methanol; it is recommended to degas methanol and butanol before use, by slow bubbling of argon for 10 minutes. The eluted luciferin is evaporated to dryness under reduced pressure. This material can be easily crystallized (see below). The yield is 20–40 mg from 1 kg of frozen specimens (250 g dry weight).

3.1.5 Properties of Cypridina Luciferin

Crystals. *Cypridina* luciferin crystallizes easily when a strong acid is added to a concentrated solution of luciferin in methanol or water, and it seems to be the only practical way to crystallize this compound. The crystals formed by the addition of hydrochloric acid to a methanolic solution of luciferin are dark red needles (Shimomura *et al.*, 1957) and the molecular formula of this state is $C_{22}H_{29}O_2N_7 \cdot 2HCl$ (M_r 496.44), containing one mole of crystallization water (Kishi *et al.*, 1966a,b). Colorless crystals can be obtained by adding hydrobromic acid and acetone to a concentrated methanolic solution of *Cypridina* luciferin; the molecular formula in this state is $C_{22}H_{27}ON_7 \cdot 2HBr$ (M_r 567.33).

Chemical structure. The structure of the free base of *Cypridina* luciferin ($C_{22}H_{27}ON_7$, M_r 405.50) was determined by Kishi *et al.* (1966a,b) as shown below (A); its *sec*-butyl group is in the same configuration as in L-isoleucine. The structure of oxyluciferin reported by the same authors contained an error, and the structure was corrected later as shown in Fig. 3.1.8 (McCapra and Chang, 1967; Stone, 1968).

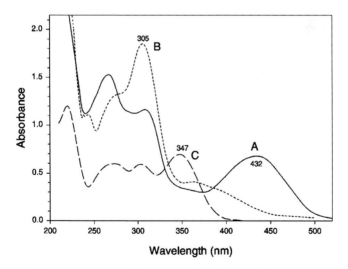

A.Yellow   B.Colorless

Cypridina luciferin

Thus, oxyluciferin has a molecular formula of $C_{21}H_{27}ON_7 \cdot 2HCl$. The total synthesis of *Cypridina* luciferin has been accomplished (Kishi *et al.*, 1966c; Inoue *et al.*, 1969; Karpetsky and White, 1971; Nakamura *et al.*, 2000).

Properties. *Cypridina* luciferin is soluble in water, methanol and other alcoholic solvents, but not in most aprotic solvents. The ultraviolet absorption spectra of luciferin and oxyluciferin are shown in Fig. 3.1.3. Luciferin in neutral solutions is yellow (λ_{max} 432 nm;

Fig. 3.1.3 Absorption spectra of *Cypridina* luciferin dihydrobromide (70 µM) in methanol (A), after addition of 1% volume of 1 N HCl (B), and oxyluciferin dihydrochloride (43 µM) in methanol (C).

ε 9,000 in methanol), and structure A is assigned to the luciferin in this state. Luciferin becomes almost colorless in methanol containing acid, and structure B is assigned to the colorless state of luciferin (Kishi *et al.*, 1966a,b). In aqueous media, luciferin is weakly fluorescent in orange (λ_{max} 540 nm). Crystals of the dihydrobromide salt are colorless, whereas those of the dihydrochloride salt are dark red.

Cypridina luciferin is chemiluminescent and extremely unstable. The compound is rapidly decomposed by oxidation under a variety of conditions, with or without light emission. Due to its extremely high affinity for molecular oxygen, luciferin in neutral and alkaline aqueous media is quickly oxidized with a trace of oxygen, even at an oxygen pressure of < 0.1 mm Hg. Oxidation by $K_3Fe(CN)_6$ (potassium ferricyanide) is instantaneous and quantitative, and the reaction is nonluminescent. This oxidant causes one-electron oxidation of luciferin, yielding a red substance that was originally named luciferin R (Shimomura *et al.*, 1957; Goto, 1968). Luciferin R does not emit significant light in the presence of *Cypridina* luciferase, but it does after treatment with a reducing agent, such as sodium borohydride and sodium dithionite. Luciferin R was found to be a dimer, consisting of two *Cypridina* luciferin molecules bound at their 5-position (Toya *et al.*, 1985). Auto-oxidation and PbO_2-oxidation of luciferin yield *Cypridina* luciferinol having the following structure (Toya *et al.*, 1983). This compound is also inactive in the luciferase-catalyzed luminescence reaction, but becomes active after treatment with sodium borohydride or sodium dithionite.

Cypridina luciferinol

Chemiluminescence. *Cypridina* luciferin emits light in various organic solvents in the presence or absence of a base. The most efficient chemiluminescence was observed in 2-methoxyethyl ether (diglyme) containing a very small amount of acetate buffer (pH 5.6), which gives a quantum yield of more than 0.03 (Goto, 1968; 10% of the quantum yield with luciferase).

It should be noted that *Cypridina* luciferin emits a fairly strong chemiluminescence in aqueous solutions in the presence of various lipids and surfactants, even in the complete absence of luciferase. The luminescence is especially conspicuous with cationic surfactants (such as hexadecyltrimethylammonium bromide) and certain emulsion materials (such as egg yolk and mayonnaise). Certain metal ions (especially Fe^{2+}) and peroxides can also cause luminescence of the luciferin. Therefore, great care must be taken in the detection of *Cypridina* luciferase in biological samples with *Cypridina* luciferin.

Light-activated luminescence. Solutions of *Cypridina* luciferin are always emitting very low levels of light due to auto-oxidation, which is easily measurable with an ordinary photomultiplier apparatus. In addition to the luminescence due to auto-oxidation, *Cypridina* luciferin emits luminescence when its solution is illuminated with light. When a luciferin solution that has been under room light for a while is inserted into the measurement compartment of a luminometer, essentially a dark box, the intensity reading of the luminometer is initially high but gradually falls to a lower level in about 1 min, indicating the presence of light-activated luminescence. The mechanism of such a light-activated luminescence may involve a free radical chain reaction (Shimomura, 1993), such as shown in the schemes below.

$$Ln^{\bullet} + O_2 \rightarrow LnOO^{\bullet}$$
$$LnOO^{\bullet} + LnH \rightarrow Ln^{\bullet} + LnOO^{-} + H^{+}$$

Under illumination, some of the luciferin molecules (LnH) that absorbed photons are changed into free radicals ($Ln^{\bullet}$), probably at carbon-2 of the imidazopyrazinone ring. The free radical instantly binds with an oxygen molecule to form a peroxide radical ($LnOO^{\bullet}$), an extremely fast reaction ($k = 10^9\,M^{-1}\,s^{-1}$; Pryor, 1976). The peroxide radical formed reacts with a luciferin molecule, generating a new free radical of luciferin and a luciferin peroxide anion ($LnOO^{-}$), the former of which is involved in the chain reaction and the latter is used in the light emitting reaction (Fig. 3.1.8). When the illumination stops, the chain reaction is gradually terminated by self-quenching of free radicals or by other quenchers, and the intensity level falls to the steady state level corresponding to auto-oxidation.

3.1.6   Oxyluciferin and Etioluciferin

The bioluminescence reaction of *Cypridina* luciferin catalyzed by *Cypridina* luciferase produces oxyluciferin (initially called oxyluciferin A). When the luciferin is left standing for a couple of days in an aqueous solution containing ammonia, it becomes oxyluciferin and etioluciferin (initially called oxyluciferin B; Shimomura *et al.*, 1957). Oxyluciferin can be converted into etioluciferin by luciferase (a slow reaction) and also by acid hydrolysis. Both oxyluciferin and etioluciferin are purified by chromatography on a column of cellulose powder using a mixed solvent of ethyl acetate-ethanol-water (5:2:3), and they are crystallized as the yellowish needles of dihydrochloride salts. Ultraviolet absorption of oxyluciferin ($C_{21}H_{27}N_7O \cdot 2HCl$) in methanol showed 4 peaks (Fig. 3.1.3): 220 nm (ε 28,000), 271 nm (ε 14,000), 302 nm (ε 13,500) and 347 nm (ε 15,400). Etioluciferin ($C_{16}H_{19}N_7 \cdot 2HCl$) showed 3 peaks in methanol containing 0.1 N HCl: 223 nm (ε 26,000), 306 nm (ε 20,000) and 410 nm (ε 5,000); and also 3 peaks in methanol containing 0.1 N NaOH: 227 nm (ε 24,000), 273 nm (ε 18,000) and 365 nm (ε 7,600) (Kishi *et al.*, 1966a,b).

3.1.7   Purification and Molecular Properties of Cypridina Luciferase

Purification of luciferase. *Cypridina* luciferase is more stable than many other luciferases, except that this enzyme is rapidly inactivated at acidity below pH 5.0. The dried specimens that have been stored for over 50 years at room temperature (sometimes exceeding 30°C) still possess strong luciferase activity that can be extracted and purified. Preparations of highly purified luciferase have been obtained by various methods (McElroy and Chase, 1951; Shimomura *et al.*, 1961, 1969; Tsuji and Sowinski, 1961; Stone, 1968; Tsuji *et al.*, 1974; Thompson *et al.*, 1989); the purification methods employed include fractional precipitation with ammonium sulfate or acetone, anion-exchange chromatography, size-exclusion chromatography, and acrylamide gel electrophoresis. A single step purification method has been reported for the luciferase that was ejected from live organisms into seawater by electrical stimulation (Kobayashi *et al.*, 2000).

Molecular weight. The molecular weight of *C. hilgendorfii* luciferase reported in the past varies considerably across a range of 50,000–80,000 (Chase and Langridge, 1960; Shimomura *et al.*, 1961, 1969; Tsuji and Sowinski, 1961; Tsuji *et al.*, 1974); it appears most likely to be 60,000–70,000. The luciferase is an acidic protein with an isoelectric point of 4.35 (Shimomura *et al.*, 1961). The absorption spectrum of luciferase is that of a simple protein without any prosthetic group, showing a peak at 280 nm. Absorbance value at 280 nm of a 0.1% luciferase solution is approximately 0.96 (Shimomura *et al.*, 1969).

Nucleotide studies. The cloning of the cDNA of *C. hilgendorfii* luciferase and the expression of the cDNA in a mammalian cell system was reported by Thompson *et al.* (1989). The primary structure of the luciferase deduced from the nucleotide sequence consists of 555 amino acid residues in a single polypeptide chain of M_r 62,171. The protein does not contain an EF-hand type calcium-binding site, and its N terminal is blocked; the N terminal amino acid is not positively identified, but a tyrosine has been suggested. Despite a high content of cysteine residues (34 residues), no free sulfhydryl group has been found in the protein (Shimomura *et al.*, 1961; Tsuji *et al.*, 1974). The cDNA of the luciferase of *Pyrocypris* (*Cypridina*) *noctiluca* has been also cloned and characterized (Nakajima *et al.*, 2004). According to the nucleotide sequence, the primary structure of this luciferase consists of 553 amino acid residues with a molecular mass of 61,415 Da.

Inhibitors. Many common enzyme inhibitors show little or no effect on the activity of *Cypridina* luciferase in the luminescence reaction (Tsuji *et al.*, 1974). However, EDTA strongly inhibits the bioluminescence reaction, showing a peculiar relationship between the

EDTA concentration and the degree of inhibition (Shimomura *et al.*, 1961). In a 30 mM sodium phosphate buffer containing 60 mM NaCl, pH 7.0, inhibition by EDTA is very strong (84%) even at an EDTA concentration as low as 25 µM, but the inhibition does not increase beyond 90% even at an EDTA concentration of 5 mM. The inhibition is completely reversed by the addition of a sufficient amount of Ca^{2+} to bind EDTA.

Why does EDTA cause only 90% inhibition, leaving 10% of the activity intact? Buffer solutions usually contain $0.1 \sim 1$ µM of contaminating Ca^{2+} when special precaution is not taken, and this concentration is much greater than the molar concentration of luciferase used in the experiments. Thus, one of the possibilities would be that Ca^{2+} interacts with the molecule of luciferase and can increase the activity of luciferase about 10 times, in spite of the fact that the molecule of luciferase lacks the Ca^{2+} binding site of EF-hand type (Thompson *et al.*, 1989). Another possibility would be that EDTA interacts directly with the molecules of luciferase, to cause the inhibition. The question remains unresolved.

3.1.8   *Luciferin-luciferase Luminescence Reaction*

Luminescence spectrum. *Cypridina* luciferin emits blue light in neutral aqueous solution in the presence of *Cypridina* luciferase and molecular oxygen (Fig. 3.1.4). The peak wavelength of the luminescence shifts between 448 nm and 463 nm depending on the buffer composition. Generally, the peak is at the red side of the range with a low ionic strength buffer, shifting significantly to the blue side on the addition of chloride salts (such as 0.1–1 M NaCl). The influence of pH is insignificant. A luminescence peak is found at 460 nm in 20 mM MES buffer, pH 6.5, and the peak shifts to 452 nm when 0.3 M NaCl is added to the solution. The natural bioluminescence of *C. hilgendorfii* occurs in seawater (containing about 0.4 M NaCl), and the emission peak is found at about 452 nm.

Reaction rate. The luminescence reaction of *Cypridina* luciferin catalyzed by *Cypridina* luciferase normally follows the first-order

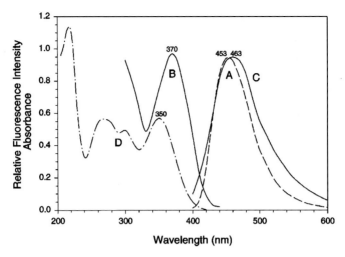

Fig. 3.1.4 Bioluminescence spectrum of *Cypridina* luciferin catalyzed by *Cypridina* luciferase (**A**), the fluorescence excitation spectrum of oxyluciferin in the presence of luciferase (**B**), the fluorescence emission spectrum of the same solution as B (**C**), and the absorption spectrum of oxyluciferin (**D**). The fluorescence of oxyluciferin alone and luciferase alone are negligibly weak. Measurement conditions: **A**, luciferin (1 µg/ml) plus a trace amount of luciferase in 20 mM sodium phosphate buffer, pH 7.2, containing 0.2 M NaCl; **B** and **C**, oxyluciferin (20 µM) plus luciferase (0.2 mg/ml) in 20 mM sodium phosphate buffer, pH 7.2, containing 0.2 M NaCl; **D**, oxyluciferin (41 µM) in 20 mM Tris-HCl buffer, pH 7.6, containing 0.2 M NaCl. All are at 20°C.

kinetics. However, when the concentration of luciferase is very low compared with that of luciferin, the initial part of the reaction can be considered as zero-order. The rate of reaction, represented by the luminescence intensity, is influenced by several factors, such as the concentrations of luciferin and luciferase, the type and concentration of salt used, and the pH and temperature of the reaction medium (Shimomura *et al.*, 1961; Shimomura and Johnson, 1970a). The effects of salts are complex, but it appears that some salt is essential for the luminescence reaction. In the case of chloride salts, the effects of NaCl, KCl, NH_4Cl and $MgCl_2$ are generally similar, and luciferase is most active with a salt concentration between 30 mM and 50 mM (Fig. 3.1.5). With $CaCl_2$, the maximum activity of luciferase is found at about 0.1 M. Luciferase is quite stable over a pH range of 5–10, and luciferase activity is optimum at pH 7.7 (Fig. 3.1.6). The optimum temperature of

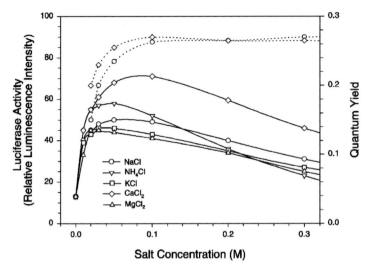

Fig. 3.1.5   Effects of salt concentration on the activity of *Cypridina* luciferase (solid lines) and quantum yield (dotted lines). In the activity measurement, *Cypridina* luciferin (1 µg/ml) was luminesced with a trace amount of luciferase in 2.5 mM HEPES buffer, pH 7.5, containing a salt to be tested, at 20°C. In the measurement of quantum yield, luciferin (1 µg/ml) was luminesced with luciferase (20 µg/ml) in 20 mM sodium phosphate buffer (for the NaCl data) or MES buffer (for the $CaCl_2$ data), pH 6.7.

luminescence intensity is found at 30°C (Fig. 3.1.7), and Michaelis constant for luciferin is reported to be 0.52 µM at 25°C (Shimomura *et al.*, 1961).

Reaction mechanism. The *Cypridina* bioluminescence reaction is believed to proceed according to the scheme shown in Fig. 3.1.8. The imidazopyrazinone part of luciferin (pK_a 8.35) is negatively charged when the luciferin is bound to luciferase (Intermediate a). Thus luciferin is easily oxygenated by O_2 at position C2, forming a peroxide anion (Intermediate b). The peroxide cyclizes, forming a dioxetanone (Intermediate c), which instantly decomposes by a concerted splitting of the 4-membered ring into CO_2 plus an amide anion of oxyluciferin in its excited state. Light is emitted when the excited state falls to its ground state (d). The mechanism involving dioxetanone was originally proposed for the chemiluminescence reaction

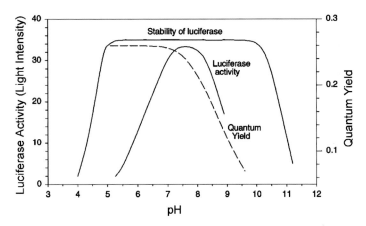

Fig. 3.1.6   Effects of pH on the activity and stability of *Cypridina* luciferase (solid lines) and the quantum yield of *Cypridina* luciferin (dashed line). In the measurements of activity and quantum yield, luciferin (1 µg/ml) was luminesced in the presence of luciferase (a trace amount for the activity measurement; 20 µg/ml for the quantum yield) in 20 mM buffer solutions of various pH containing 0.1 M NaCl, at 20°C. In the stability measurement, luciferase (a trace amount) was left standing in 0.1 ml of the buffer solutions of various pH for 30 min at 20°C, then the activity was measured by adding 1 ml of 50 mM sodium phosphate buffer, pH 6.5, containing 0.1 M NaCl and 1 µg of luciferin, at 20°C. The activity and stability data are taken from Shimomura *et al.*, 1961, with permission from John Wiley & Sons Ltd.

of luciferin by McCapra and Chang (1967), and confirmed in the bioluminescence by ^{18}O-labeling experiments (Shimomura and Johnson, 1971; see also Section 1.1.8).

Light-emitter of *Cypridina* bioluminescence. Oxyluciferin is practically non-fluorescent in aqueous solutions, but becomes brightly fluorescent when bound to *Cypridina* luciferase, and the spectrum of this fluorescence (λ_{max} 463–465 nm) closely matches the spectrum of the bioluminescence (λ_{max} 453–455 nm) (Fig. 3.1.4). The spectral correspondence suggests that the luciferase-bound oxyluciferin is the light-emitter of bioluminescence (cf. Shimomura *et al.*, 1969). The small difference in these spectra probably represents an environmental change of oxyluciferin at the binding site of luciferase after the light emission. It appears that the environment of oxyluciferin at the binding site of luciferase is hydrophobic and sufficiently polarizing to promote the formation of the amide anion.

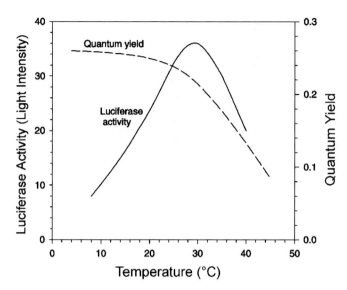

Fig. 3.1.7   Effects of temperature on the activity of *Cypridina* luciferase (solid line) and the quantum yield of *Cypridina* luciferin (dashed line). Luciferin (1 µg/ml) was luminesced in the presence of luciferase (a trace amount for the activity measurement; 20 µg/ml for the quantum yield) in 50 mM sodium phosphate buffer, pH 6.8, containing 0.1 M NaCl.

Luciferase turnover. The luciferase-catalyzed light-emitting reaction that forms oxyluciferin is fast, but the hydrolysis reaction of oxyluciferin into etioluciferin by luciferase is slow. The turnover rate (catalytic center activity) of luciferase was reported to be about 30/s for the luminescence reaction, and 0.03/s for the hydrolysis of oxyluciferin (Shimomura *et al.*, 1969).

Side reaction. The luminescence reaction of *Cypridina* luciferin catalyzed by luciferase involves a side reaction (Fig. 3.1.8). In the luminescence reaction, 85–90% of luciferin is converted into oxyluciferin and CO_2 accompanied by light emission, whereas 10–15% of luciferin is converted directly into etioluciferin plus a keto-acid without light emission (Shimomura and Johnson, 1971). In the chemiluminescence reactions of *Cypridina* luciferin in organic solvents (such as diglyme, acetone, pyridine and DMSO), the proportion of the dark side reaction

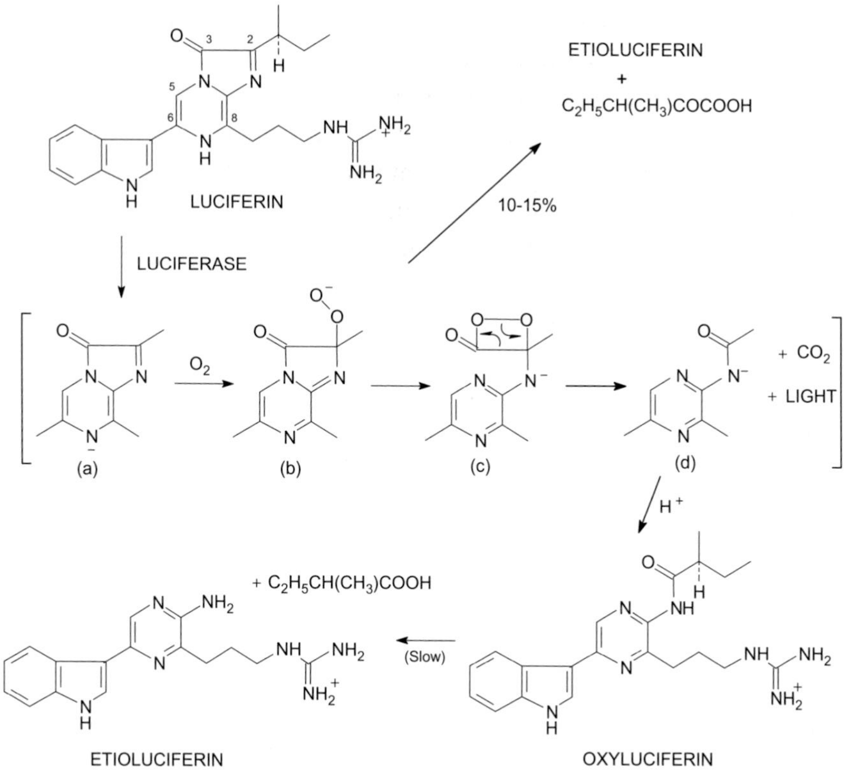

Fig. 3.1.8   A diagram showing the reactions involved in the luciferase-catalyzed luminescence of *Cypridina* luciferin.

increases to 35–70%. The mechanism of this side reaction, which may involve the heterolysis of the O-O group of Intermediate b without the formation of dioxetanone, remains to be clarified.

3.1.9   *Quantum Yield*

Johnson *et al.* (1962) measured the quantum yield of *Cypridina* luciferin in the luciferase-catalyzed reaction for the first time, using a photomultiplier calibrated with two kinds of standard lamps. The measurement gave a value of 0.28 ± 0.04 at 4°C in 50 mM sodium phosphate buffer, pH 6.5, containing 0.3 M NaCl. The quantum yield value was later corrected to 0.296 ± 0.044 using a revised molecular weight of luciferin (Shimomura and Johnson, 1970a). A study has been carried out to examine the conditions that might affect the quantum yield of luciferin (Shimomura and Johnson, 1970a). The results revealed that the effect of salts is strong, and the highest quantum yield is obtained with NaCl or $CaCl_2$. In the case of NaCl, the quantum yield is very low when NaCl is not added, and steeply increases with an increase in the concentration of NaCl, reaching a plateau at about 0.1 M, then stays at the same level up to 0.5 M; the effect of $CaCl_2$ is similar to that of NaCl except that the plateau is reached at a concentration slightly lower than 0.1 M (Fig. 3.1.5). The quantum yield of luciferin steeply decreases when the pH of the medium is raised beyond 7.0, probably due to the loss of luciferin by nonluminescent auto-oxidation (Fig. 3.1.6). The quantum yield is also affected by the medium temperature. The quantum yield is optimum near 0°C, and decreases only slightly with rising temperature up to about 20°C; however, a further rise in temperature causes a steep decrease (Fig. 3.1.7).

In the measurement of the quantum yield, the concentrations of luciferin and luciferase are very important. If the amount of one of these components is insufficient, the loss of luciferin by auto-oxidation becomes large, resulting in a quantum yield that is significantly lower than the true value. Specifically, a very low concentration of luciferin is highly susceptible to fast auto-oxidation, and thus results in a decrease of quantum yield. Also, a low concentration of luciferase extends the reaction time, thus increasing the loss of luciferin due to auto-oxidation, resulting in a lower quantum yield. Thus, in order to measure the bioluminescence quantum yield of *Cypridina* luciferin with reasonable accuracy, it is recommended to use at least 0.1 µg/ml of luciferin and 10 µg/ml of luciferase, in addition to three other conditions discussed above: a salt concentration (NaCl or $CaCl_2$) higher than 0.1 M; a medium pH between 6 and 7; and a medium temperature lower than 20°C. Some examples of the quantum yield of *Cypridina* luciferin in bioluminescence and chemiluminescence reactions are shown below (Table 3.1).

Table 3.1 Quantum Yields of *Cypridina* Luciferin in Bioluminescence and Chemiluminescence

Reaction Mixture (1 ml)	Quantum Yield
Luciferin (1 μg) plus *C. hilgendorfii* luciferase (80 μg) in 20 mM sodium phosphate buffer containing 0.3 M NaCl, pH 6.5, at 10°C	0.30
Same as above, but in 99% D_2O instead of H_2O	0.33
Luciferin (unspecified amount) in diglyme containing 5 μl of 1 M sodium acetate buffer, pH 5.6, at room temperature	0.03*
Luciferin (1 μg) in acetonitrile containing 5 μl of 0.3 M sodium acetate buffer, pH 5.6, at 25°C	0.02
Luciferin (1 μg) in dimethylacetamide, at 25°C	0.005

*From the data of Goto *et al.*, 1969.

3.2 Euphausiids *Euphausia pacifica* and *Meganyctiphanes norvegica*

Euphausiids are commonly called krill. Two species of euphausiid, *E. pacifica* and *M. norvegica* (Fig. 3.2.1), are distributed widely and abundantly, the former in the North Pacific and the latter in the North Atlantic. *E. pacifica* is commercially harvested off the coasts of Japan and British Columbia, Canada, for use in aquaculture and also as human food. Mature *E. pacifica* reach a length of about 20 mm, and *M. norvegica* 40 mm. They emit very bright blue pinpoints of light from 10 highly developed photophores, consisting of one pair on the eyestalks, two pairs on the thoracic segments, and four single organs on the abdominal segments.

Efforts by Harvey (1931) to find luciferin and luciferase in euphausiids were unsuccessful, but J. D. Doyle (1966, personal communication) succeeded in obtaining the extracts of luciferin and luciferase from the photophores of *M. norvegica* and *Thysanoessa raschii*.

3.2.1 Involvement of the Fluorescent Compound F and Protein P

Isolation of F and P. The first attempt to isolate and purify the substances responsible for the light emission of *M. norvegica* was made by Shimomura and Johnson (1967). They isolated two substances, a protein (P) and a fluorescent compound (F), which produce a blue light

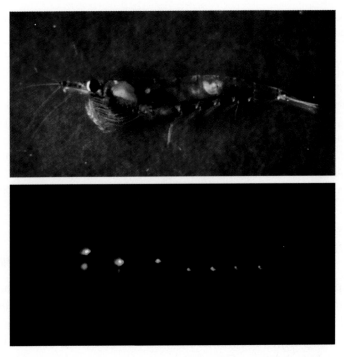

Fig. 3.2.1   The krill *Meganyctiphanes norvegica*, in daylight (*top*) and in the dark (*bottom*).

when mixed together in aqueous solutions in the presence of molecular oxygen. The emitted light is spectrally indistinguishable from the fluorescence of F. The luminescence reaction is extremely sensitive to the pH of the medium; an increase of pH from 7.4 to 7.6 results in an increase of luminescence intensity from 5% to 80% of the maximum intensity (Fig. 3.2.2). Compound F is soluble in water and alcohols, but insoluble in most organic solvents. The solutions of F in water and ethanol are very rapidly oxidized in air even at 0°C; they can be stored at −80°C for 1–2 days. Compound F is also inactivated by acids; it is so sensitive to acidity that it cannot be stored with dry ice (CO_2) unless it is sealed in a glass vial. The protein P is also unstable and quickly loses its activity when stored at 0°C, or even at −80°C.

   In the purification of F and P, the speed of purification must overcome that of inactivation due to the extremely unstable nature of these

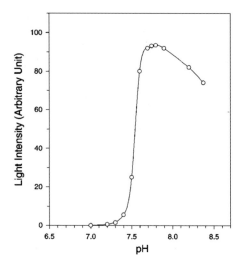

Fig. 3.2.2   Influence of pH on the initial light intensity of *euphausiid* luminescence when the fluorescent compound F and protein P are mixed in 25 mM sodium phosphate buffers of various pH values, each containing 1 M NaCl, at near 0°C. Both F and P were obtained from *Meganyctiphanes norvegica*. From Shimomura and Johnson, 1967, with permission from the American Chemical Society.

substances. Therefore, a great deal of effort was made to refine and speed up the purification of these substances, mainly by trial and error. It should be noted, however, that both F and P in the organisms before extraction are quite stable at −75°C, and also when stored in dry ice.

In our report on the bioluminescence of *Meganyctiphanes* (Shimomura and Johnson, 1967), the extremely unstable nature of the substance P caused us to interpret the functions of P and F incorrectly, the former as a photoprotein and the latter as a catalyst, as pointed out by Hastings (1968). The error was corrected 28 years later (Shimomura, 1995a), F being unambiguously shown to be a luciferin and P, a luciferase, on the basis that the quantum yield of F is about 0.6 at 0°C, while P can be recycled many times in the luminescence reaction.

Assay of the luminescence activities of F and P. A test tube containing 1 ml of 20 mM Tris-HCl buffer containing 30 mM calcium chloride, 0.01% BSA and 1 mM 2-mercaptoethanol (pH 7.5 at 24°C,

8.1 at 0°C) is cooled at 0°C, and then a solution of F ($<7\,\mu$l; in 50% ethanol) is added. The luminescence reaction is started by the injection of P ($<30\,\mu$l; in 20 mM sodium arsenate containing 0.5 M KCl, pH 6.5). For assaying F, integrated total light is measured; for assaying P, light intensity is measured. In the assay of P, 5–7 μl of a solution of F (A_{388} 5–6) is routinely used.

3.2.2   Fluorescent Compound F

Extraction and purification. An extremely unstable fluorescent substance, F, can be extracted and purified from frozen specimens of *E. pacifica* (from Victoria, British Columbia, Canada) and *M. norvegica* (from Bergen, Norway, and Millport, Scotland). To minimize the decomposition of F, all processes are carried out under an argon atmosphere at the lowest temperature appropriate for each step, and the solutions are kept neutral or slightly alkaline. All solvents should be degassed by bubbling argon gas before use. The following is our final version that can be completed in less than 10 hours; the basic principle of the method is the same as in the original procedure (Shimomura and Johnson, 1967; Nakamura *et al.*, 1988).

Extraction: Frozen krill (85 g) is briefly homogenized with 60% ethanol (220 ml) at about 0°C, and centrifuged. The supernatant is rapidly concentrated under reduced pressure in a 2-liter flask at about 40°C (using a rotary evaporator, a mechanical vacuum pump, and a large condensate trap immersed in dry ice/acetone) to a volume of 15–20 ml. After the addition of 30 ml of cold ethanol, the solution is temporarily stored at −30°C. Materials similarly prepared from 6 batches (510 g krill in total) are combined, centrifuged, and the supernatant is concentrated to 30 ml, and then mixed with 70 ml of ethanol. Compound F in the solution is extracted with 120 ml of *n*-butanol.

Alumina chromatography: The butanol extract containing F is poured onto a column of alumina (2.6 × 9 cm; Woelem basic alumina Grade 1), and F adsorbed is eluted with 50% ethanol containing 0.6% ammonium hydroxide. The fractions containing F are combined (a trace of Tris is added) and concentrated to about 5 ml, and mixed with 40 ml of ethanol. The precipitate formed is removed by centrifugation, and then the supernatant is diluted with 140 ml of 50% ethanol.

DEAE-cellulose chromatography: The 50% ethanol solution is poured onto a column of DEAE cellulose (2.6 × 10 cm), and F adsorbed at the top is eluted with 20 mM Tris buffer, pH 7.5, containing 0.2 M NaCl. A low pressure of argon gas is applied to accelerate the flow rate. The fractions containing F are combined, concentrated, and desalted using ethanol.

HPLC: The compound F obtained is further purified on a column of TSK DEAE-5PW, using 40% aqueous acetonitrile containing 85 mM NaCl and 3 mM $NaHCO_3$. The final yield of F is about 2 mg (purity 80–90%) from 500 g of frozen specimens. A small amount of 2-mercaptoethanol added at the last desalting step may significantly improve the purity of the final product.

Properties and chemical structure of F. A solution of F shows a characteristic absorption peak at 388 nm in 50% ethanol or at 390 nm in aqueous solutions (Fig. 3.2.3; Shimomura and Johnson, 1967; Shimomura, 1995a). The fluorescence emission spectrum shows a peak at 476 nm (Fig. 3.2.4; Shimomura and Johnson, 1968a). Compound F is rapidly inactivated in the presence of a trace of oxygen, and also by weak acidity, around pH 4 or the acidity caused by dry ice. The rate of inactivation of F by oxidation in 50% ethanol is 75%/day at 0°C in air, and 2%/day at −75°C under argon.

$K_3Fe(CN)_6$ *oxidation*: Compound F is stoichiometrically inactivated by oxidation with $K_3Fe(CN)_6$ (Shimomura and Johnson, 1967); thus, it is possible to estimate the molecular extinction coefficient (ε) of the 388–390 nm absorption peak by titrating F with $K_3Fe(CN)_6$. The ε value obtained by the titration in 50% ethanol was 15,400 (assuming the reaction to be one-electron oxidation) or 30,800 (assuming two-electron oxidation). Two other methods of lesser precision were used to determine the true ε value: 1) the dry weight of the ethyl acetate extract of an acidified solution of F gave an ε value of 14,100; 2) the comparison of NMR signal intensities gave a value of 11,400 ± 2,000 in water (H. Nakamura, Y. Oba, and A. Murai, 1995, personal

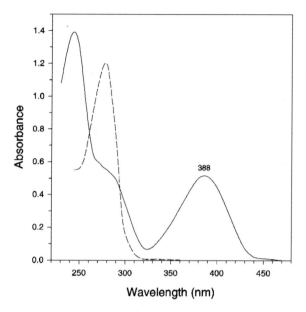

Fig. 3.2.3 Absorption spectra of the fluorescent compound F in 50% ethanol (solid line) and the protein P in 20 mM sodium arsenate, pH 6.5, containing 0.5 M KCl (dashed line). Both F and P were obtained from *Meganyctiphanes norvegica*. From Shimomura, 1995a, with permission from John Wiley & Sons Ltd.

communication). These data conclusively show that the oxidation of F with $K_3Fe(CN)_6$ is a one-electron oxidation, and that the ε value for the 388 nm peak in 50% ethanol is 15,400 (Shimomura, 1995a). The one-electron oxidation is not surprising because the oxidation of *Cypridina* luciferin with $K_3Fe(CN)_6$ is also one-electron oxidation (Shimomura, 1960; Goto, 1968).

Chromic acid oxidation: Compound F yields various pyrrole derivatives by chromic acid oxidation and alkali treatment (Shimomura, 1980), providing important information concerning the chemical structure of F. A mild chromic acid oxidation of F ($CrO_3/KHSO_4/H_2O$, room temp.) yielded 3-methyl-4-vinylmaleimide (structure **1** in Fig. 3.2.5) and an aldehyde (structure **2**), whereas vigorous chromic acid oxidation ($CrO_3/2N\ H_2SO_4$, 90°C) gave hematinic acid (structure **3**) (Shimomura, 1980). Under basic condition (NaOH/MeOH, 65°C), a pyrromethanone (structure **4**) was obtained

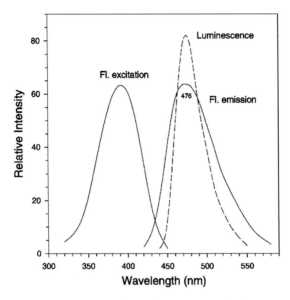

Fig. 3.2.4   Fluorescence spectra of F (solid lines), and the bioluminescence spectrum of F plus P, in 20 mM Tris-HCl, pH 7.6, containing 0.15 M NaCl, at 4°C. Both F and P were obtained from *Meganyctiphanes norvegica*. From Shimomura and Johnson, 1968a.

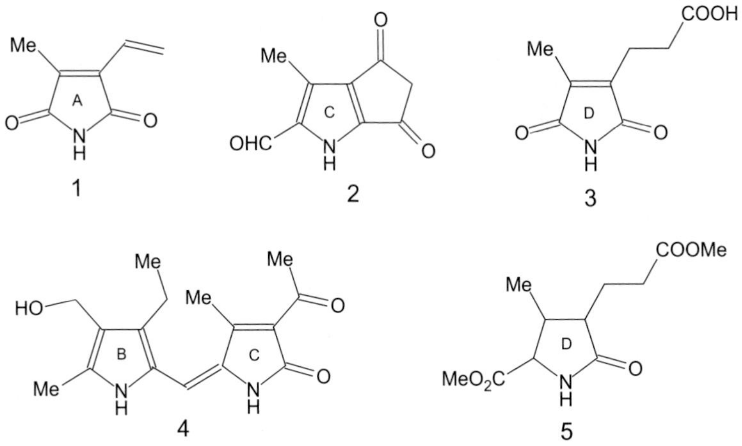

Fig. 3.2.5   Chemical structures of the compounds derived from F.

(Nakamura *et al.*, 1988). These results suggest that F has the structure of a bile pigment type tetrapyrrole.

Structure of Oxy-F: Compound F is extremely unstable and is difficult to obtain at a level of purity suitable for NMR studies. However, an oxidation product, Oxy-F, formed when F is left standing at $-20°C$, is considerably more stable than F and can be purified to a sufficiently high level of purity. Oxy-F is nonfluorescent and shows absorption maxima at 237 nm and 275 nm (shoulder). The high-resolution FAB mass spectrum indicated the molecular formula of Oxy-F to be $C_{33}H_{38}O_9N_4Na_2$ [m/z 703.2363 $(M+Na)^+$ and 681.2483 $(M+H)^+$]. The ^1H and ^{13}C NMR data allowed the assignment of structure 7 to oxy-F (Fig. 3.2.6; Nakamura *et al.*, 1988).

Structure of F: Although F has never been obtained in a completely pure state, the FAB mass spectral data of F [m/z 687 $(M+Na)^+$ and 665 $(M+H)^+$], and the comparison of the ^1H and ^{13}C NMR spectra of F with those of Oxy-F, suggested structure 6 for this compound. To confirm this structure, F was subjected to ozonolysis, followed by diazomethane treatment. The expected diester 5 was successfully isolated, indicating that 6 is indeed the structure of compound F (Nakamura *et al.*, 1988). The structure of the luminescence reaction product of F is considered to be 8 on the basis of comparison with the dinoflagellate luminescence system (see Chapter 8).

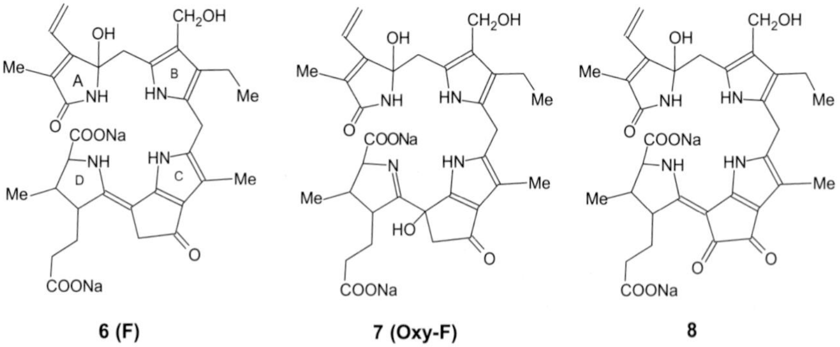

6 (F)   **7 (Oxy-F)**   **8**

Fig. 3.2.6 Chemical structures of compound F (6; the euphausiid luciferin), a product obtained when F was left standing at –20°C for 2 weeks (7), and the product of bioluminescence reaction (8).

3.2.3 Protein P

Extraction and purification. Protein P was extracted and purified from *E. pacifica* and *M. norvegica* at near 0°C (Shimomura, 1995a). The outline of the procedure used is as follows.

Frozen krill (65 g) is homogenized in 200 ml of cold water, and centrifuged. The supernatant is adjusted to pH 5.75, and centrifuged again. The clear supernatant obtained is partially saturated with $(NH_4)_2SO_4$ (50 g), and the resulting clouded mixture is centrifuged, discarding the supernatant. The pellets are dissolved in 8 ml of a pH 6.5 buffer containing 20 mM sodium arsenate and 20 mM KCl, and chromatographed on a column of Sephacryl S-300 (2.6 × 33 cm; Pharmacia) using the same buffer (flow rate 100 ml/hr). The eluate fractions containing the first 50% of the activity were discarded. The rest of the active fractions are combined, diluted with an equal volume of the pH 6.5 buffer, and added onto an affinity chromatography column of biliverdin-Sepharose 4B (prepared from biliverdin and Pharmacia EAH Sepharose 4B; 1.6 × 8 cm) that has been packed with the pH 6.5 buffer. The activity of P is adsorbed at the top part of the column. The column is first washed with 0.25 M KCl/20 mM sodium arsenate, at pH 6.5. Then, P is eluted by a linear increase of KCl concentration from 0.25 M to 0.65 M, at a flow rate of about 1 ml/min. The eluate fractions of nearly constant specific activity (luminescence activity/A_{280}) are combined and used in the experiments. The total time required is about 5 hours.

Properties of the affinity purified protein P. The properties of the proteins P obtained from *M. norvegica* and *E. pacifica* are similar (Shimomura, 1995a). The molecular weights estimated by size-exclusion chromatography on Sephacryl S-400 and Ultrogel AcA 22 (IBF) are both approximately 600,000. In the presence of 0.02% SDS, P dissociates into two inactive molecular species (roughly 85,000 and 19,000). A solution of purified P, A_{280} 1.0, contained 0.86 mg/ml of protein. The specific luminescence activity measured immediately after purification was 4.8–5.0 × 10^{12} photons/s per 1 ml of A_{280} 1.0 solution. The protein can be stored with liquid nitrogen without any

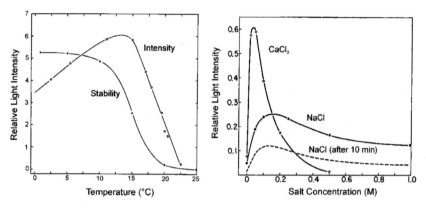

Fig. 3.2.7 Left panel: Effects of temperature on the luminescence intensity and stability of the protein P from *Meganyctiphanes*. The initial light intensity was measured with F plus P in 5 ml of 20 mM Tris-HCl/0.15 M NaCl, pH 7.5, at various temperatures. In the stability test, P was kept at the indicated temperature for 10 min, then mixed with 5 ml of 25 mM Tris-HCl/1 M NaCl, pH 7.59, containing F, to measure initial light intensity. Right panel: Effect of the concentration of salts on the light intensity of the luminescence of F plus P, in 25 mM Tris-HCl, pH 7.6, at near 0°C. In the case of NaCl, the light intensity decreased to about a half after 10 min. From Shimomura and Johnson, 1967, with permission from the American Chemical Society.

loss of activity. The rates of the spontaneous inactivation of *E. pacifica* luciferase and *M. norvegica* luciferase were 67% per day and 45% per day, respectively, at 0°C in 20 mM sodium arsenate/0.5 M KCl, at pH 6.5. The rate of inactivation increases steeply when the temperature is raised above 10°C (Fig. 3.2.7, left panel; Shimomura and Johnson, 1967).

3.2.4   Luminescence Reaction

The fluorescent compound F, a luciferin, emits blue light (λ_{max} 476 nm; Fig. 3.2.4) in the presence of molecular oxygen and the protein P, a luciferase. In the luminescence reaction, F is changed into an oxidized form (structure 8, Fig. 3.2.6). The luminescence reaction is highly sensitive to pH, with a narrow optimal range around pH 7.8 (Fig. 3.2.2); the optimum salt concentration is 0.15 M for NaCl and 0.04 M for $CaCl_2$ (Fig. 3.2.7, right panel; Shimomura and Johnson, 1967). The protein P is a slow-working luciferase, with a turnover number at 0°C of only about 30/hr for the initial 20 min and 20/hr after 1 hour (Shimomura, 1995a). Total light emitted during a 50-hour period indicated that the bioluminescence quantum yield of F is about 0.6 at 0°C, and P recycles many times (Shimomura, 1995a). The luminescence intensity produced by the reaction of F and P is the brightest at near 14°C, and decreases to about a half the level when cooled to 0°C or warmed up to 18°C (Fig. 3.2.7, left panel; Shimomura and Johnson, 1967). The product of luminescence reaction (8) is nonfluorescent; thus, its excited state is unlikely to be the light emitter. On the other hand, the peaks of the bioluminescence spectrum and the fluorescence emission spectrum of F almost coincide, although the latter spectrum is somewhat broader than the former (Fig. 3.2.4). The close agreement of the emission peaks suggests that the bioluminescence light emitter is probably an intermediate (prior to the formation of new carbonyl in 8) that has a chromophore virtually identical to that of F. The detailed reaction mechanism remains to be clarified (for a possible chemical mechanism, see Section 8.2.7).

The *in vivo* luminescence of euphausiids. The photophores of captured live specimens emit light spontaneously, or by mechanical stimulation, or by the action of 5-hydroxytryptamine. According to Herring and Locket (1978), the live specimens of *Nematoscelis megalops* emit a luminescence with an emission maximum at 463 nm, whereas a filtered homogenate luminesced with a maximum at 472 nm; the cause of the difference has not been clarified. Emission spectra recorded from homogenates of many species of euphausiids are sharp and show maxima in a narrow range from 467 to 473 nm (FWHM 44–53 nm) (Herring, 1983). In the *in vivo* luminescence of certain euphausiids, a pigment other than compound F is possibly involved. For example, the homogenate of *Thysanoessa raschii* and the terminal glows of *E. pacifica* and *M. norvegica* showed a bimodal emission spectrum, with a sharp primary peak at 476 nm and a second, lower peak between 520 and 540 nm (Boden and Kampa, 1959, 1964; Kampa and Boden, 1956).

3.3   The Decapod Shrimp *Oplophorus gracilirostris*

Certain genera of decapod shrimps, such as *Heterocarpus* and *Oplophorus*, emit strikingly bright light from luminous clouds projected into seawater. In the case of *O. gracilirostris* (body weight 3 g), the shrimp possesses secretory glands at the base of the antennae and legs that eject a luminous secretion, as well as photophores on the legs. The luminescence reaction takes place when coelenterazine, the luciferin (Inoue and Kakoi, 1976), is oxidized by molecular oxygen in the presence of *Oplophorus* luciferase (Shimomura *et al.*, 1978). The assay methods of coelenterazine and the luciferase are described in Section C5 of the Appendix.

3.3.1   *Oplophorus Luciferase*

Extraction and purification (Shimomura *et al.*, 1978). Dr. Yata Haneda and Dr. Shoji Inoue kindly provided us with specimens of *O. gracilirostris*, which had been picked out one by one from large commercial catches of *Sergestes lucens* netted in Suruga Bay, Japan. The material was air-dried or freeze-dried before shipment to the U.S.

Dried shrimp was ground, defatted with benzene, and then extracted with cold water. The luciferase extracted was purified first by a batch adsorption onto DEAE cellulose (elution with 0.4 M NaCl), followed by gel filtration on a column of Sephadex G-150, anion-exchange chromatography on a column of DEAE-cellulose (gradient elution 0.05–0.5 M NaCl), and gel filtration on a column of Ultrogel AcA 34. The specific activity of the purified luciferase was 1.7×10^{15} photons $\cdot$ s^{-1} mg^{-1}, and the yield in terms of luciferase activity was about 28%.

The protein can be further purified by hydrophobic interaction chromatography on a column of Butyl Sepharose 4 Fast Flow (Pharmacia; elution with decreasing concentration of $(NH_4)_2SO_4$ starting at 1.5 M), and gel filtration on a column of Superdex 200 Prep (Pharmacia; Inouye *et al.*, 2000).

Molecular properties of *Oplophorus* luciferase. The molecular weight of the natural form of *Oplophorus* luciferase is most likely to be about 106,000 (Inouye *et al.*, 2000). The luciferase has a tendency to aggregate and lose activity in acidic solutions (pH < 7) and also when precipitated with ammonium sulfate. During SDS-PAGE analysis, the molecules of native *Oplophorus* luciferase are split into two proteins with molecular masses 19 kDa and 35 kDa. The cDNAs encoding these two proteins were cloned (Inouye *et al.*, 2000). The expression of the cDNAs in bacterial and mammalian cells indicated that the 19 kDa protein, not the 35 kDa protein, is capable of catalyzing the luminescent oxidation of coelenterazine. The 19 kDa protein consists of 169 amino acid residues with a calculated molecular mass of 18,689.50 and a pI of 4.70.

The native luciferase having a molecular weight of 106,000 probably consists of two units of the functional 19 kDa protein and two units of the 35 kDa protein. The value of $A_{280,1cm}$ for a solution containing 1 mg/ml of the native luciferase is calculated to be about 0.9 from the inferred amino acid sequence. The function of the 35 kDa protein remains unclear, although it might have a role in the stabilization of the 19 kDa protein.

3.3.2   Coelenterazine-luciferase Reaction

The bioluminescence reaction of *Oplophorus* is a typical luciferin-luciferase reaction that requires only three components: luciferin (coelenterazine), luciferase and molecular oxygen. The luminescence spectrum shows a peak at about 454 nm (Fig. 3.3.1). The luminescence is significantly affected by pH, salt concentration, and temperature. A certain level of ionic strength (salt) is necessary for the activity of the luciferase. In the case of NaCl, at least 0.05–0.1 M of the salt is needed for a moderate rate of light emission, and about 0.5 M for the maximum light intensity.

pH Effect: The luminescence intensity is optimum at about pH 9, and the intensity decreases steeply at both sides of the optimum

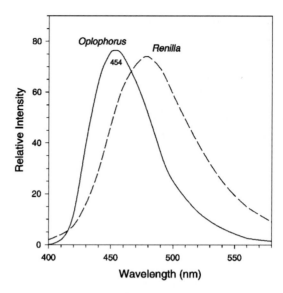

Fig. 3.3.1 Luminescence spectrum of coelenterazine catalyzed by the luciferase of the decapod *Oplophorus* in 15 mM Tris-HCl, pH 8.3, containing 50 mM NaCl (solid line). For comparison, the luminescence catalyzed by the luciferase of the anthozoan sea pansy *Renilla* is shown with dashed line (in 25 mM Tris-HCl, pH 7.5, containing 0.1 M NaCl).

(Fig. 3.3.2). The total light elicitable (or the quantum yield of coelenterazine) is nearly constant in a wide range of pH, 6.0 to 10.0, and falls off at both ends.

Heat stability: The *Oplophorus* luminescence system is more thermostable than several other known bioluminescence systems; the most stable system presently known is that of *Periphylla* (Section 4.5). The luminescence of the *Oplophorus* system is optimum at about 40°C in reference to light intensity (Fig. 3.3.3; Shimomura *et al.*, 1978). The quantum yield of coelenterazine is nearly constant from 0°C to 20°C, decreasing slightly while the temperature is increased up to 50°C (Fig. 3.3.3); at temperatures above 50°C, the inactivation of luciferase becomes too rapid to obtain reliable data of quantum yield. In contrast, in the bioluminescence systems of *Cypridina*, *Latia*, *Chaetopterus*, luminous bacteria and aequorin, the relative quantum yields decrease steeply when the temperature is raised, and become almost zero at a temperature near 40–50°C (Shimomura *et al.*, 1978).

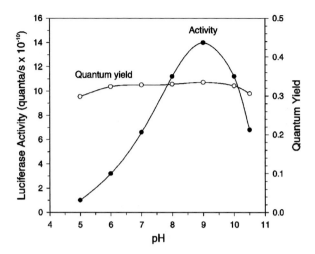

Fig. 3.3.2 Influence of pH on the activity of luciferase (•) and the quantum yield of coelenterazine (○) in the bioluminescence of *Oplophorus*. The measurements were made with coelenterazine (4.5 μg) and luciferase (0.02 μg) for the former, and coelenterazine (0.1 μg) and luciferase (100 μg) for the latter, in 5 ml of 10 mM buffer solutions at 24°C. The buffer solutions used: sodium acetate (pH 5.0), sodium phosphate (pH 6.0–7.5), Tris-HCl (pH 7.5–9.1), and sodium carbonate (pH 9.5–10.5), all containing 50 mM NaCl. Replotted from Shimomura *et al.*, 1978, with permission from the American Chemical Society.

Quantum yield and luciferase activity: The quantum yield of coelenterazine in the luminescence reaction catalyzed by *Oplophorus* luciferase was 0.34 when measured in 15 mM Tris-HCl buffer, pH 8.3, containing 0.05 M NaCl at 22°C (Shimomura *et al.*, 1978). The specific activity of pure luciferase in the presence of a large excess of coelenterazine (0.9 μg/ml) in the same buffer at 23°C was 1.75×10^{15} photons $\cdot$ s^{-1} mg^{-1} (Shimomura *et al.*, 1978). Based on these data and the molecular weight of luciferase (106,000), the turnover number of luciferase is calculated at 55/min.

Mechanism of luminescence reaction. The chemical reaction of *Oplophorus* bioluminescence can be represented by the following simplified scheme:

$$\text{Coelenterazine} + O_2 \xrightarrow{\text{luciferase}} \text{Coelenteramide} + CO_2 + \text{Light}$$

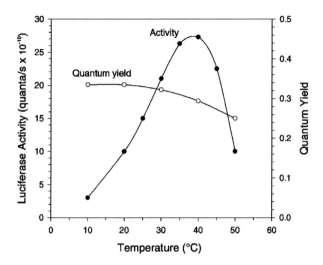

Fig. 3.3.3   Effects of temperature on the activities of luciferase (•) and the quantum yields of coelenterazine (o) in the *Oplophorus* bioluminescence reaction. The activity was measured with coelenterazine (4.5 µg) and luciferase (0.05 µg), and the quantum yields with coelenterazine (0.2 µg) and luciferase (200 µg), in 5 ml of 15 mM Tris-HCl buffer, pH 8.3 (at 25°C), containing 50 mM NaCl. Coelenterazine was first added to the buffer solution at the designated temperature, then the luminescence reaction was started by a rapid injection of 0.1 ml of luciferase solution. Replotted from Shimomura *et al.*, 1978, with permission from the American Chemical Society.

The product coelenteramide is not noticeably fluorescent in aqueous solutions, but is highly fluorescent in organic solvents and also when the compound is in the hydrophobic environment of a protein. When coelenterazine is luminesced in the presence of *Oplophorus* luciferase, the solution after luminescence (the spent solution) is not fluorescent, presumably due to the dissociation of coelenteramide from the luciferase that provided a hydrophobic environment at the time of light emission. An analogous situation exists in the bioluminescence system of *Renilla* (Hori *et al.*, 1973).

The luminescence reaction of coelenterazine is initiated by the peroxidation of coelenterazine at its C2 carbon by molecular oxygen (Fig. 3.3.4). Then, the peroxidized coelenterazine decomposes into coelenteramide plus CO_2, producing the energy needed for the light emission. For the mechanism of the decomposition of peroxide that produces the energy, two different pathways can be considered.

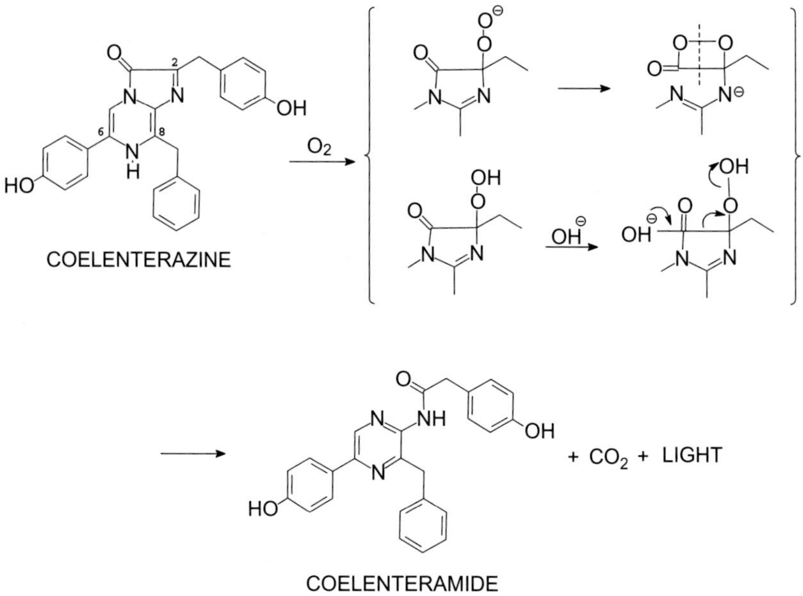

COELENTERAZINE

COELENTERAMIDE

Fig. 3.3.4 Reaction mechanism of the coelenterazine bioluminescence showing two possible routes of peroxide decomposition, the dioxetanone pathway (upper route) and linear decomposition pathway (lower route). The *Oplophorus* bioluminescence takes place via the dioxetanone pathway. The light emitter is considered to be the amide-anion of coelenteramide (see Section 5.4).

One is the concerted decomposition of a dioxetanone structure that is proposed for the chemiluminescence and bioluminescence of both firefly luciferin (Hopkins *et al.*, 1967; McCapra *et al.*, 1968; Shimomura *et al.*, 1977) and *Cypridina* luciferin (McCapra and Chang, 1967; Shimomura and Johnson, 1971). The other is the "linear decomposition mechanism" that has been proposed for the bioluminescence reaction of fireflies by DeLuca and Dempsey (1970), but not substantiated. In the case of the *Oplophorus* bioluminescence, investigation of the reaction pathway by ^{18}O-labeling experiments has shown that one O atom of the product CO_2 derives from molecular oxygen, indicating that the dioxetanone pathway takes place in this bioluminescence system as well (Shimomura *et al.*, 1978). It appears that the involvement of a dioxetane intermediate is quite widespread in bioluminescence.

3.4   Copepoda

Marine copepods are all very small (usually 0.5–2 mm long) and most of them are free-swimming planktonic forms. They exist in enormous numbers, playing an extremely important role in marine food chains. Bioluminescent species are found in the orders Cyclopoida and Calanoida. The former contains the luminous genus *Oncaea*, and the latter contains many luminous genera, including *Metridia*, *Pleuromamma*, and *Gaussia*. Light is produced as a secretion from epidermal glands in the limbs or body (Herring, 1978a). Harvey (1952) confirmed that the bioluminescence of copepods requires molecular oxygen by an experiment carried out in pure hydrogen, but he had negative results in his efforts to demonstrate a luciferin-luciferase reaction, which is puzzling.

Campbell and Herring (1990) examined eight species of copepods, and found that all of them contain coelenterazine (the luciferin) and a luciferase. In the *Euaugaptilus* species, over 90% of the luciferase was found in their "legs," and over 40% of the total coelenterazine was found in the bodies.

The luciferases have been cloned in the cases of *Gaussia* (Ballou *et al.*, 2000; Verhaegen and Christpoulos, 2002) and *Metridia longa* (Golz *et al.*, 2002; Markova *et al.*, 2004). According to the symposium abstract by Ballou *et al.* (2000), the recombinant *Gaussia princeps* luciferase has a molecular weight of 19,900. The luminescence reaction of coelenterazine catalyzed by the recombinant *Gaussia* luciferase shows a broad pH optimum with a peak at 7.7 for the emission of blue light (λ_{max} 470 nm), and the luminescence activity of the luciferase is strongly dependent on the concentration of monovalent cations. Analysis of the gene indicates the presence of a secretory signal sequence that functions in both prokaryotes and eukaryotes. The enzyme is exceptionally resistant to exposure to heat, and also to acidic and basic conditions. The luciferase has a tendency to self-aggregate into inactive forms, like many other coelenterazine luciferases.

According to Markova *et al.* (2004), the cDNA encoding the luciferase of *Metridia longa* was cloned and sequenced. The luciferase is a 219-amino acid protein with a molecular weight of 23,885.

The protein contains an N-terminal signal peptide of 17 amino acid residues for secretion. The luminescence reaction of coelenterazine catalyzed by the recombinant luciferase shows a luminescence emission maximum at 485 nm, whereas the luminescence catalyzed by the native luciferase shows a maximum at 480 nm.

To this author, it seems unfortunate that the cloning of luciferases preceded the isolation of natural luciferases in the study of the luminous copepod, drawing interest away from the future study of natural copepod luciferases.

CHAPTER
4

THE JELLYFISH *AEQUOREA* AND OTHER LUMINOUS COELENTERATES

The organisms of the phylum Cnidaria (or Coelenterata) are characterized by apparent radial symmetry of body forms and simple structures consisting of two layers of cells surrounding an internal space (coelenteron) which also functions as a digestive cavity. Coelenterates occur in various forms, such as plant-shaped hydroids, sea pens and sea anemones, as well as floating jellyfishes and siphonophores. They are predominantly sessile — attached by a base — at least during one stage of their life cycles. Many species of coelenterates are bioluminescent, except those of sea anemones and corals. In all cases as far as known, their bioluminescence is caused by coelenterazine or its derivatives existing in their light organs. The phylum Cnidaria comprises three classes: Hydrozoa, Scyphozoa and Anthozoa, which are briefly explained below.

Hydrozoa. Hydrozoans may be in the form of medusae or polyps and may occur singly or in colonies. Although some hydrozoans show only the medusoid form, most species possess a polynoid stage in their life cycles. This class includes the luminescent genera *Aequorea, Mitrocoma (Halistaura), Obelia* and *Phialidium*. Various siphonophoran genera are bioluminescent, but little is known about the chemical nature of their luminescence. Most bioluminescent hydrozoans contain a photoprotein that emits light in the presence of Ca^{2+} (Shimomura *et al.*, 1962, 1963a,b; Morin and Hastings, 1971a; Levine and Ward, 1982), often accompanied by a green fluorescent protein with the exception of siphonophores (Herring, 1983). These hydrozoan photoproteins contain a moiety of coelenterazine in the molecules. There are many species of *Aequorea* and *Obelia*, and possibly all of them are bioluminescent. They are distributed very widely throughout the world.

Scyphozoa. Scyphozoans are most frequently referred to as jellyfish, and generally larger in size than hydrozoans. In this class, the medusa form is dominant in the life cycle; the polyp form is restricted to a small larval stage. The class Scyphozoa includes the luminescent genera *Pelagia, Periphylla* and *Atolla*. In the case of *Periphylla*, light is emitted by a luciferin-luciferase reaction in the presence of oxygen (Shimomura and Flood, 1998), in which the luciferin is coelenterazine. In the case of *Pelagia*, however, Harvey (1926b) noted that the extract of *Pelagia noctiluca* emitted a bright light in the absence of oxygen, suggesting a possible involvement of a photoprotein instead of a luciferin-luciferase reaction. Moreover, the extraction of a Ca^{2+}-sensitive photoprotein from *Pelagia* has been reported (Morin and Hastings, 1971a; Morin and Reynolds, 1972), indicating that the bioluminescence system of *Pelagia* differs from that of *Periphylla*. Scyphozoans do not possess a green fluorescent protein.

Anthozoa. Anthozoans are plant-shaped polyps, either solitary or colonial, completely lacking the medusoid stage. They are found along coastal waters and include the luminescent genera *Renilla* (the sea pansies), *Cavernularia* (the sea cactuses), and *Ptilosarcus* and *Pennatula* (the sea pens). Bioluminescent anthozoans emit light by a luciferin-luciferase reaction that involves coelenterazine as the luciferin (Shimomura and Johnson, 1975b). Most bioluminescent anthozoans contain a green fluorescent protein.

Organisms in the phylum Ctenophora, commonly known as comb jellies, are included at the end of this chapter because of their close relationship to coelenterates.

4.1   The Hydrozoan Medusa *Aequorea aequorea*

Introductory information. This species was first named *Medusa aequorea* (Forskal, 1775), and renamed *Aequorea forskalea* (Peron and Lesueur, 1809). However, H. B. Bigelow pointed out that, since *Medusa* is no longer used while the genus name *Aequorea* has long been recognized, this species should be named *Aequorea aequorea* Forskal by right of priority (Johnson and Snook, 1927). Kramp (1965) maintained Bigelow's view. In recent years, however, the name *Aequorea victoria* (Murbach and Shearer, 1902; Arai and Brinckmann-Voss, 1980) has been frequently used to indicate this species. The author considers the latter name as a local synonym for the *Aequorea aequorea* species in the areas of British Columbia and Puget Sound (Shimomura, 1998, 2006).

A mature specimen of *Aequorea aequorea* (Fig. 4.1.1) looks like a transparent, hemispherical umbrella or, sometimes, a saucer. This species is highly variable in form and color, especially in the number of radial canals and tentacles (Mayer, 1910; Russel, 1953; Kramp, 1965). The species is distributed very widely — the coasts of the Atlantic from Norway to South Africa and Cape Cod to Florida, the Mediterranean, the northeastern Pacific, the Iranian Gulf, and the east coast of Australia (Kramp, 1959, 1965, 1968). Mature specimens in the vicinity of Friday Harbor, Washington, measure 7–10 cm in diameter and weigh about 50 g. The light organs, consisting of about 200 tiny granules, are distributed evenly along the edge of the umbrella, making a full circle. Soaking a specimen in a dilute KCl solution in a darkroom causes the light organs to luminesce, displaying a ring of green light in the darkness. If a specimen is soaked in distilled water, a dimmer green ring first observed gradually changes into blue with the progress of cytolysis of the cells. Under an ultraviolet light, a specimen of live jellyfish

Fig. 4.1.1   The hydrozoan jellyfish *Aequorea aequorea*.

shows a bright ring of green fluorescence, similar to the luminescence caused by KCl.

Until 1988, this species of jellyfish was extremely abundant in the area of the San Juan Islands and the Strait of Georgia. A stream of scattered floating jellyfish, riding on the tide current, was a common sight on a summer day. Sometimes the surface of the water was nearly completely covered by a dense population of jellyfish. The jellyfish were a nuisance to salmon fishermen, annoying them by clogging their nets. However, the abundant jellyfish were a great advantage to us scientists who needed them, and we have done most of our studies of the *Aequorea* bioluminescence with the specimens collected at the Friday Harbor Laboratories, University of Washington, and the vicinity of Friday Harbor. We collected 50,000–80,000 specimens almost every summer between 1966 and 1980. Mysteriously, however, this species in the area suddenly disappeared sometime around 1990 for unknown reason, making it difficult to collect even a few specimens of *Aequorea* (Mills, 2001).

Collection of Aequorea. Specimens of *A. aequorea* are individually collected with a net specially designed for that purpose, to prevent damage to the light organs located along the edge of the umbrella. The collecting net is made with an oval-shaped wire frame (30 cm long) attached to a 1.8 meter-long handle (bamboo or wood), and a piece of plastic window screen. The plastic screen is sewn on to the wire frame using dental floss, making an almost flat net. Specimens at 1.5-meter depth, as well as those at the surface, can be easily caught on the net. The jellyfish is then slid into a bucket containing seawater; it is important to avoid unnecessary sliding of the specimens on the net, to minimize the damage to the light organs. The specimens can be kept for one day in a tank with running seawater.

4.1.1 *History of the Biochemical Study of Aequorea Bioluminescence*

Harvey (1921) is probably the first person who studied the luminescent substance of jellyfish *Aequorea*. He cut off the margin of the umbrella containing light organs from a score of jellyfish with a pair of scissors, making thin strips. Then, he squeezed the strips through four layers of cheesecloth and obtained a turbid luminescing liquid "squeezate," which glowed for some hours. After the luminescence ceased, an addition of fresh water or cytolytic agents such as saponin to the squeezate revived luminescence, but isotonic cane sugar solution or seawater did not cause the emission of light. Harvey thought that the light was emitted when photogenic cells and granules were dissolved. He also found that, when the luminous materials of *Aequorea*, *Mitrocoma* and *Phialidium* are dried over $CaCl_2$, the dried materials give a bright light when moistened again. However, his attempts to demonstrate the presence of luciferin and luciferase in *Aequorea* and *Mitrocoma* were unsuccessful.

Harvey (1952) noted: "R. S. Anderson has tested *Aequorea* at Friday Harbor and found that his hydromedusan will luminesce under strict anaerobic conditions." It was an extremely important observation retrospectively. The ability to luminesce in the absence of oxygen had been also observed with *Medusa hemisphaerica* by Macartney (1810) and with radiolarians, ctenophores, and *Pelagia* by Harvey (1926b).

In the summer of 1961, F. H. Johnson and the author tried to obtain cell-free solutions of the light-emitting substance of *Aequorea*. First, we tried to extract luciferin and luciferase, but all our efforts failed. Thus, we changed our strategy and tried to extract the light-emitting substance, whatever it might be. Our basic idea was to extract the substance under the conditions that reversibly inhibit luminescence; therefore, we tried to find out such conditions. After an exhaustive effort, we discovered an unexpectedly simple method (see Shimomura, 1995b): the luminescence could be reversibly inhibited and the substance could be extracted with a buffer solution of pH 4.0. The cell-free pH 4.0 extract was very dimly luminescent, but the light intensity increased more than 10 times upon neutralization. The same pH 4.0 extract, however, emitted a bright flash of light when mixed with seawater, or a neutral solution of $CaCl_2$, revealing that Ca^{2+} is an essential requirement in the luminescence reaction of *Aequorea*. Based on this information, we devised the extraction method of the light-emitting substance utilizing the Ca^{2+}-chelator EDTA. The purification of the extracted material led to the discovery of the first example of photoprotein, aequorin (Shimomura *et al.*, 1962). A detailed account of the discovery of aequorin has been described elsewhere (Shimomura, 1995, 2005a,b).

4.1.2 *Extraction and Purification of Aequorin*

Material for extraction. In specimens of *Aequorea*, aequorin exists in tiny granular light organs located along the outer margin of the umbrella. When the margin containing the light organs is cut off from the umbrella and only the resulting strip (2–3 mm width) is used, about 99% of the unnecessary body mass can be eliminated. Thus, we used only those strips for the extraction of aequorin. The strips can be kept in cold seawater (4–5°C) up to 12 hours without any loss of active aequorin. For the best final yield of purified aequorin, it is important to make the strips as narrow as possible without damaging the light organs. Because cutting the strips with a pair of scissors is time-consuming and impractical for a large number of jellyfish, a jellyfish-cutting machine was constructed utilizing a meat cutter blade (Fig. 4.1.2; Johnson, 1970). With this machine, one can easily cut 600

Aequorea cutting machine by Johnson

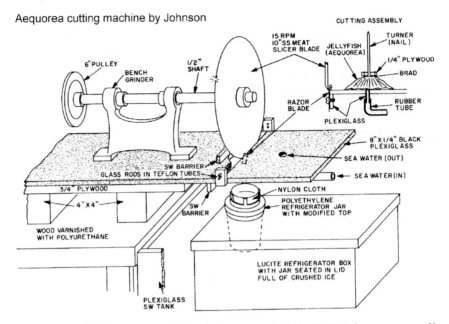

Fig. 4.1.2 Jellyfish cutting machine designed and built by F. H. Johnson to cut off the thin strings of tissue containing light organs of *Aequorea* (from Johnson, 1970). A jellyfish is placed on the seawater outlet of the black Plexiglas platform. The umbrella is spread by holding and turning the jellyfish with a turner that has spikes (brads), and then the specimen is moved toward the intersecting point of rotating meat cutter blade and razor blade to cut off the outer edge of the umbrella at a width of 2–3 mm. The excised strip drops into a container that is chilled with ice. The meat cutter blade is rotated by a motor using a pulley and belt.

or more specimens per hour, which is 10 times faster than using scissors. A simplified version of the jellyfish cutter has been developed by Blinks *et al*. (1978).

Extraction of aequorin. The method originally used by Shimomura *et al*. (1962) was considerably modified and improved by 1970 to allow the extraction of a large number of specimens (Johnson and Shimomura, 1978). The strips containing light organs prepared from 500 mature specimens (7.5–10 cm in diameter) are kept in 800 ml of chilled seawater. The seawater is drained through a piece of Dacron gauze, and the strips are added into a 2-liter flask containing 1 liter of cold 50 mM EDTA, pH 6.0–6.5, saturated with ammonium sulfate, which causes the strips to shrink. The flask is vigorously shaken for about one minute to dislodge the particles containing light organs from the strips. If the shrunken strips form a wad, it should be cut into smaller pieces, and the flask shaken again. Then, the luminescence activity of the fluid is assayed to confirm that nearly all of the particles containing active material are dislodged. The dislodgment of the particles (fluorescent in green) can be also visually monitored under a long-wave ultraviolet light.

The mixture is squeezed through a piece of Dacron gauze (50–100 mesh). The tissue mass retained on the gauze is discarded. The clouded fluid containing the particles of active material (10^{13}–10^{14} photons/ml) is mixed with 80 ml of analytical grade Celite powder and filtered on Whatman No. 3 paper (precoated with a thin layer of Celite) on a 15-cm Büchner funnel, with aspirator suction. The filter cake containing the particles of active material is transferred into a flask containing 400 ml of cold 50 mM EDTA, pH 6.5, and the flask is shaken to cytolyze the particles. The clouded fluid containing active material is filtered on Whatman No. 4 paper layered on No. 3 paper on a 18.5 cm Büchner funnel (precoated with Celite) attached to a chilled 2-liter suction flask containing 280 g of $(NH_4)_2SO_4$, with aspirator suction. The filter cake is then washed with 80 ml of cold 50 mM EDTA, pH 6.5, in the same setting. The contents of the suction flask are stirred occasionally for one day in a refrigerator to complete the precipitation of crude aequorin (and GFP). The precipitate can be collected by centrifugation at 0°C (12 min. at 17,000 g), or by filtration on paper with the aid of Celite. The crude aequorin at this stage can be kept in a freezer for years without any loss of activity.

Blinks used a different method to dislodge the particles of active material from the strips cut off from the jellyfish (Blinks *et al.*, 1978). The strips are shaken in cold seawater, and the particles dislodged are harvested by filtration on a Büchner funnel with the aid of Celite. The filter cake is first washed with 50 mM EDTA, pH 8.0, containing $(NH_4)_2SO_4$ at 75% saturation, to remove seawater. Then, the particles are cytolyzed and aequorin is extracted *in situ* by washing the filter cake with cold 50 mM EDTA, pH 8.0. The filtrate is clear and slightly greenish. The active matter in the filtrate is precipitated by saturation with $(NH_4)_2SO_4$. The advantage of Blinks' method is that it is simpler, and results in purer extracts than using the Shimomura-Johnson method. The disadvantage is that the yield is somewhat lower, and also that the isoforms composition of aequorin obtained does not represent the original composition due to the uneven loss of the isoforms in this procedure. Thus, Blinks' method is recommended when the iso-form composition of the product is unimportant.

Assay of aequorin. The assay of aequorin is simple. To a vial containing a small amount of aequorin sample (1–100 µl), 1 ml of 10 mM calcium acetate solution is injected, measuring the total amount of light emitted. The amount of the total light is proportional to the amount of aequorin in the sample.

Purification of aequorin. The purification method of aequorin reported by Shimomura *et al.* (1962) was essentially the repetition of column chromatography on DEAE-cellulose, the only usable, efficient chromatographic adsorbent available at the time. Since then, various different types of chromatographic media have been developed, and the purification method has been steadily improved.

The methods and techniques presently available for the purification of aequorin are summarized below. In the description, all buffers are pH 6.5–8, and chromatography is performed at 0–5°C.

1. Size-exclusion chromatography (gel-filtration). As the gel material, Sephadex G-75 and 100 (Pharmacia) and Ultrogel AcA 44 and 54 (IBF Biotechnics) were frequently used. Superdex 200 Prep available in recent years is an excellent medium. Gel-filtration is performed in a buffer containing 5–10 mM EDTA and 0.2 M NaCl. When the concentration of aequorin is low (≤ 1 mg/ml), aequorin is eluted as a protein of molecular weight about 21,000. However, aequorin has a tendency to reversibly aggregate at higher concentrations (≥ 5 mg/ml), especially in the presence of ammonium sulfate. Thus, in a buffer containing 1 M $(NH_4)_2SO_4$, aequorin is eluted in a range corresponding to molecular weights of 50,000 to 100,000. Therefore, it is possible to purify aequorin by gel-filtration only (Shimomura and Johnson, 1976): First, aequorin is chromatographed at a low sample concentration

(1 mg/ml) on a gel-filtration column using a buffer containing 5–10 mM EDTA and 0.2 M NaCl. Then the aequorin fraction obtained is concentrated and re-chromatographed on a column of smaller diameter using a buffer containing 1 M $(NH_4)_2SO_4$. The fractions corresponding to the molecular weight of 50,000–100,000 contain virtually pure aequorin.

2. *Anion exchange chromatography.* Although DEAE cellulose is still useful today, several types of newer media, such as Q Sepharose and DEAE Sepharose (both from Pharmacia) are now preferentially used in the purification of aequorin. In anion-exchange chromatography, aequorin is adsorbed onto a column with a buffer containing 5–10 mM EDTA and a low concentration of salt (< 0.1 M), and then the adsorbed aequorin is eluted with a higher concentration of salt. Natural aequorin elutes as a broad band due to its heterogeneous nature. For HPLC, TSK DEAE-5PW is an excellent column for the resolution of aequorin iso-forms (Shimomura, 1986a).

3. *Hydrophobic interaction chromatography.* Butyl Sepharose 4 Fast Flow (Pharmacia) is an excellent medium for purifying aequorin. Aequorin is adsorbed on the column using a buffer containing 5–10 mM EDTA and 1.8 M $(NH_4)_2SO_4$, and then eluted stepwise with buffer solutions containing decreasing concentrations of $(NH_4)_2SO_4$. Aequorin is eluted at an ammonium sulfate concentration between 1 M and 0.5 M. Because apoaequorin is eluted only at an ammonium sulfate concentration lower than 0.1 M, aequorin is cleanly separated from apoaequorin. Utilizing this fact, it is possible to prepare a sample of pure aequorin using a single column of Butyl Sepharose 4, as described below.

A sample of aequorin (purity $> 80\%$) is first luminesced by adding a sufficient amount of Ca^{2+}. To the spent luminescence solution, ammonium sulfate is dissolved to a concentration of 1 M, and then the solution is added onto a column of Butyl Sepharose 4. The apoaequorin adsorbed on the column is eluted stepwise with buffer solutions containing decreasing concentrations of $(NH_4)_2SO_4$ starting from 1 M. Apoaequorin is eluted at a $(NH_4)_2SO_4$ concentration lower than 0.1 M. The apoaequorin eluted is regenerated with coelenterazine in the presence of 5 mM EDTA and 2 mM 2-mercaptoethanol (an overnight reaction at 3–5°C; see Section 4.1.5). To the solution of regenerated aequorin, ammonium sulfate is dissolved to a concentration of 1.8 M, and the solution is added onto the same Butyl Sepharose 4 column prepared with a buffer solution containing 1.8 M $(NH_4)_2SO_4$ and 5 mM EDTA. Aequorin adsorbed on the column is eluted stepwise with buffer solutions containing decreasing concentrations of $(NH_4)_2SO_4$. The purity of aequorin obtained should be virtually 100%.

Freeze-drying of aequorin. The process of freeze-drying always results in some loss in the luminescence activity of aequorin. Therefore, aequorin should not be dried if a fully active aequorin is required. The loss is usually 10% or more. The loss can be somewhat lessened by adjusting the buffer composition; the use of 100 mM KCl and some sugar (50–100 mM) seems to be beneficial. The buffer composition used at the author's laboratory is as follows: 100 mM KCl, 50 mM glucose, 3 mM HEPES, 3 mM Bis-Tris, and at least 0.05 mM EDTA, pH 7.0.

4.1.3   Properties of Aequorin

General properties. Aequorin (M_r about 21,000) is an unusual protein that holds a large amount of energy. The energy can be released in the form of light through an intramolecular reaction triggered by calcium ions. The Ca^{2+}-triggered reaction takes place regardless of the presence or absence of molecular oxygen, resulting in the decomposition of the protein into coelenteramide, CO_2 and apoaequorin, accompanied by the emission of light (λ_{max} 465 nm). The molecule of aequorin is conformationally highly rigid and stable, but the molecule of apoaequorin is less rigid and changeable according to various observations, such as the fluorescence anisotropy of tryptophan residues and the resistance to papain digestion (La and Shimomura, 1982).

Although aequorin is non-fluorescent, the spent solution after luminescence is brightly fluorescent in blue due to the presence of coelenteramide. Although pure coelenteramide is poorly fluorescent in aqueous solutions, it becomes strongly fluorescent in a hydrophobic environment. In the presence of Ca^{2+}, coelenteramide

and apoaequorin form a fluorescent complex, of which the dissociation constant is 7.1×10^{-6} M at pH 7.4 at 25°C (Morise et al., 1974;
based on M_r 21,000 for aequorin). The blue fluorescent complex is
often referred to as BFP, and its fluorescence emission (λ_{max} 470 nm)
closely matches the bioluminescence emission of aequorin (λ_{max} about
465 nm; Shimomura and Johnson, 1970b).

According to Charbonneau et al. (1985), aequorin is a single chain
peptide consisting of 189 amino acid residues, with an unblocked
amino terminal. The molecule contains three cysteine residues and
three EF-hand Ca^{2+}-binding domains. The absorption spectra of
aequorin and BFP are shown in Fig. 4.1.3, together with the luminescence spectrum of aequorin and the fluorescence spectrum of BFP.

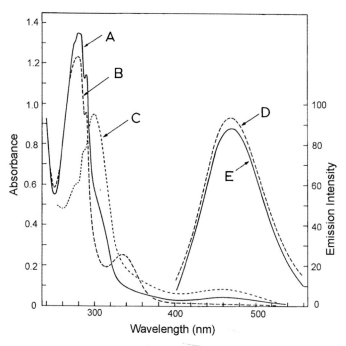

Fig. 4.1.3 Absorption spectra of aequorin (**A**), spent solution of aequorin after
Ca^{2+}-triggered luminescence (**B**), and the chromophore of aequorin (**C**). Fluorescence emission spectrum of the spent solution of aequorin after Ca^{2+}-triggered bioluminescence, excited at 340 nm (**D**). Luminescence spectrum of aequorin triggered
with Ca^{2+} (**E**). Curve C is a differential spectrum between aequorin and the protein
residue (Shimomura et al., 1974b); protein concentration: 0.5 mg/ml for **A** and **B**,
1.0 mg/ml for **C**. From Shimomura and Johnson, 1976.

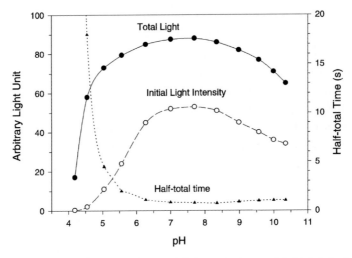

Fig. 4.1.4 Influence of pH on the total light emission and initial light intensity of aequorin. Buffer solutions containing 0.1 mM calcium acetate, 0.1 M NaCl, and 10 mM sodium acetate (for pH < 7) or 10 mM Tris-HCl (for pH > 7) were adjusted to various pH with acetic acid or NaOH, and then 2 ml of the solution was added to 3 μl of aequorin solution containing 1 mM EDTA to elicit luminescence, at 22°C. The data shown are a revision of Fig. 9 in Shimomura *et al.*, 1962. The half-total time is the time required to emit 50% of total light.

A concentrated solution of aequorin is yellowish, due to its weak absorption at 460 nm.

The luminescence of aequorin in terms of total light is efficient in a wide range of pH, from 4.5 to beyond 10 (Fig. 4.1.4). The light intensity is also optimum in a broad pH range of 7–8.5. The time course of the aequorin luminescence reaction is roughly the first order in the presence of various concentrations of Ca^{2+} (Fig. 4.1.5; Shimomura *et al.*, 1963b).

Some of the properties of aequorin are listed in Table 4.1.1.

Heterogeneity. Natural aequorin is not a homogeneous protein; it is a mixture of many isoforms having isoelectric points ranging from 4.2 to 4.9 (Blinks and Harrer, 1975). The isoform composition may vary to some extent by the purification method employed, due to uneven loss of isoforms during purification. Consequently, the properties of each preparation of aequorin may also vary. By anion-exchange

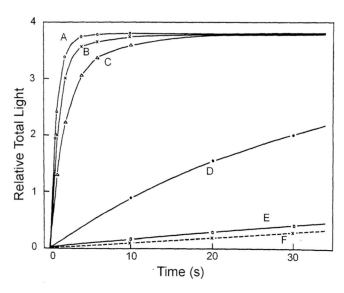

Fig. 4.1.5 The time course of aequorin luminescence measured with various concentrations of Ca^{2+}. Calcium acetate solution (5 ml) was added to 10 µl of aequorin solution to give the final Ca^{2+} concentrations of 10^{-2} M (**A**), 10^{-4} M (**B**), 10^{-5} M (**C**), 10^{-6} M (**D**), and 10^{-7} M (**E**) at 25°C. The dashed line (**F**) represents the light emitted following the addition of deionized distilled water that had been redistilled in quartz. The concentration of EDTA derived from the aequorin sample was 10^{-7} M (final conc.). From Shimomura *et al.*, 1963b, with permission from John Wiley & Sons Ltd.

HPLC, about one dozen of the isoforms, aequorins A, B, C, –J, were isolated (Shimomura, 1986a; Shimomura *et al.*, 1990). An example of HPLC separation of isoforms is shown in Fig. 4.1.6, and a comparison of the properties of the isoforms is given in Table 4.1.2.

Specificity to Ca^{2+}. Several kinds of cations other than Ca^{2+} can elicit the light emission of aequorin (Shimomura and Johnson, 1973d). Some lanthanide ions (such as La^{3+} and Y^{3+}) can trigger the luminescence as efficiently as Ca^{2+}. In addition, Sr^{2+}, Pb^{2+} and Cd^{2+} cause significant levels of luminescence. Cu^{2+} and Co^{2+} give slight luminescence only in slightly alkaline solutions (pH 8.0), but Be^{2+}, Ba^{2+}, Mn^{2+}, Fe^{2+}, Fe^{3+} and Ni^{2+} do not elicit any light from aequorin. In the test of biological systems, however, aequorin is considered to be highly specific to Ca^{2+} because the occurrence of a significant amount of metal cations other than Ca^{2+} is unlikely. In *in vitro*

Table 4.1.1   Main Properties of Natural Aequorin and its Ca^{2+}-triggered Luminescence

Molecular weight	20,000–22,000[a,b,c]
Isoelectric points	4.2–4.9[d]
$A_{1\%,1cm}$ at 280 nm	27–30[c,e] (18.0 for apoaequorin[e])
$A_{1\%,1cm}$ at 460 nm	0.81[c,f]
Solubility in aqueous buffer	> 30 mg/ml[a]
Content of functional group	1 coelenterazine group/molecule[a]
Half-life in EDTA solution	7 days at 25°C[a]
Luminescence emission maximum	465 ± 5 nm
Luminescence activity	4.3–4.5 × 10^{15} photons/mg at 25°C[e,g]
Quantum yield	0.175 at 15°C[a]
	0.15–0.16 at 25°C[c,g]
	0.11 at 35°C[a]
Number of Ca^{2+} needed for luminescence	2 Ca^{2+}/molecule[i]
Detection limit of Ca^{2+}	1 nM Ca^{2+} in low ionic strength buffers[j]
	100 nM in the presence of 0.15 M KCl [j]
Rate constant of luminescence	
Rise	100–300 s^{-1} at 20–25°C[k,l]
Decay	1.0–1.2 s^{-1} at 20–25°C[k,l]

a. Shimomura and Johnson, 1979a   g. Shimomura and Johnson, 1976
b. Charbonneau et al., 1985   h. Shimomura and Johnson, 1970b
c. Shimomura, 1986a   i. Shimomura, 1995c
d. Blinks and Harrer, 1975   j. Shimomura and Shimomura, 1984
e. Shimomura and Johnson, 1969   k. Hastings et al., 1969
f. Shimomura and Johnson, 1975c   l. Loschen and Chance, 1971

tests, the aequorin luminescence caused by all metal ions, except Ca^{2+}, Sr^{2+} and lanthanides, can be completely masked by including 1 mM sodium diethyldithiocarbamate in the test solution (Shimomura and Johnson, 1975d).

Activators and inhibitors. The total amount of light emitted in Ca^{2+}-triggered luminescence is increased by certain alcohols: for example, 10% by 2 mM n-hexanol, 30% by 2 mM n-heptanol, and 18% by saturated n-octanol (Shimomura et al., 1962; Neering and Fryer, 1986). The mechanism of the activation is unclear. No other types of activation is known.

The luminescence is strongly inhibited by bisulfite and p-dimethylaminobenzaldehyde even at micromolar concentrations of the reagents; these reagents destroy the ability of aequorin to

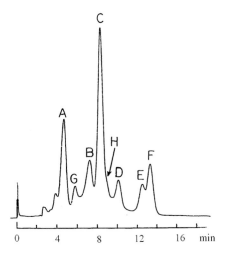

Fig. 4.1.6   HPLC analysis of a sample of purified natural aequorin on a TSK DEAE-5PW column (0.75 × 7.5 cm) eluted with 10 mM MOPS, pH 7.1, containing 2 mM EDTA and sodium acetate. The concentration of sodium acetate was increased linearly from 0.25 M to 0.34 M in 14 min after the injection of the sample. Full-scale 0.02 A. Flow rate 1 ml/min. Reproduced with permission, from Shimomura, 1986a. © the Biochemical Society.

luminesce, thus decreasing the total light emission. High concentrations (> 50 mM) of inorganic salts, such as NaCl and KCl, and millimolar concentrations of Mg^{2+}, decrease the light intensity, probably by competing with Ca^{2+}; thus, they are weakly inhibitory. Inhibition by the calcium-chelators EDTA and EGTA are discussed below.

The effects of calcium chelators. EDTA and EGTA affect the Ca^{2+}-triggered luminescence of aequorin in two ways (Shimomura and Shimomura, 1982; Ridgway and Snow, 1983). Firstly, the chelators remove free Ca^{2+} in solutions, causing the suppression of light emission. Secondly, the free (unchelated) form of the chelators interacts (binds) directly with the molecules of aequorin, causing an inhibition of luminescence (Shimomura and Shimomura, 1984). The second type of inhibition is significant in solutions of low ionic strengths, but relatively small in the presence of a high concentration of KCl (0.1 M or more); the decrease in inhibition is presumably due to the preexisting inhibition by KCl. Equilibrium dialysis of 0.1 mM aequorin in

Table 4.1.2   Properties of the Isoforms of Aequorin (Shimomura, 1986a)

	Aequorin					
	A	B	C	D	E	F
Molecular weight	22,800	22,300	21,300	21,900	20,400	20,100
A (1%, 1 cm) at 280 nm	2.92	3.10	3.02	2.92	3.01	3.00
Luminescence maximum (nm)	460	466	468	468	472	472
Luminescence activity (10^{15} photons/mg)	4.35	4.46	4.95	4.76	4.82	5.16
Quantum yield	0.165	0.165	0.175	0.174	0.163	0.172
First-order reaction rate constant (s^{-1})	0.95	1.21	1.17	1.13	1.30	1.33
Median Ca^{2+} sensitivity (pCa)*						
Low ionic strength	6.91	7.20	7.15	7.32	7.23	7.28
With 0.1 M KCl	5.81	5.96	6.00	6.25	6.01	6.11
Estimated content (%)	15.9	12.0	31.8	8.2	6.3	9.4

*Median sensitivity: The value of pCa ($-\log [Ca^{2+}]$) at which the initial maximum intensity is equal to $(I_0 I_{max})^{0.5}$, where I_0 is the intensity in the absence of Ca^{2+} and I_{max} is the intensity in 0.01 M Ca^{2+} (Shimomura and Shimomura, 1985).

3 mM MOPS/50 mM NaCl, pH 7.0, containing 5 mM, 1 mM and 0.2 mM ^{14}C-EDTA resulted in the binding of 2.3, 1.9 and 1.0 mol EDTA/mol aequorin, respectively, clearly showing the affinity between EDTA and aequorin (Shimomura and Shimomura, 1982).

The interaction between aequorin and a chelator must be carefully considered when estimating Ca^{2+} concentrations with aequorin in a calcium buffer containing EDTA or EGTA. This is particularly crucial when using a common calcium buffer system that contains a constant total concentration of a chelator in the buffer solutions of various Ca^{2+} concentrations; in such a buffer system, a buffer of lower Ca^{2+} concentration contains a higher concentration of the free form of the chelator, resulting in an increased inhibition.

Relationship between Ca^{2+} concentration and luminescence intensity. In the measurement of Ca^{2+} concentration with aequorin, the calibration of the relationship between Ca^{2+} concentration and luminescence intensity is essential. However, the application of this relationship is complicated by the chelator used to set the Ca^{2+} concentration, for the reason noted above. To minimize the complication, we used only a minimum amount of EDTA to protect aequorin in the measurements to obtain the relationship between Ca^{2+}-concentration and light intensity, and plotted the data as shown in Fig. 4.1.7 (Shimomura and Johnson, 1976). The concentration of EDTA was 2×10^{-8} M, and the total concentration of salt was also decreased to 2 mM to minimize the contaminating Ca^{2+} from the salt. In this plot, a middle portion of the curve, between Ca^{2+}-concentrations of 3×10^{-8} M and 3×10^{-7} M, is a straight line with a slope of 2.0, indicating that the intensity is proportional to the square of Ca^{2+} concentration.

The slope of 2.0 was also reported by Azzi and Chance (1969), Ashley (1970), and Baker *et al.* (1971), although several values between 2.0 and 4.0 were also reported by other investigators. The slope of 2 indicates that one aequorin molecule needs to be bound with two Ca^{2+} for the emission of light to take place; the requirement of two Ca^{2+} was confirmed later by the titration of aequorin with calcium ions (Shimomura, 1995c; Shimomura and Inouye, 1996). In Fig. 4.1.7,

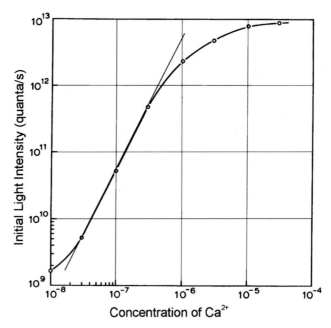

Fig. 4.1.7 Relationship between the concentration of Ca^{2+} and the initial maximum intensity of luminescence when 2.5 ml of 2 mM sodium acetate (ultrapure grade) containing the indicated amount of calcium acetate was added to 5 μl of aequorin stock solution, at 25°C. The aequorin stock solution contained 0.7 mg of aequorin in 1 ml of 2 mM sodium acetate containing 10^{-5} M EDTA. When no Ca^{2+} was added the maximum intensity was 1.1×10^9 quanta/s. From Shimomura and Johnson, 1976.

the inflexion at high Ca^{2+} concentration indicates the saturation of the Ca^{2+}-binding site of aequorin, and the inflexion at low Ca^{2+} concentration is due to the Ca^{2+}-independent luminescence that is indigenous to the aequorin molecule (aequorin is always emitting a very low level of light, regardless of Ca^{2+}). In the presence of added KCl, the curve shifts to the right, indicating that the apparent Ca^{2+}-sensitivity of aequorin is decreased. The examples showing the effect of KCl concentration in the presence of 1 mM EDTA and 1mM CDTA (*trans*-1,2-diaminocyclohexane-N,N,N′,N′-tetraacetic acid) were given by Blinks *et al.* (1982, p. 25).

Effects of various Ca^{2+} chelators. The effects of various Ca^{2+} chelators on the luminescence of aequorin are compared in the

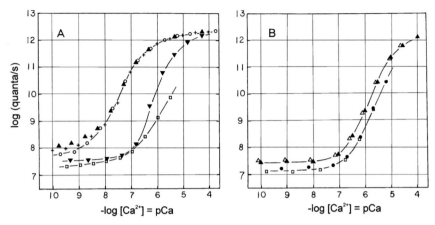

Fig. 4.1.8 Influence of various calcium chelators on the relationship between Ca^{2+} concentration and the luminescence intensity of aequorin, at 23–25°C: (*panel A*) in low-ionic strength buffers (I < 0.005) and (*panel B*) with 150 mM KCl added. Buffer solutions (3 ml) of various Ca^{2+} concentrations, pH 7.05, made with or without a calcium buffer was added to 2 µl of 10 µM aequorin solution containing 10 µM EDTA. The calcium buffer was composed of the free form of a chelator (1 or 2 mM) and various concentrations of the Ca^{2+}-chelator (1:1) complex to set the Ca^{2+} concentrations (the concentration of free chelator was constant at all Ca^{2+} concentrations). The curves shown are obtained with: 1 mM MOPS (▲), 1 mM glycylglycine (+), 1 mM citrate (○), 1 mM EDTA plus 2 mM MOPS (□), 1 mM EGTA plus 2 mM MOPS (•), 2 mM NTA plus 2 mM MOPS (▼), and 2 mM ADA plus 2 mM MOPS (△). In the chelator-free buffers, MOPS and glycylglycine, Ca^{2+} concentrations were set by the concentration of calcium acetate. Reproduced with permission, from Shimomura and Shimomura, 1984. © the Biochemical Society.

presence and absence of 150 mM KCl (Shimomura and Shimomura, 1984), as shown in Fig. 4.1.8. In obtaining these data, the concentrations of the inhibitory free forms of chelators were kept constant (1 mM or 2 mM) at all Ca^{2+} concentrations measured, to keep the inhibitory effects constant in the range of Ca^{2+} concentration studied (Shimomura and Shimomura, 1982).

In low ionic strength solutions (I < 0.005; panel A), the curves obtained with the pH buffers MOPS and glycylglycine and also with a weak Ca^{2+}-chelator citric acid are nearly superimposable on each other, whereas those obtained with the buffers that contain strong chelators EDTA and NTA are shifted almost 2 pCa units to the right. The curve for NTA is somewhat atypical; probably the Ca^{2+}-NTA ratio of the Ca^{2+}-NTA complex used was not exactly 1.0. In the presence of 150 mM KCl (panel B), all curves obtained with the pH buffers and weak Ca^{2+}-chelators (including ADA) coincide, but they are markedly shifted to the right as a group, whereas the curves obtained with EDTA and EGTA are only slightly affected. Consequently, there is a difference of only 0.2–0.3 pCa unit between the curves obtained in the presence of 1 mM free EDTA and in the absence of the chelator, showing a significant decrease in the effects of free EDTA (and EGTA) in the presence of 150 mM KCl.

Luminescence of aequorin in the absence of Ca^{2+}. All solutions of aequorin, as well as freeze-dried aequorin, are always luminescing spontaneously and constantly even in the absence of Ca^{2+} (and in the presence of a large excess of EDTA), resulting in a gradual decrease in aequorin's capability of total light emission. The intensity of this type of luminescence is quite low at 0°C, though it can easily be measured with a luminometer. The intensity is highly temperature-dependent and increases steeply when the temperature is raised (Shimomura and Johnson, 1979a). The temperature-dependent luminescence is also observed with freeze-dried aequorin suspension in certain organic solvents, such as toluene, acetone and diglyme. The quantum yield of the spontaneous luminescence of dried aequorin when warmed (< 50°C), alone or with an organic solvent, is generally in the range of 0.003–0.005, whereas that of aequorin in aqueous solution is considerably less (about 0.001 at 43°C).

Aqueous solutions of aequorin also emit light upon the addition of various thiol-modification reagents, such as p-quinone, Br_2, I_2, N-bromosuccinimide, N-ethylmaleimide, iodoacetic acid, and p-hydroxymercuribenzoate (Shimomura et al., 1974b). The luminescence is weak and long-lasting (~ 1 hour). The quantum yield varies with the conditions, but seldom exceeds 0.02 at 23–25°C. The luminescence is presumably due to destabilization of the functional moiety caused by the modification of thiol and other groups on the aequorin molecule.

Stability of aequorin. Information on the stability of aequorin is important when using this photoprotein as a calcium probe. As already noted, aequorin is always emitting a low level of luminescence, thus deteriorating itself. The stability of aequorin in aqueous solution containing EDTA or EGTA varies widely depending the temperature, pH and the concentration of salts. It is most important to keep the temperature as low as possible. The half-life of the luminescence activity of aequorin in 10 mM EDTA, pH 6.5, is about 7 days at 25°C (Table 4.1.1), and 60–150 days at 2–4°C; it appears that the rate of the inactivation of aequorin decreases gradually and significantly when a sample is kept undisturbed at the same temperature in darkness.

Stability of aequorin in a solution is dramatically increased by the addition of ammonium sulfate. Thus, when solutions of aequorin in 5 mM HEPES, pH 7.2, containing 5 mM EDTA plus 0.5, 1.0 or 2.0 M $(NH_4)_2SO_4$ were left standing at 20–22°C for 80 days in darkness, they retained 65%, 90%, and 95% of their original activity, respectively. Under similar conditions, aequorin solutions in pH 6.5 buffer containing 10 μM EDTA and 0.5 M or 1.0 M $(NH_4)_2SO_4$ retained 4% and 90% of the original activity, respectively.

Freeze-dried aequorin is also quite stable, but the process of drying always causes a loss of luminescence activity (see the last part of Section 4.1.2). All forms of aequorin are satisfactorily stable for many years at −50°C or below, but all rapidly deteriorate at temperatures above 30–35°C. A solution of aequorin should be stored frozen whenever possible because repeated freeze-thaw cycles cause little harm to aequorin activity.

4.1.4   *Discovery of the Coelenterazine Moiety in Aequorin*

AF-350. The chemical study of aequorin was started in the late 1960s. First, we tried to isolate the functional chromophore from the protein, but every trial to extract the chromophore triggered an intramolecular reaction of aequorin, resulting in a change or breakdown of the original chromophore. Thus, we shifted our target from the original chromophore to any significant fragment of the chromophore that is split off from the aequorin molecule. In 1969, we found that the denaturation of aequorin with urea in the presence of 2-mercaptoethanol results in the formation of a blue fluorescent substance, which can be extracted with *n*-butanol

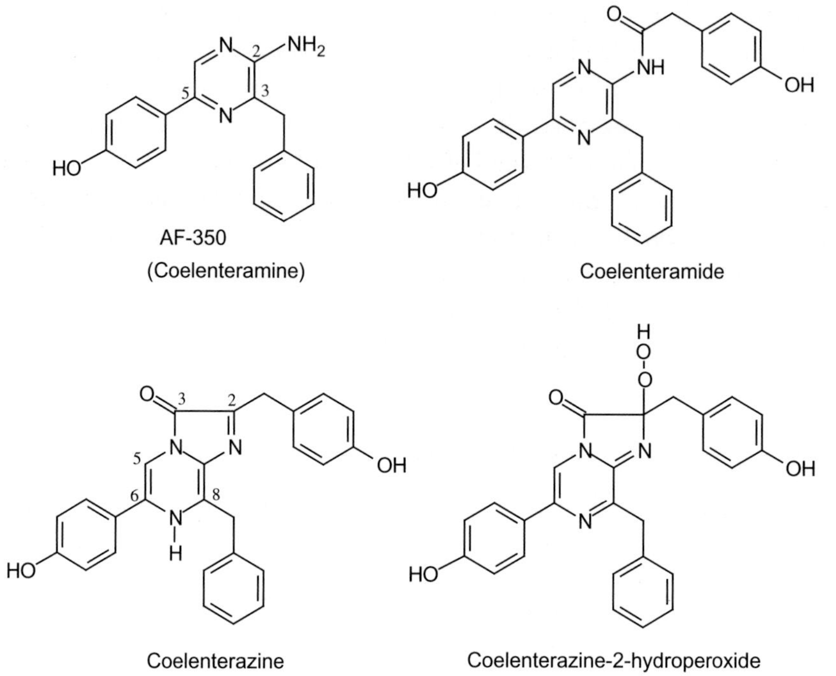

Fig. 4.1.9   Chemical structures of coelenterazine and related compounds.

(Shimomura and Johnson, 1969). The blue fluorescent substance showed an ultraviolet absorption maximum at 350 nm (in ethanol); thus, it was named AF-350 (renamed coelenteramine later). We obtained about 1 mg of AF-350 from 125 mg of pure aequorin that had been extracted and purified from 2.5 tons of *Aequorea*, and then determined its structure as shown in Fig. 4.1.9 (Shimomura and Johnson, 1972).

Coelenteramide and coelenterazine. The structure of AF-350 contains the same aminopyrazine skeleton as in *Cypridina* etioluciferin and oxyluciferin (Fig. 3.1.8), suggesting that the bioluminescence reaction of aequorin might resemble that of *Cypridina* luciferin. To investigate such a possibility, we prepared the reaction product of aequorin luminescence by adding Ca^{2+} to a solution of aequorin. The product solution (blue fluorescent) was made acidic, and extracted with *n*-butanol. A blue fluorescent compound extracted was purified and its structure was determined. The compound was a derivative of AF-350, coelenteramide, shown in Fig. 4.1.9 (Shimomura and Johnson, 1973c). This result strengthened the close relationship between the luminescence reactions of aequorin and *Cypridina* luciferin, and implied that aequorin luminescence might involve coelenterazine as the luciferin or the light-emitting principle of aequorin (Shimomura *et al.*, 1974b). Although coelenterazine was a hypothetical compound at this point, it was actually isolated a year later from the luminous squid *Watasenia* and chemically synthesized (Inoue *et al.*, 1975).

4.1.5   *Regeneration of Aequorin from Apoaequorin*

A solution of aequorin was luminesced with an excess amount of Ca^{2+}. The blue fluorescent solution obtained was passed through a column of Sephadex G-25 prepared with 10 mM Tris-HCl buffer containing 10 mM EDTA, pH 7.5, to remove coelenteramide and Ca^{2+}. The eluted protein, apoaequorin, was regenerated into aequorin by incubation with synthetic coelenterazine and 2 mM 2-mercaptoethanol at 5°C for 3 hours (Shimomura and Johnson, 1975c). The regenerated aequorin was indistinguishable from the original aequorin in all properties tested. Thus, the existence of a coelenterazine moiety in aequorin molecule was positively confirmed. In molecules of aequorin, coelenterazine exists in a form of peroxide (Fig. 4.1.9), as indicated by various lines of chemical evidence (Shimomura and Johnson, 1978) and ^{13}C-NMR data (Musicki *et al.*, 1986). A general scheme of the luminescence and regeneration of aequorin is shown in Fig. 4.1.10.

During the process of regeneration, apoaequorin in a somewhat unfolded conformation is converted into a rigid conformation around a core of coelenterazine. It appears that coelenterazine has a strong capability to refold the unfolded apoaequorin molecules. In fact, apoaequorin denatured under various drastic conditions, such as treatment with 1 M HCl, 1 M NaOH, and 6 M urea, or heating at 95°C, can be regenerated into aequorin with yields over 50% (Shimomura and Shimomura, 1981).

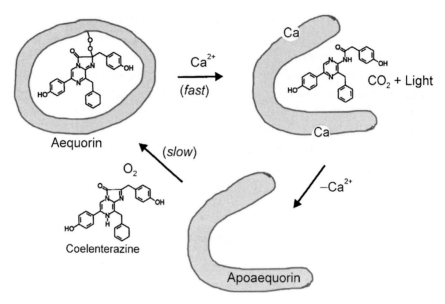

Fig. 4.1.10 A diagram illustrating the luminescence reaction and regeneration of aequorin. From Shimomura, 2005b, with permission from World Scientific Publishing Co.

Requirements in aequorin regeneration. In the regeneration reaction, apoaequorin, coelenterazine and molecular oxygen are essential requirements, and Ca^{2+} must be absent. The role of 2-mercaptoethanol (or DTT) is auxiliary; it increases the yield of regenerated aequorin, possibly by protecting the SH groups of cysteine residues in the apoaequorin molecule. In fact, when all three cysteine residues of recombinant apoaequorin had been replaced with serine residues by a site-specific mutagenesis, the modified aequorin regenerated from this apoaequorin was fully active in the luminescence reaction, although the rate of the regeneration was significantly lower (Kurose *et al.*, 1989). Thus, the cysteine residues probably play a role in the regeneration reaction, but not in the luminescence reaction.

The regeneration of native-type aequorin is not hindered by the presence of coelenteramide. Therefore, apoaequorin in a spent solution of luminescence can be regenerated into aequorin by simply adding an excess amount of EDTA to chelate the existing Ca^{2+}, together with coelenterazine and some 2-mercaptoethanol, followed by incubation at 0–5°C for several hours. If the regeneration reaction is performed in the presence of Ca^{2+} (without adding EDTA), the regenerated aequorin will react with Ca^{2+} as it forms, resulting in a weak long-lasting luminescence that continues until all coelenterazine is consumed. In this case, apoaequorin acts as a luciferase; however, it is a very sluggish enzyme having a turnover number of 1–2 per hour.

Yield of regenerated aequorin. Regeneration of aequorin is a slow reaction, which takes 25–30 minutes for a half completion, and about 3 hours for a 90% completion at 5°C; the reaction is essentially complete in 12 hours. The concentration of apoaequorin has important effects on both the rate of regeneration and the yield of regenerated aequorin, as shown in Fig. 4.1.11. The relationship

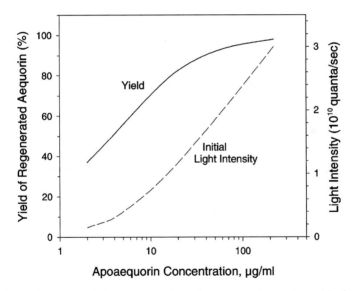

Fig. 4.1.11 Influence of the concentration of apoaequorin on the yield of regenerated aequorin after 12 h at 4°C (solid line), and on the initial light intensity of the apoaequorin-catalyzed luminescence of coelenterazine (dashed line). The regenerated aequorin was measured with a 10 µl portion of a reaction mixture (0.5 ml) made with 10 mM Tris-HCl, pH 7.5, containing 1 mM EDTA, 5 mM 2-mercaptoethanol, 10 µl of methanolic 0.6 mM coelenterazine, and various amounts of apoaequorin. The luminescence activity of apoaequorin was measured in 2 ml of 10 mM Tris-HCl, pH 7.5, containing 0.5 M NaCl, 2 mM $CaCl_2$, 2 mM 2-mercaptoethanol, 10 µl of methanolic 0.2 mM coelenterazine, and various amounts of apoaequorin. Reproduced with permission, from Shimomura and Shimomura, 1981. © the Biochemical Society.

between apoaequorin concentration and the reaction rate (light intensity) is not linear; therefore, a calibration curve would be needed to assay apoaequorin on the basis of its luciferase activity catalyzing the luminescence of coelenterazine. The data also indicate that more than 0.05 mg/ml of apoaequorin would be needed to achieve a near quantitative regeneration. The yield of aequorin regeneration with low concentrations of apoaequorin can be significantly increased by including 0.1% BSA in the regeneration buffer.

4.1.6   Recombinant Aequorin

The cloning and expression of apoaequorin cDNA were accomplished by two independent groups in 1985–1986. One of them analyzed the cDNA clone AQ440 they obtained, and reported that apoaequorin is composed of 189 amino acid residues (M_r 21,400) with a NH_2-terminal valine and a COOH-terminal proline (Inouye et al., 1985, 1986); the results are consistent with the amino acid sequence data of native aequorin reported by Charbonneau et al. (1985). The other group reported that the cDNA AEQ1 they obtained had contained the entire protein coding region of 196 amino acid residues, which included 7 additional residues attached to the N-terminus, and the apoaequorin expressed in E. coli showed a molecular weight of 20,600 (Prasher et al., 1985, 1987). In our HPLC analysis, the recombinant aequorin from the former group matched well with the recombinant aequorin sample purchased from Molecular Probes, which is believed to be originated from the latter group, but did not exactly match with any of the iso-forms of natural aequorin (Shimomura et al., 1990).

Overexpression of apoaequorin (Inouye et al., 1989, 1991). To produce a large quantity of apoaequorin, an apoaequorin expression plasmid piP-HE containing the signal peptide coding sequence of the outer membrane protein A (ompA) of E. coli (Fig. 4.1.12) was constructed and expressed in E. coli. The expressed apoaequorin was secreted into the periplasmic space of bacterial cells and culture medium. The cleaving of ompA took place during secretion; thus the

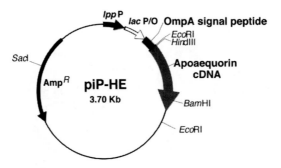

Fig. 4.1.12   Apoaequorin expression plasmid piP-HE (from Inouye *et al.*, 1989, and Shimomura and Inouye, 1999). Reproduced with permission from Elsevier.

secreted apoaequorin contained 191 amino acid residues, two residues more than the 189 residues that were previously deduced from the cDNA sequence. The increase of the two residues is caused by the replacement of the NH_2-terminus valine with Ala-Asn-Ser-, due to the use of a linker sequence between the ompA signal sequence and the apoaequorin sequence. The apoaequorin secreted into the culture medium is purified by acid precipitation and DEAE-cellulose chromatography, yielding 180 mg of purified apoaequorin from one liter of the supernatant of culture medium.

An improved method of producing recombinant aequorin was devised based on the fact that the expression of the peak amount of apoaequorin in bacterial cells occurs several hours before the secretion into culture medium (Shimomura and Inouye, 1999). The cells containing apoaequorin in the periplasmic space, before secretion, are extracted under a very mild condition and, at the same time, converted into aequorin. The purification of the extract by two steps of column chromatography yields a high-purity preparation of recombinant aequorin.

X-ray structure of aequorin (Head *et al.*, 2000). X-ray crystallography was performed with the recombinant aequorin prepared by the improved method described above. The crystals of recombinant aequorin were grown in a high concentration of ammonium sulfate. The results revealed that aequorin is a globular molecule containing a

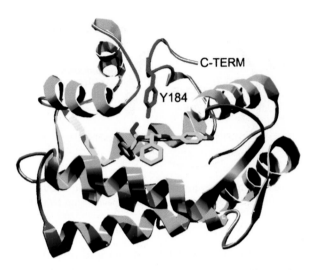

Fig. 4.1.13   A ribbon representation of the crystal structure of recombinant aequorin molecule showing the secondary structure elements in the protein. Alpha-helices are denoted in cyan, beta-sheet in yellow, loops in magenta; coelenterazine (yellow) and the side chain of tyrosine 184 are shown as stick representations. From Head *et al.*, 2000, with permission from Macmillan Publishers.

hydrophobic core cavity that accommodates the ligand coelenterazine-2-hydroperoxide (Fig. 4.1.13). The hydroperoxide is stabilized by hydrogen bonds involving the protein residues. The core cavity accommodating coelenterazine moiety is situated in the center of the protein, and it is closed to a spherical probe of 1.4 Å radius; there is no solvent access to the coelenterazine ligand from the surface. Lining the cavity are the side chains of 21 hydrophobic residues, 3 tyrosines, 3 histidines, 2 threonines and 1 lysine. The molecule contains four helix-loop-helix EF-hand domains, of which three (E-F hands I, III and IV) can bind Ca^{2+}. According to Toma *et al.* (2004), the loop of the E-F hand IV is clearly deformed. Thus, it appears most likely that the binding of the E-F hands I and III with Ca^{2+} triggers the luminescence of aequorin.

4.1.7   Semisynthetic Aequorins

The core cavity of the aequorin molecule can accommodate various synthetic analogues of coelenterazine in place of coelenterazine. The replacement of the coelenterazine moiety of aequorin can be performed by a simple process. First, aequorin is luminesced by adding Ca^{2+}. Then the apoaequorin produced is regenerated with an analogue of coelenterazine in the presence of EDTA, 2-mercaptoethanol (or DTT) and molecular oxygen (Shimomura *et al.*, 1988, 1989, 1990, 1993a). The product is called a semisynthetic aequorin, and identified with an italic prefix (see Tables 4.1.3 and 4.1.4). Semisynthetic aequorins can be prepared from both native aequorin and recombinant aequorin, using various synthetic analogues of coelenterazine. About 50 kinds of semisynthetic aequorins were prepared and their characteristics investigated. Some of the characteristics of semisynthetic aequorins are significantly different from those of natural aequorin, particularly in the sensitivity to Ca^{2+}, the rate of the luminescence reaction, and the rise time of luminescence, which are summarized in Tables 4.1.3 and 4.1.4. The relationship between Ca^{2+} concentration and the initial light intensity of some semisynthetic aequorins are shown in Fig. 4.1.14.

Structural limitations of coelenterazine analogues. Aequorin can accommodate a surprisingly wide range of coelenterazine analogues in its core cavity. Some structural limitations of coelenterazine analogues for producing active semisynthetic aequorins are given below (Shimomura *et al.*, 1988, 1989; Ohmiya *et al.*, 1993a).

R^1 group: This group must be an aromatic ring. The original *p*-hydroxyphenyl group can be replaced by a bulky naphthyl group, but not by an adamantane group. The substitution of the original phenolic OH (*p*) with F or H increases the relative intensity (thus Ca^{2+} sensitivity) of the semisynthetic aequorin produced, but the substitution of OH with bulky I, Br or Cl, or the substitution of the phenyl group with a naphthyl group decreases the sensitivity, probably by interfering with the conformational changes of protein structure that follow the binding of Ca^{2+} to aequorin (Toma *et al.*, 2004). The linker CH_2 group cannot be removed, although it can be replaced by $(CH_2)_2$ or $(CH_2)_3$ (Qi *et al.*, 1991). Interestingly, the ratio of the total light amounts from the semisynthetic aequorins with 2-$CH_2C_6H_5$, 2-$(CH_2)_2C_6H_5$, and 2-$(CH_2)_3C_6H_5$ groups are 1.0:0.09:0.32, showing that $(CH_2)_2$ linker is less favorable than $(CH_2)_3$ linker.

Table 4.1.3 Properties of Semisynthetic Aequorins prepared from a single batch of Natural Aequorin measured at 23–25°C (Shimomura, 1991a)

No. (Prefix)	Structural Modification of Coelenterazine[a]	Luminescence Max. (nm)	Relative Total Light Amount[b]	Relative Intensity in 10^{-6} or 10^{-7} M Ca^{2+}[c]	Half-total Light Time (s)[d]
1	None	465	1.00	1.00	M
2 (b)	R^1: C_6H_5	464	0.82	10	M
3 (f)	R^1: $C_6H_4F(p)$	473	0.80	18	M
4 (f2)	R^1: $C_6H_3F_2(m,p)$	470	0.68	24	M
5 (f5)	R^1: C_6F_5	465	0.14	25	M
6 (cl)	R^1: $C_6H_4Cl(p)$	463	0.85	0.19	4
7 (br)	R^1: $C^6H_4Br(p)$	473	0.80	0.37	2.4
8 (i)	R^1: $C_6H_4I(p)$	476	0.70	0.03	8
9 (n)	R^1: β-naphthyl	467	0.26	0.01	5
10 (b)	R^2: n-butyl	448	0.79	12	M
11 (ip)	R^2: isopropyl	441	0.54	47	1
12 (cp)	R^2: cyclopentyl	442	0.95	15	F
13 (ch)	R^2: cyclohexyl	452	1.00	4.2	F
14 (chm8)	R^2: cyclohexylmethyl	442	0.77	2	M
15 (m8)	R^2: benzyl	444	0.60	11	M
16 (f8)	R^2: p-fluorophenyl	466	0.70	2.8	M
17 (fb)	R^1: $C_6H_4F(p)$, R^2: n-butyl	460	0.37	73	M
18 (fip)	R^1: $C_6H_4F(p)$, R^2: isopropyl	449	0.16	140	M
19 (bcp)	R^1: C_6H_5, R^2: cyclopentyl	444	0.67	190	F

(Continued)

Table 4.1.3 (Continued)

No. (Prefix)	Structural Modification of Coelenterazine[a]	Luminescence Max. (nm)	Relative Total Light Amount[b]	Relative Intensity in 10^{-6} or 10^{-7} M Ca^{2+}[c]	Half-total Light Time (s)[d]
20 (*fcp*)	R^1: $C_6H_4F(p)$, R^2: cyclopentyl	452	0.57	135	M
21 (*bch*)	R^1: C_6H_5, R^2: cyclohexyl	450	0.70	45	M
22 (*fch*)	R^1: $C_6H_4F(p)$, R^2: cyclohexyl	461	0.70	66	F
23 (*m5*)	R^4: methyl	438	0.28	0.33	M
23A (*m*)	R^5: CH_2	460	0.01	~1	F
24 (*e*)	R^5: CH_2CH_2	405, 465	0.50	4	F
25 (*eb*)	R^1: C_6H_5, R^5: CH_2CH_2	405, 465	0.47	9	F
26 (*ef*)	R^1: $C_6H_4F(p)$ R^5: CH_2CH_2	410, 475	0.50	35	F
27 (*ecb*)	R^2: cyclohexyl, R^5: CH_2CH_2	400	0.46	4	F
28 (*ebch*)	R^1: C_6H_5, R^2: cyclohexyl, R^5: CH_2CH_2	400	0.30	14	F
29 (*efch*)	R^1: C_6H_4F, R^2: cyclohexyl, R^5: CH_2CH_2	400	0.30	55	F
30 (*ecp*)	R^2: cyclopentyl, R^5: CH_2CH_2	400, 440	0.29	11	F

(Continued)

Table 4.1.3   (Continued)

No. (Prefix)	Structural Modification of Coelenterazine[a]	Luminescence Max. (nm)	Relative Total Light Amount[b]	Relative Intensity in 10^{-6} or 10^{-7} M Ca^{2+}[c]	Half-total Light Time (s)[d]
31 (ebcp)	R^1: C$_6$H$_5$, R^2: cyclopentyl, R^5: CH$_2$CH$_2$	400, 440	0.29	50	F
32 (efcp)	R^1: C$_6$H$_4$F, R^2: cyclopentyl, R^5: CH$_2$CH$_2$	400, 440	0.24	172	F
33 (v)	R^5: CH=CH	485	0.035	~10	70
34 (a)	R^3: NH$_2$	460	0.08		F

Structure labels: R^1 (-C$_6$H$_4$OH-p), R^2 (-C$_6$H$_5$), R^3 (-OH), R^4, R^5

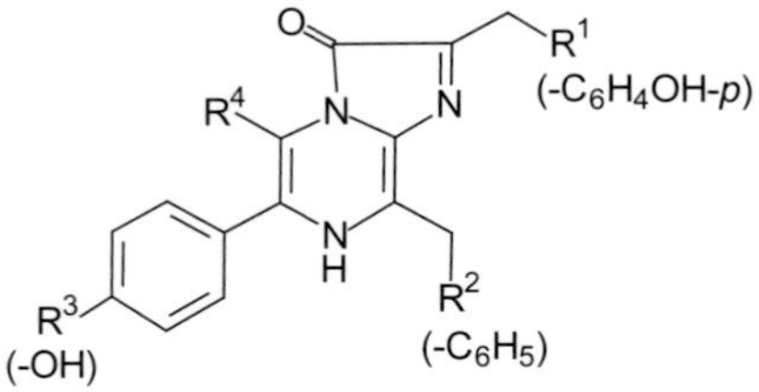

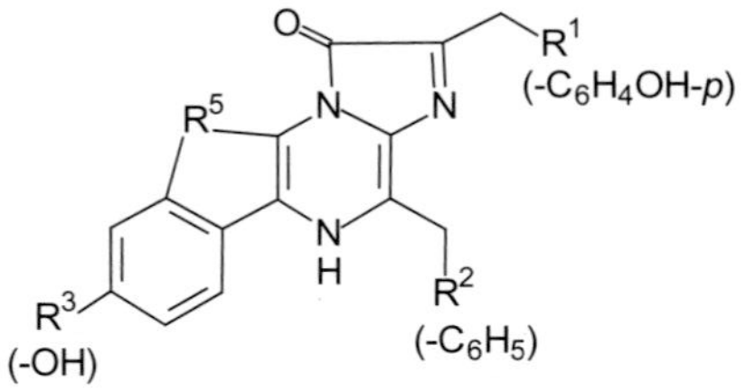

a. This column shows only the changes from the original structures of coelenterazine and *e*-coelenterazine. The original groups are shown in the parentheses in the structures shown above.

b. The ratio of the total light amount emitted from a semisynthetic aequorin to that of natural aequorin, on the basis of equal protein amounts.

c. The ratio of the luminescence intensity of semisynthetic aequorin to that of natural aequorin, with 10^{-7} M Ca^{2+} when the value is 1 or larger, and with 10^{-6} M Ca^{2+} when the value is less than 1, on the basis of equal protein amounts, measured in 1 mM EGTA calcium buffer containing 100 mM KCl and 2 mM MOPS (pH 7.0). The value represents the sensitivity to Ca^{2+}.

d. The time required to emit 50% of the total light in 10 mM calcium acetate at 25°C: F, 0.15–0.3 s; M, 0.4–0.8 s (the half-rise time of luminescence: F, 2–4 ms, all others, 6–20 ms).

Table 4.1.4   Semisynthetic Aequorins derived from Recombinant Aequorin (Shimomura *et al.*, 1993a)

No. (Prefix)	Structural Modification of Coelenterazine[a]	Luminescence Max. (nm)	Relative Total Light Amount[b]	Relative Intensity at 10^{-6} or 10^{-7} M Ca^{2+c}	Half-total Light Time (s)[d]
1	None	466	1.00	1.00	M
2 (h)	R¹: C_6H_5	466	0.75	16	M
3 (f)	R¹: $C_6H_4F(p)$	472	0.80	20	M
4 (f2)	R¹: $C_6H_3F_2(m,p)$	470	0.80	30	M
6 (cl)	R¹: $C_6H_4Cl(p)$	464	0.92	0.6	5
9 (n)	R¹: β-naphthyl	468	0.25	0.15	5
9' (nJ)[e]	R¹: β-naphthyl	467	0.30	0.07	5
12 (cp)	R²: cyclopentyl	442	0.63	28	F
13 (ch)	R²: cyclohexyl	453	1.00	15	F
17 (fb)	R¹: $C_6H_4F(p)$, R²: *n*-butyl	460	0.20	1100	2
19 (hcp)	R¹: C_6H_5, R²: cyclopentyl	445	0.65	500	F
21 (hch)	R¹: C_6H_5, R²: cyclohexyl	450	0.52	80	F
22 (fcb)	R¹: C_6H_4F, R²: cyclohexyl	462	0.43	73	M
23 (m5)	R⁴: methyl	440	0.37	2	M
24 (e)	R⁵: CH_2CH_2	405, 472	0.50	6	F
26 (ef)	R¹: $C_6H_4F(p)$, R⁵: CH_2CH_2	405, 470	0.35	40	F
27 (ech)	R²: cyclohexyl, R⁵: CH_2CH_2	402, 440	0.40	8	F
Fluorescein-labeled[f]		528	1.00	2	M

a, b, c, and d. See footnotes of Table 4.1.3.
e. Prepared from aequorin isoform J.
f. Fluorescein was chemically bound to apoaequorin, followed by regeneration with unmodified coelenterazine.

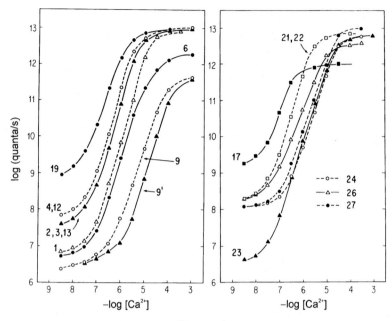

Fig. 4.1.14   Relationship between Ca^{2+} concentration and the initial light intensity of various recombinant semisynthetic aequorins and n-aequorin J (a semisynthetic natural aequorin made from isoform J). The curve number corresponds to the number of semisynthetic aequorin used in Table 4.1.4. A sample aequorin (3 μg) was in 3 ml of calcium-buffer solution containing 1 mM total EGTA, 100 mM KCl, 1 mM Mg^{2+} and 1 mM MOPS (pH 7.0), at 23–24°C. From Shimomura *et al.*, 1993a, with permission from Elsevier.

R^2 group: This group must be larger than the ethyl group. No aromaticity is needed at this position. Replacement of the original phenyl group with a cyclohexyl or cyclopentyl group increases the sensitivity to Ca^{2+} and shortens the rise time without a significant decrease in the total light emission. Unlike the aromatic phenyl group, these aliphatic groups do not interact hydrophobically with the surrounding amino acid residues, giving these groups more conformational freedom, and consequently promoting the light-emitting reaction (Toma *et al.*, 2004). A bulky adamantane group can be accommodated at this position. A carboxyl group attached to the end of a substituent shows an adverse effect.

R^3 group: A p-hydroxyphenyl or p-aminophenyl group at this position appears to be an essential requirement for the formation of active aequorin. It cannot be replaced by a naphthol group. An introduction of CH_3 or OH group to the α-positions of the phenyl group is detrimental to the luminescence reaction, presumably by obstructing free rotation of the phenyl ring around its bond axis with the imidazopyrazinone ring.

Sensitivity to Ca^{2+}. Aequorin is a highly sensitive indicator for detecting and measuring Ca^{2+}, and it has been used as a valuable probe for studying intracellular calcium. In comparing the sensitivities of various kinds of semisynthetic aequorin, it would be desirable to standardize the sensitivity. A simple and practical method to grade the Ca^{2+} sensitivity would be to compare the luminescence intensity of a semisynthetic aequorin with that of native-type aequorin under identical conditions, though it is by no means a perfect way. In this method, the initial light intensities of both aequorins are measured in a buffer solution of a certain Ca^{2+} concentration that would give their intensity readings in the steep slope region of the log (light intensity)$-$log $[Ca^{2+}]$ curve (see Figs. 4.1.7 and 4.1.14), and the sensitivity is expressed by the ratio of the light intensities of a semisynthetic aequorin to that of native-type aequorin. This method is used in Tables 4.1.3 and 4.1.4. With this method, a very wide range of Ca^{2+}-sensitivities is found with various natural and recombinant semisynthetic aequorins.

The second procedure is to measure the luminescence intensities at various Ca^{2+} concentrations and plot log (light intensity) against $-$log $[Ca^{2+}]$ for each aequorin. Examples of this method are shown in Fig. 4.1.14. This method provides more detailed information on the sensitivity of each aequorin. Generally, an increase in Ca^{2+} sensitivity shifts the curve to the left.

The third method uses a term "median sensitivity" to express the sensitivity (Shimomura and Shimomura, 1985). Median sensitivity is the pCa value ($-$log $[Ca^{2+}]$) at which the initial light intensity is equal to the square root of $I_0 I_{max}$, where I_0 is the initial intensity measured with no Ca^{2+} added and I_{max} is the initial intensity measured with 10 mM Ca^{2+}. Thus, at the median sensitivity on pCa scale, the initial light intensity value is at the midpoint of I_0 and I_{max} on log (light intensity) scale. Experimentally, the median sensitivity value can be obtained relatively easily, by measuring initial light intensities only at 2–3 different Ca^{2+} concentrations in addition to the values of I_0 and I_{max}. This method is used in Table 4.1.2. Due to the square law relationship between light intensity and Ca^{2+}-concentration, a difference of 1.0 pCa in the median sensitivity values corresponds to approximately 100 in the relative luminescence intensity in the first method.

e-Aequorins. *e*-Aequorins contain a ligand of *e*-coelenterazine or its analogues. In the structure of *e*-coelenterazines, the C5 of the imidazopyrazinone is connected to the α-C atom of the 6-(*p*-hydroxyphenyl) group with an ethylene linker (CH_2-CH_2); thus the planes of the two ring systems are restrained and kept at a certain angle. The luminescence reactions of *e*-aequorins are fast, with a half-rise time of 2–4 ms and a half-total light (or half-decay) time of 0.15–0.3 s. *e*-Type semisynthetic aequorins differ significantly from other aequorins also in certain other properties. For example, *e*-type aequorins show bimodal luminescence spectra with peaks at 400–405 nm and 440–475 nm (Tables 4.1.3 and 4.1.4). The ratio of the two peaks varies not only by the type of aequorins but also by the measurement conditions such as pH and Ca^{2+} concentration (Shimomura *et al.*, 1988, 1989). The fluorescence emission spectra of the spent solutions of *e*-type aequorins are different from their spectra of Ca^{2+}-triggered luminescence (Fig. 4.1.15), showing the fluorescence emission maxima at about 420 nm (Shimomura, 1995d). This must indicate a change in the protein environment surrounding the coelenteramide moiety after the emission of light.

One of the most significant differences between *e*-aequorins and ordinary aequorins is seen in their regeneration reaction. The rates of the regeneration of aequorins vary widely by the structures of coelenterazine analogues, as well as by the presence or absence of Ca^{2+} (Shimomura *et al.*, 1993a; Shimomura, 1995d), as shown in Table 4.1.5. In the absence of Ca^{2+}, regenerated aequorin accumulates in the reaction medium during the course of regeneration reaction. In the presence of Ca^{2+}, the regenerated aequorin spontaneously reacts

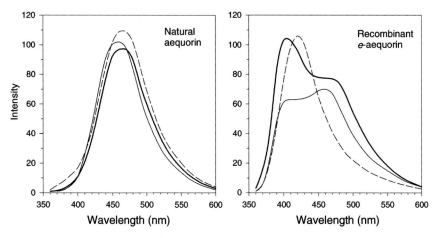

Fig. 4.1.15 Comparison of the luminescence and fluorescence emission spectra of natural aequorin (*left panel*) and recombinant *e*-aequorin (*right panel*): the luminescence spectra of Ca^{2+}-triggered reaction (dark solid lines), the fluorescence emission spectra of the spent solution containing 2 mM Ca^{2+} (dashed lines), and the luminescence spectra of the spent solution after addition of coelenterazine (light solid lines). Reproduced with permission, from Shimomura, 1995d. © the Biochemical Society.

with Ca^{2+}, causing a weak, steady luminescence until coelenterazine is exhausted. The data in Table 4.1.5 indicate that *e*-aequorin is scarcely regenerated in the presence of Ca^{2+}, a distinct difference from the ordinary types of aequorins (except *cl*-aequorin which is regenerated rather slowly). It appears that the regeneration of *e*-type aequorins from apoaequorin and *e*-coelenterazines is inhibited by Ca^{2+}.

Preparation of semisynthetic aequorins. The best yield of semisynthetic aequorins can be obtained by using the apoaequorin prepared from aequorin immediately before use. Apoaequorin stored, even for 2–3 days, or recombinant apoaequorin isolated from a bacterial culture will give a significantly lower yield due to their somewhat unfolded molecular conformation. Fresh apoaequorin prepared by the Ca^{2+}-triggered luminescence reaction appears to have the conformation best suited for regeneration.

To prepare apoaequorin from aequorin, 10 mM calcium acetate is added dropwise to a solution of aequorin (1–2 mg/ml) containing less than 2–3 mM EDTA until its ability to luminesce is completely

Table 4.1.5   Relative Rates of the Regeneration of Semi-synthetic Aequorins in the absence and presence of Ca^{2+} at 24°C (Shimomura, 1995d)

Coelenterazine Analogue used*	Initial Rate of Regeneration (in relative light units/s)	
	Without Ca^{2+} (with EDTA)	In the presence of Ca^{2+}
(Coelenterazine)	7.5	2.7
h	1.3	4.5
f	1.2	4.0
cl	5.4	0.5
ch	6.6	12.0
hch	1.7	21.8
hcp	1.2	14.2
fch	1.3	9.0
e	12.4	0.13
eh	2.4	0.04
ehch	1.8	0.07
efcp	1.4	0.13

The initial rate is shown in terms of the total light emitting capacity of regenerated aequorin in arbitrary units, when 40 µg of apoaequorin in 0.5 ml of pH 7.5 buffer is allowed to react with 2.5 µg of coelenterazine analogues. One unit corresponds to approx. 2.2×10^{11} quanta.

*Shown with a prefix that identifies the structure of coelenterazine analogue. See Table 4.1.3.

exhausted. This will take 10–20 min. Because the binding of Ca^{2+} with aequorin and EDTA produces acidity, the aequorin solution used should have contained at least 10 mM of a pH buffer. If the aequorin solution contains a high concentration of EDTA (>5 mM), it is recommended to decrease EDTA to 0.1–1 mM by gel filtration before adding Ca^{2+}. After the luminescence is completed, the apoaequorin in the spent solution can be precipitated with ammonium sulfate and centrifuged. The pellet is dissolved in a regeneration buffer: 10 mM HEPES, pH 7.6, containing 2–5 mM EDTA, 0.2–0.5 M KCl, NaCl or $(NH_4)_2SO_4$, and 2–5 mM 2-mercaptoethanol (or 1–2 mM DTT). Alternatively, the spent solution can be passed through a gel filtration column (such as Sephadex G-25 Fine) that has been prepared with the regeneration buffer. The gel filtration method should be chosen when coelenteramide hinders the formation of semisynthetic aequorin.

The regeneration is started by the addition of a solution of coelenterazine analogue to the regeneration buffer containing apoaequorin at 0–5°C. The amount of the coelenterazine analogue added should be about 10–20% excess of apoaequorin on molar basis. The progress of regeneration is periodically monitored using a small portion of the reaction mixture. The formation of a semisynthetic aequorin is often slower than that of natural aequorin, and the regeneration is usually completed in a period of 6 hours to 2 days.

Coelenterazine analogues are easily soluble in methanol like coelenterazine. When methanol is used, however, the methanol concentration in the regeneration mixture should not exceed 5%. If the use of methanol must be avoided, dissolve the coelenterazine analogue in water with the help of a trace amount of 1 M NaOH. However, coelenterazines in alkaline condition are extremely unstable. Therefore, the solution must be made rapidly in argon atmosphere and added at once to the regeneration mixture containing apoaequorin.

When regeneration is completed, excess coelenterazine and any low molecular weight substances (such as 2-mercaptoethanol) can be removed from regenerated aequorins by gel filtration. A technique of gel filtration suitable for a small volume of sample (0.1–0.2 ml) has been described (Shimomura, 1991a).

4.1.8 The In Vivo Luminescence of Aequorea

Live specimen of *Aequorea aequorea* emits green light (λ_{max} 508–509 nm) whereas the photoprotein aequorin isolated from the same jellyfish emits blue light (λ_{max} 460–465 nm). The difference is due to the presence of a green fluorescent protein (GFP) in the photogenic cells of *Aequorea*. Although a green fluorescence substance in *Aequorea* had been found by Davenport and Nicol (1955), it was isolated and identified as a protein for the first time during the purification of aequorin in 1961 (Shimomura *et al.*, 1962; Johnson *et al.*, 1962). Morise *et al.* (1974) isolated three major isoforms of GFP, and partially characterized the proteins; further characterization was reported by Prendergast and Mann (1978).

Properties of GFP. *Aequorea* GFP is a relatively stable protein with a molecular weight of about 27,000 (Shimomura, 1979). The protein is resistant to denaturation by proteases (Roth and Ward, 1983; Roth, 1985) and detergents (González *et al.*, 1997). *Aequorea* GFP can be easily crystallized by dialysis against pure water (Morise *et al.*, 1974). The crystals are thin needles (2–3 μm wide, 50–100 μm long) with a hollow core along the axis of the crystal, and show the striking anisotropy of fluorescence emission (Inoue *et al.*, 2002). The fluorescence emission spectrum shows a sharp peak with a maximum at 508–509 nm (Fig. 4.1.16). The fluorescence quantum yield is approximately 0.8 (Morise *et al.*, 1974; Kurian *et al.*, 1994).

According to Ward (1998), *Aequorea* GFP exists as the monomer at low concentrations (< 0.5 mg/ml), and the dimeric form becomes predominant at concentrations higher than 5 mg/ml. The absorption spectrum of the monomer shows three peaks: 280 nm (ε 22,000), 397 nm (ε 27,600), and 475 nm (ε 14,000), of which the last peak

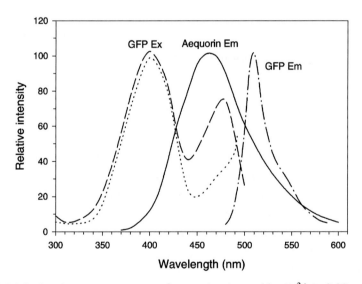

Fig. 4.1.16 Luminescence spectrum of aequorin triggered by Ca^{2+} (solid line; λ_{max} 465 nm), and the fluorescence spectra of *Aequorea* GFP: excitation (dashed line; λ_{max} 400 nm and 477 nm) and emission (dash-dot line; λ_{max} 509 nm). The dotted line is the fluorescence excitation spectrum of GFP in the light organs, showing that 480 nm excitation peak is almost missing — an evidence showing that GFP in light organs exists in an aggregated form having a very low ε value at 480 nm.

is greatly decreased when dimerized (ε 3,000); thus, the ratio of A_{475}/A_{400} is variable by the concentration of GFP, and increases by dilution (Morise *et al.*, 1974). When the molecule of *Aequorea* GFP is fused with the aequorin molecule, the A_{475} value is significantly increased at the expense of the A_{400} value (Gorokhovatsky *et al.*, 2004).

Structure of GFP and its chromophore. To study the chromophore of GFP, a sample of GFP was denatured by heating it at 90°C. It was digested with papain, and then a peptide containing the fluorophore was isolated and purified from the digested mixture. The structural study of the peptide has indicated that the chromophore of GFP is an imidazolone derivative shown below (Shimomura, 1979). This chromophore structure was confirmed later by Cody *et al.* (1993) in a hexapeptide isolated from GFP. It is intriguing that the structure of the GFP chromophore is a part of the structure of coelenterazine.

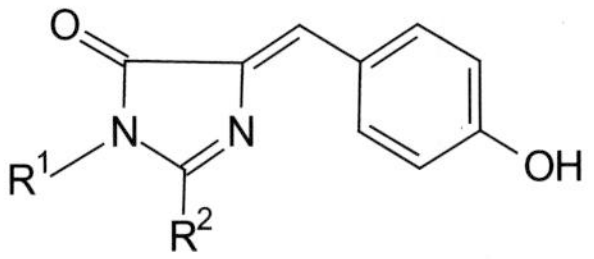

R^1, R^2 = peptide chain

The gene of *Aequorea* GFP was cloned by Prasher *et al.* (1992), and expressed in *E. coli* and *Caenorhabditis elegans* by Chalfie *et al.* (1994) and in *E. coli* by Inouye and Tsuji (1994a). The X-ray structure of recombinant GFP was solved by Ormö *et al.* (1996) and Yang *et al.* (1996, 1997). The protein is in the shape of a cylinder consisting of 11 strands of β-sheets and an α-helix inside (which contains the chromophore), with short helical segments on the ends of the cylinder. Thus the chromophore is sealed and protected from the outside medium.

Energy transfer. Because the light emitted from the jellyfish *Aequorea* is green, there must be an energy transfer from the blue-light emitter of aequorin to GFP (Johnson *et al.*, 1962), and Morin and Hastings (1971b) have suggested the involvement of Förster-type resonance mechanism. For resonance energy transfer to take place, the molecular distance between aequorin and GFP must be short enough (less than about 100 Å) and there must be a significant spectral overlap between the luminescence spectrum of aequorin and the absorption spectrum of GFP. The latter requirement is apparently satisfied in the case of *Aequorea* (Fig. 4.1.16). Morise *et al.* (1974) have demonstrated that the Förster-type resonance energy transfer can take place when both proteins, aequorin and GFP, are coadsorbed on fine particles of DEAE ion-exchanger. It is clear that the coadsorption of the proteins made the molecular distance between aequorin and GFP short enough for the occurrence of the energy transfer. A graphic illustration of the mechanism is shown in Fig. 4.1.17.

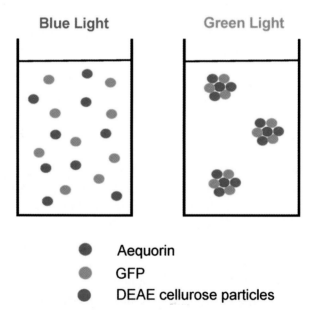

Fig. 4.1.17   Graphic illustration of Förster-type resonance energy transfer from aequorin to *Aequorea* GFP. In the vessel at left, a solution contains the molecules of aequorin and GFP randomly distributed in a low ionic strength buffer. The vessel at right contains a solution identical with the left, except that it contains some particles of DEAE cellulose. In the solution at right, the molecules of aequorin and GFP are coadsorbed on the surface of DEAE particles. Upon an addition of Ca^{2+}, the solution at left emits blue light from aequorin (λ_{max} 465 nm), and the solution at right emits green light from GFP (λ_{max} 509 nm).

The photogenic cells of *A. aequorea* contain very high concentrations of both GFP and aequorin, estimated at 5% each (Morise *et al.*, 1974), or more likely 2.5% GFP and several times higher concentration of aequorin (Cutler, 1995). Cutler (1995) and Cutler and Ward (1993, 1997) have reported a simulation of the *in vivo* luminescence, in which they have mixed very high concentrations of aequorin and GFP (10–20 mg/ml each) in a capillary tube of 0.2 mm diameter, observing the emission of a green light as the evidence for the occurrence of the resonance energy transfer.

4.2   The Hydroid *Obelia* (Hydrozoan)

Bioluminescence of hydroids has been known and studied since the early 18th century (Harvey, 1952). However, the biochemical research on the bioluminescence of hydroids began only after the finding by Morin and Hastings (1971a) that the luminescence of the *Obelia* species is due to the presence of Ca^{2+}-sensitive photoproteins. They isolated and partially purified the photoproteins from four species of *Obelia*. The isolated photoproteins emitted blue light (λ_{max} 475 nm) when Ca^{2+} is added, differing from the green luminescence (λ_{max} 508 nm) emitted from the live specimens. Thus, the bioluminescence of hydroids resembles that of the jellyfish *Aequorea*, suggesting the presence of a green fluorescent protein in the light organs (see Section 4.7).

4.2.1   *Natural Obelins*

The photoproteins isolated from the *Obelia* species were named obelins, and their luminescence characteristics have been investigated with the materials obtained from *O. geniculata* (Campbell, 1974), and *O. australis* and *O. geniculata* (Stephenson and Sutherland, 1981). According to the latter paper, the SDS-PAGE analysis of the purified obelin showed a small number of protein bands (probably isoforms) corresponding to the molecular weights of 20,000 to 30,000, with a major band at 24,000; gel-filtration gave a molecular weight of approx. 21,000. The extraction and properties of obelins have also been reported by Vysotski *et al.* (1989, 1990, 1991, 1993, 1995).

Various characteristics of obelins, including the sensitivities to Ca^{2+} and other metal ions, are generally similar to those of aequorin although there are some distinct differences. It is reported that Cd^{2+} triggers the luminescence of obelin (Trofimov *et al.*, 1994; Bondar *et al.*, 1995), but the finding is not new in the photoproteins since Cd^{2+} has been known to elicit the luminescence of aequorin (Shimomura and Johnson, 1973d). Obelin, like aequorin, may possess an affinity for the free (unchelated) form of a Ca^{2+}-chelator, such as EDTA and EGTA. Such an affinity might cause an inhibition of luminescence, as in the case of aequorin (see Section 4.1.3). The chemical mechanism of obelin luminescence is considered to be identical with that of aequorin.

4.2.2   *Recombinant Obelin*

A cDNA encoding apoobelin was obtained from *O. longissima* and sequenced (Illarionov *et al.*, 1995). The deduced amino acid sequence of the apoobelin consists of 195 amino acid residues, with a calculated molecular mass of about 22.2 kDa, closely matching the apoproteins of other Ca^{2+}-sensitive photoproteins such as aequorin from the jellyfish *Aequorea* (Inouye *et al.*, 1985; Prasher *et al.*, 1985) and clytin from the jellyfish *Phialidium gregarium* (Inouye and Tsuji, 1993). To obtain recombinant apoobelin, the cDNA encoding apoobelin was expressed in *E. coli* (Illarionov *et al.*, 2000). The recombinant apoobelin produced was purified and converted into obelin by incubation with coelenterazine in the presence of molecular oxygen and 2-mercaptoethanol or dithioerythritol, as in the case of aequorin.

Spectral properties. The absorption spectra of recombinant obelins obtained from *O. geniculata* and *O. longissima* are nearly identical showing two peaks, 280 nm and 460 nm (Markova *et al.*, 2002), resembling the data of aequorin. However, the emission spectra of these obelins in Ca^{2+}-triggered luminescence are significantly different from each other (Fig. 4.2.1; Markova *et al.*, 2002), showing their peaks at 495 nm (*O. geniculata*) and 485 nm (*O. longissima*); the peaks are considerably red-shifted from the peaks of natural

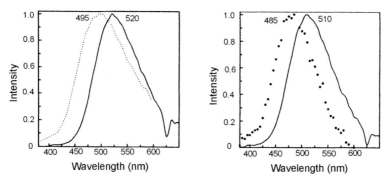

Fig. 4.2.1   Luminescence spectra of the Ca^{2+}-triggered light emission of recombinant obelins (dotted lines), and the fluorescence emission spectra of their spent solution after luminescence (solid lines). *Left*: obelin derived from *O. geniculata*; *right*: obelin derived from *O. longissima*. Reproduced from Markova *et al.*, 2002, with permission from the American Chemical Society.

obelins (475 nm for both *O. geniculata* and *O. longissima*; Morin and Reynolds, 1969; Morin and Hastings, 1971a,b) and aequorin (460–465 nm). With regard to their fluorescence characteristics, recombinant obelins and aequorin are clearly different after they have been luminesced with Ca^{2+}. Spent solutions of the recombinant obelins from *O. geniculata* and *O. longissima* show fluorescence emission maxima at 520 nm and 510 nm, respectively (Fig. 4.2.1; Markova *et al.*, 2002), in contrast to the spent solution of aequorin (both natural and recombinant) that shows a fluorescence emission maximum at about 460 nm, coinciding with the luminescence emission maximum of Ca^{2+}-triggered bioluminescence. In the case of obelins, the marked differences between the fluorescence emission maxima (520 nm and 510 nm) and the maxima of Ca^{2+}-triggered luminescence (495 nm and 485 nm) must indicate a significant change in the environment of the light emitter, coelenteramide, after light is emitted.

Concerning the Ca^{2+}-triggered luminescence of obelin, Deng *et al.* (2001) reported an interesting observation (Fig. 4.2.2): the luminescence of the recombinant obelin from *O. longissima* is blue (λ_{max} 475 nm), whereas a mutant of this obelin, in which the amino acid residue-92 has been changed from tryptophan to phenylalanine, emits

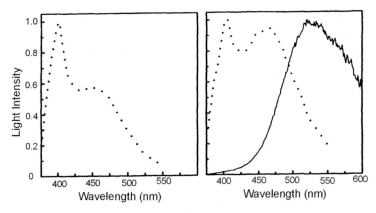

Fig. 4.2.2 *Left panel*: Uncorrected Ca^{2+}-triggered bioluminescence spectrum of W92F obelin derived from *O. longissima*. *Right panel*: Corrected bioluminescence spectrum of the same obelin (dotted line), and the fluorescence emission spectrum of the spent solution after luminescence (solid line). From Deng *et al.*, 2001, with permission of the Federation of the European Biochemical Societies.

a luminescence shifted to violet (λ_{max} 400 nm). The 400 nm emitting species must be the neutral form of coelenteramide, and the 475 nm emitting species is probably the amide anion. It appears that the protein conformation of the mutant enhanced the protonation of the amide anion emitting species.

X-ray structure of obelin. The first crystallographic study of recombinant obelin was reported with the material crystallized using sodium citrate (Vysotski *et al.*, 1999). The complete X-ray structures of recombinant obelins were solved with the proteins derived from *O. longissima* (Liu *et al.*, 2000, 2003; Deng *et al.*, 2002a) and *O. geniculata* (Deng *et al.*, 2002b). Obelins have compact globular structures containing four helix-turn-helix EF-hand motifs, of which three can bind Ca^{2+}, and their peroxidized coelenterazine ligands are located in the hydrophobic core cavity of the proteins, as in the case of aequorin.

Deng *et al.* (2004a,b) prepared the crystals of the spent obelin (W92F mutant from *O. longissima*) that had been luminesced with Ca^{2+}, and successfully obtained the X-ray structure of apoobelin as an important information in elucidating the mechanism of the luminescence reaction.

Obelin as a Ca^{2+}-indicator. Obelin is a useful Ca^{2+} indicator over a Ca^{2+}-concentration range between $10^{-6.5}$ and $10^{-3.5}$ M (Stephenson and Sutherland, 1981), in contrast to aequorin that has been shown to be sensitive over a range between $10^{-7.5}$ and $10^{-4.5}$ M under similar conditions (Blinks *et al.*, 1978). Recombinant obelin might be superior to aequorin in the following two aspects (Illarionov *et al.*, 2000; Markova *et al.*, 2002). Firstly, obelin reacts with Ca^{2+} faster than aequorin in the luminescence reaction; the rate constant for the rise of obelin luminescence is about $450\,s^{-1}$, and that of aequorin luminescence is $115\,s^{-1}$. Secondly, the inhibitory effect of Mg^{2+} on the luminescence of obelin is less than half of that on the luminescence of aequorin. However, the possible inhibition of the luminescence reaction of obelin by the free (unchelated) form of a chelator, such as EDTA or EGTA, has not been investigated. It should be pointed out that the Ca^{2+}-sensitivity and response speed of aequorin can be modified across a wide range by replacing the coelenterazine moiety in the protein with various analogues, and that would also be applicable in the case of obelin.

4.3   The Hydrozoan Medusa *Phialidium gregarium*

The small jellyfish *Phialidium gregarium* (diameter 15–20 mm) used to be abundant at Friday Harbor, Washington, in summer and autumn until about 1990. Levine and Ward (1982) isolated and purified a Ca^{2+}-sensitive photoprotein from this jellyfish and named it phialidin. They extracted the photoprotein from whole specimens with an EDTA-containing buffer. The photoprotein extract was precipitated with ammonium sulfate, and purified by the following methods: gel-filtration (BioGel P-150, minus 400 mesh), anion-exchange chromatography (DEAE Bio-Gel A), and gel-filtration (Sephadex G-75, superfine).

According to Levine and Ward (1982), the molecular weight of phialidin is 23,000, and the luminescence intensity of Ca^{2+}-triggered luminescence is optimal in a broad range of pH (6.0–9.5; Fig. 4.3.1) in resemblance to aequorin. The rate constant for the decay of luminescence was $0.67\,s^{-1}$ in a pH 7.0 buffer at 22°C. The sensitivity to

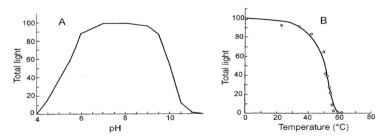

Fig. 4.3.1 Effect of pH on the total light emission of phialidin (**A**), and the temperature stability profiles of phialidin (minute open circles) and aequorin (solid line) (**B**). In A, each buffer contained 0.1 M $CaCl_2$ plus 0.1 M Tris, glycine or sodium acetate, the pH being adjusted with NaOH or HCl. In B, the photoprotein samples in 10 mM Tris-EDTA buffer solution, pH 8.0, were maintained at a test temperature for 10 min, and immediately cooled in an ice water bath. Then total luminescence activity was measured by injecting 1 ml of 0.1 M $CaCl_2$/Tris-HCl, pH 7.0, to 10 μl of the test solution. From Levine and Ward (1982), with permission from Elsevier.

Ca^{2+} is slightly lower than that of aequorin, indicating that the photoprotein is suited for measuring higher concentrations of Ca^{2+} than measurable with aequorin (Shimomura and Shimomura, 1985).

Inouye and Tsuji (1993) cloned the cDNA encoding the photoprotein of the hydroid *Clytia gregaria*. The sequence analysis of the cDNA indicated that the photoprotein clytin consists of 189 amino acid residues, with a calculated molecular weight of 21,605, and has three EF-hand structures to bind Ca^{2+} ions. According to their description, however, the organisms used were apparently a jellyfish commonly called *Phialidium gregarium*, which is thought to be the medusa stage of the hydroid *Clytia gregaria*. To minimize confusion, the present author suggests using the name "phialidin" for the native photoprotein, restricting the name "clytin" for the recombinant photoprotein originated from *Phialidium*.

4.4   Other Bioluminescent Hydrozoans

The hydrozoan medusa *Halistaura* (*Mitrocoma*) *cellularia* (**Shimomura et al., 1963a**). This species looks like *Phialidium gregarium* in shape, but is much larger in size (diameter 5–6 cm). They are thin, very fragile, disk-shaped medusae. Like *Aequorea*, this species was very abundant in the Friday Harbor region in the state of Washington until about 1990. The medusa has granular light organs along the outer edge of the body making a full circle, like *Aequorea*. The light organs are brightly fluorescent in green. In extracting the photoprotein, the morphology and fragility of this medusa made it difficult to cut off the outer margin containing the light organs, as was done with *Aequorea*. Thus, whole specimens of *Halistaura* were homogenized with a blender in cold saturated ammonium sulfate solution containing EDTA. Then the homogenate was filtered on a Büchner funnel with the aid of a small amount of Celite. The photoprotein in the filter cake was extracted with a cold EDTA solution, and purified by repeated anion-exchange chromatography on DEAE cellulose (better chromatography media are available today).

The purified photoprotein, designated halistaurin, resembled aequorin in various aspects. Its absorption spectrum showed a narrow shoulder at 290 nm and a slight bulge at 310 nm, in addition to the 280 nm protein peak. After the Ca^{2+}-triggered luminescence reaction (emission λ_{max} 460 nm), the absorption bulge at 310 nm shifted to 335 nm. The product after luminescence showed a fluorescence emission peak at 465 nm (excitation peak at 340 nm). An addition of $NaHSO_3$ diminished the total light-emitting capability without affecting the rate of the luminescence reaction, indicating that $NaHSO_3$ destroys the luminescence function of the photoprotein. The reaction rate of the Ca^{2+}-triggered luminescence of halistaurin is slower than that of aequorin (less than half).

The cDNA for this photoprotein has been cloned and expressed in *E. coli*, and the recombinant protein obtained was named mitrocomin (Fagan *et al.*, 1993). Mitrocomin consists of 190 amino acid residues with a tyrosine residue at the C-terminus, and has three Ca^{2+}-binding sites.

Other hydrozoans. Various bioluminescent coelenterates have been comparatively studied in reference to the components necessary for their *in vivo* luminescence (Morin and Hastings, 1971a; Cormier *et al.*, 1973; Shimomura and Johnson, 1975b, 1979b).

4.5   The Scyphozoans *Pelagia* and *Periphylla*

4.5.1   *Pelagia noctiluca*

Luminous scyphozoans include historically famous *Pelagia*, in addition to *Periphylla* and *Atolla*. Harvey (1952) states in his book: All the great naturalists of the 16th and 17th centuries speak of the luminous "Plumo marinus," referring to *Pelagia noctiluca*, and repeat Pliny's story of the slime of this jellyfish which adheres to fingers and covers objects with a luminous coat. Today, research reports on the scyphozoan jellyfish *Pelagia* are abundant on certain subjects, such as its distribution, population and sting toxin, but are scarce on the chemical aspects of bioluminescence. According to Harvey (1926b), *Pelagia* does not require oxygen for luminescence. His remark suggests the possible involvement of a photoprotein in the luminescence of *Pelagia*. Indeed, the extraction of a Ca^{2+}-activated photoprotein from *Pelagia* has been reported by Morin and Hastings (1971a) and Morin and Reynolds (1972) although the details are unknown. According to the latter report, the luminescence of *Pelagia* is extracellular with an emission maximum at 475 nm.

4.5.2   *Periphylla periphylla*

The luminous deep-sea medusa *Periphylla periphylla* (Fig. 4.5.1) is widely distributed in various oceans in the world. Certain Norwegian fjords are densely populated with large specimens of this species, each measuring up to 20 cm in diameter, 25 cm in height, and weighing over 600 g (Fosså, 1992). Histological studies indicated that the luminescence of *P. periphylla* is associated with two distinct sources: one is represented by minute, irregularly shaped cytoplasmic granules in the cortical layer of maturing ovarian eggs, and the other is represented by the clusters of even smaller, mostly spherical grains within the cytoplasm of highly specialized photocytes distributed throughout the exumbrellar epithelium (Herring, 1990; Flood *et al.*, 1996). The latter photocytes are highly concentrated along the margin of the medusae, which is divided into 16 evenly sized lappets (Fig. 4.5.2).

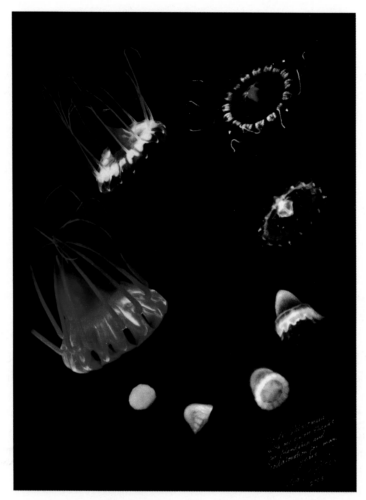

Fig. 4.5.1   Life cycle of the scyphozoan jellyfish *Periphylla periphylla*, counter-clockwise from egg (white, about 1 mm diameter) to mature adult (orange-red, about 15 cm diameter, 400 g weight). This illustration is a gift from Dr. Per Flood.

The scyphozoan *Periphylla* emits light with a luciferin-luciferase reaction using coelenterazine as the luciferin, differing from *Pelagia* in the same class and all luminous hydrozoans that luminesce with photoproteins.

Isolation of multiple forms of *Periphylla* luciferase. In agreement with the histological findings noted above, two types of luciferase, a

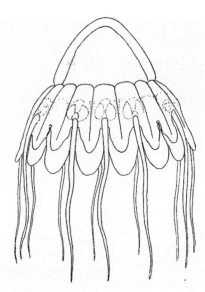

Fig. 4.5.2   A drawing of *Periphylla periphylla*. From Russell, 1970, with permission from Cambridge Unversity Press.

soluble form and an insoluble particulate form, have been isolated and purified by a laborious process (Shimomura and Flood, 1998; Shimomura *et al.*, 2001). The soluble form "luciferase L", which is responsible for the exumbrellar bioluminescence display of the medusa, was extracted from the lappets and the dome mesoglea. The insoluble aggregated luciferase, which exists in particulate forms, was obtained from the ovaries. The total luciferase activity of the insoluble form per specimen was over 100 times that of luciferase L. The size of the particulates, measured by differential filtration, was larger than 0.2 μm and smaller than 2 μm; the actual size is probably close to the low end of this range according to previous microscopic observation (Flood *et al.*, 1996). The particulate matter is highly active in catalyzing the luminescence of coelenterazine, like soluble luciferase L, but its involvement in the *in vivo* bioluminescence has not been confirmed.

The particulate matter can be solubilized with 2-mercaptoethanol, giving a mixture of luciferase oligomers with molecular masses in multiples of approximately 20 kDa. The luciferase activity is increased several times by the solubilization. The purification of the solubilized luciferases was done mainly by column chromatography on Toyopearl Ethyl 650 M (hydrophobic interaction), Superdex 200 Prep (size exclusion) and Toyopearl SP650 M (cation exchange). The isolated luciferases having the molecular masses of 20 kDa, 40 kDa, and 80 kDa were designated luciferase A, luciferase B and luciferase C respectively. Luciferase L (32 kDa) can be split by 2-mercaptoethanol into luciferase A and an inactive accessory protein (approximately 12 kDa). The accessory protein is suspected to be responsible for the unusual heat stability of luciferase L (see below).

Properties of the luciferases. According to Shimomura and Flood (1998) and Shimomura *et al.* (2001), all *Periphylla* luciferases L, A, B and C catalyze the oxidation of coelenterazine, resulting in the emission of blue light (λ_{max} 465 nm). Luciferases B (40 kDa) and C (80 kDa) are apparently the dimer and tetramer, respectively, of luciferase A (20 kDa). The presence of a salt is essential for the activity of luciferase, and the optimum salt concentration is about 1 M in the case of NaCl for all forms of luciferases. The luminescence intensity of luciferase L is maximum near 0°C, and decreases almost linearly with rising temperature, falling to zero intensity at 60°C; the luminescence intensity profiles of luciferases A, B and C show their peaks at about 30°C (Fig. 4.5.3). The Michaelis constants estimated for luciferases A, B and C with coelenterazine are all about 0.2 μM, and that for luciferase L is 1.2 μM.

Luminescence activity. The specific luminescence activities (quanta/s emitted from 1 ml of a solution of $A_{280\,nm,1\,cm}$ 1.0) of luciferases A, B and C are in a range of $1.2 \sim 4.1 \times 10^{16}$ photons/s when measured with the standard assay buffer (20 mM Tris-HCl, pH 7.8, containing 1 M NaCl, 0.05% BSA, and 0.14 μg/ml of coelenterazine, at 24°C). These are the highest specific activities of coelenterazine luciferases.

Stability. Luciferase L (32 kDa) is highly resistant to heat and its activity is little affected by boiling it for 1 min (Fig. 4.5.4); however, it is unstable in low ionic-strength solutions if some BSA is not included in the solution. Luciferases A, B and C are also highly resistant to inactivation; their luminescence activities are only slightly diminished

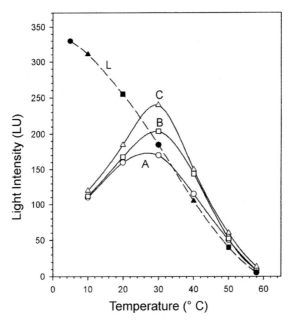

Fig. 4.5.3   Effect of temperature on the light intensity of coelenterazine catalyzed by *Periphylla* luciferases A, B, C and L, in 3 ml of 20 mM Tris-HCl, pH 7.8, containing 1 M NaCl and 0.05% BSA. The luminescence reaction was started by the addition of 10 μl of 0.1 mM methanolic coelenterazine. The amounts of luciferase used for the measurement of each point: luciferase A, 170 LU; luciferase B, 190 LU; luciferase C, 210 LU; luciferase L, 210 LU. From Shimomura *et al.*, 2001.

at pH 1 and pH 10 (Fig. 4.5.5) and are enhanced in the presence of 1–2 M guanidine hydrochloride, although they are less stable to heat than luciferase L. BSA (0.05%) is useful to prevent the spontaneous inactivation of all forms of *Periphylla* luciferase; certain cationic detergents, such as lauroylcholine chloride (LCC; 0.005%), are also effective in preventing inactivation.

Inhibitors. Luciferases A, B, C and L are all strongly inhibited by Cu^{2+}. In the case of luciferase L, light intensity is decreased by 25% with 10 μM Cu^{2+} and 85% with 30 μM Cu^{2+}. With luciferase C, the inhibition was 70% with 1 μM Cu^{2+} and 97% with 10 μM Cu^{2+}. The inhibition by Cu^{2+} is much stronger than that of the other metal ions, such as Ba^{2+}, Cd^{2+}, Zn^{2+} and Pb^{2+}, and it appears almost specific.

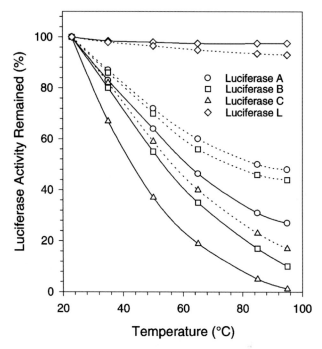

Fig. 4.5.4 Heat stability of *Periphylla* luciferases A, B, C and L in 20 mM Tris-HCl buffer (pH 7.8) containing 1 M NaCl and 0.05% BSA (solid lines) or 0.01% LCC (dotted lines). The buffer (1 ml) containing a luciferase sample was placed in a glass test tube that had been pre-equilibrated at a temperature in a water-bath. After 2 min, the test tube was briefly cooled in cold water, and then luciferase activity in 10 μl of the solution was measured in 3 ml of the pH 7.8 buffer containing 0.3 μM coelenterazine at 24°C. From Shimomura *et al.*, 2001.

Quantum yield of coelenterazine. The quantum yields of coelenterazine in the luminescence reaction catalyzed by luciferases A, B and C are all close to 0.30 at 24°C, which is one of the highest values among coelenterazine luciferases. The amount of luciferase L obtained was insufficient to measure reliable data of specific activity and quantum yield.

Luciferase in eggs. The eggs of *P. periphylla* are extremely rich with the insoluble form of luciferase, which drastically decreases upon fertilization. In the eggs, the particles containing aggregated luciferase are in the cortical layer (Flood *et al.*, 1996). The total content of

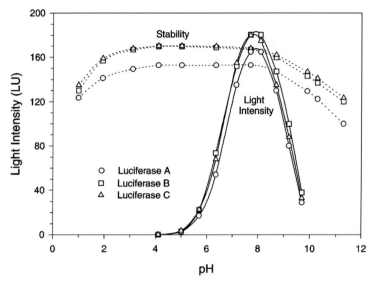

Fig. 4.5.5 Effect of pH on the luminescence of coelenterazine catalyzed by *Periphylla* luciferases **A**, **B** and **C**, and on the stability of the luciferases. The effect on light intensity (solid lines) was measured in 3 ml of 50 mM phosphate buffers, pH 4.1–7.25, and 50 mM Tris-HCl buffers, pH 7.1–9.7, all containing 1 M NaCl, 0.025% BSA, and 0.3 μM coelenterazine. To measure the stability (dotted lines), a luciferase sample (5 μl) was left standing for 30 min at room temperature in 0.1 ml of a buffer solution containing 1 M NaCl and 0.025% BSA and having a pH to be tested, and then luciferase activity in 10 μl of the solution was measured in 3 ml of 20 mM Tris-HCl, pH 7.8, containing 1 M NaCl, 0.05% BSA, and 0.3 μM coelenterazine at 24°C. The amounts of luciferases used for measuring each point were: luciferase A, 150 LU; luciferases B and C, 170 LU. One LU = 5.5×10^8 quanta/s. From Shimomura *et al.*, 2001.

luciferase in one egg is very large for its small size (1 μg or 5×10^{-11} mole/egg). Unexpectedly, the eggs contained a negligibly small amount of coelenterazine (1×10^{-14} mole/egg), although it is possible that coelenterazine had been lost during the preparation of the material.

Upon fertilization, the amount of luciferase in the eggs begins to decrease and reaches a minimum at a late embryonic or early juvenile stage, which contains about only 3% of the initial amount of luciferase. Therefore, the vigorous biosynthesis of luciferase must start at a later stage of development, considering that large adult specimens contain very large amounts of luciferase. The decrease of luciferase in the eggs is puzzling and intriguing. Why does the egg contain a large amount of luciferase in the first place? What is the function or purpose of the luciferase in the eggs?

4.6   The Anthozoan *Renilla* (Sea Pansy)

The studies of the sea pansy *Renilla* (Fig. 4.6.1) and the hydrozoan *Aequorea* have the longest history in the bioluminescence of coelenterates. Cormier (1961) demonstrated the luciferin-luciferase reaction of *Renilla* and investigated some of the properties of the luminescence reaction. The luciferase was extracted from ammonium sulfate-treated specimens with 50 mM phosphate buffer, pH 7.5, containing 1 mM glutathione, and the luciferin solution was prepared by re-extracting the residue with the same buffer and heating the extract at 100°C for one and a half minutes. The luciferin-luciferase luminescence reaction required ATP, ADP or AMP, and oxygen. Cormier soon discovered that the real requirement for the *Renilla* luciferin-luciferase reaction was 3′,5′-diphosphoadenosine that activates luciferin, instead of ATP, ADP or AMP (Cormier, 1962). Cormier and Hori (1964)

Fig. 4.6.1   Two examples of luminous anthozoans: the sea pansy *Renilla reniformis* (left) and the sea pen *Ptilosarcus gruneyi* (right).

found that luciferin can be activated by heating it in a pH 1.0 solution at 100°C for 2 minutes, instead of using 3′,5′-diphosphoadenosine. Later, Hori *et al.* (1972) found that the mechanism of the activation is the removal of a sulfate group from luciferin by hydrolysis. Thus, the substance originally termed luciferin is now called luciferyl sulfate.

Renilla **luciferin and luciferyl sulfate.** Luciferyl sulfate is extracted from the organisms with methanol. The extract is concentrated, defatted with benzene, and the aqueous layer is extracted with ethyl acetate. Luciferyl sulfate in the extract is purified by chromatography on a column of Sephadex LH-20 (Hori *et al.*, 1972; Inoue *et al.*, 1977a). Free luciferin is obtained by heating the sulfate in 0.1 N HCl at 90°C for 50 sec, followed by extraction with ethyl acetate. The luciferin of *Renilla* thus obtained was found to be identical with coelenterazine (Inoue *et al.*, 1977a; Hori *et al.*, 1977).

Renilla **luciferase.** The luciferase of *Renilla reniformis* has been purified and characterized by Karkhanis and Cormier (1971) and Matthews *et al.* (1977a). The purified luciferase has a molecular weight of 35,000, and catalyzes the luminescence reaction of coelenterazine. The luciferase-catalyzed luminescence is optimum at pH 7.4, at a temperature of 32°C, and in the presence of 0.5 M salt (such as NaCl or KCl). The luciferase has a specific activity of 1.8×10^{15} photons $\cdot$ s^{-1}mg^{-1}, and a turnover number of 111/min. The luminescence spectrum shows a maximum at 480 nm. The absorbance A_{280} of a 0.1% luciferase solution is 2.1. The luciferase has a tendency to self-aggregate, forming higher molecular weight species of lower luminescence activities.

The cDNA encoding the luciferase of *Renilla reniformis* has been obtained and expressed in *Escherichia coli* (Lorenz *et al.*, 1991). The cDNA contained an open reading frame encoding a 314-amino acid sequence. The recombinant *Renilla* luciferase obtained had a molecular weight of 34,000, and showed an emission maximum at 480 nm in the luminescence reaction of coelenterazine, in good agreement with the data of natural *Renilla* luciferase.

Quantum yield of luciferin. Various values of quantum yield have been reported for coelenterazine in the luminescence reaction catalyzed by *Renilla* luciferase: 0.055 (Matthews *et al.*, 1977a), 0.07 (Hart, *et al.*, 1979), and 0.10–0.11 (with a recombinant form; Inouye and Shimomura, 1997). The quantum yield is significantly increased in the presence of *Renilla* green fluorescent protein (GFP); see below.

The *in vivo* luminescence of *Renilla*. One specimen of *Renilla mülleri* (about 6 g when contracted) contains approximately 2 nmol of luciferyl sulfate (coelenterazine enol-sulfate) plus 0.8 nmol of coelenterazine bound to the binding protein (Shimomura and Johnson, 1979b). The sulfate is considered to be a storage form of coelenterazine, and can be converted into free coelenterazine by luciferin sulfokinase in the presence of $3'5'$-diphosphoadenosine (Cormier *et al.*, 1970) or by an acid treatment as already mentioned.

Like in many other luminous coelenterates, the *in vivo* luminescence of *Renilla* appears to be controlled by Ca^{2+}. In *Renilla*, the bioluminescence system is packaged in discrete subcellular particles termed lumisomes that contain luciferase, GFP and coelenterazine bound to a binding protein (Anderson and Cormier, 1973; Cormier, 1978). When Ca^{2+} concentration rises by nerve stimulation, Ca^{2+} binds to the binding protein, which causes the release of the free form of coelenterazine. The oxidation of the liberated coelenterazine catalyzed by luciferase results in the formation of the excited state of luciferase-bound coelenteramide. In the presence of GFP, the energy of the excited state is transferred into GFP, resulting in the emission of green light (λ_{max} 509). In the absence of GFP, blue light (λ_{max} 480 nm) would be emitted from the amide anion of coelenteramide. The mechanism is illustrated in a simplified scheme shown in Fig. 4.6.2. It should be noted that energy transfer from the excited state coelenteramide to GFP does not take place in solutions of *Renilla* luciferase containing the GFP of *Aequorea*.

The spectra of the luminescence of coelenterazine catalyzed by recombinant *Renilla* luciferase in the presence and absence of *Renilla* GFP are shown in Fig. 4.6.3 (Lorenz *et al.*, 1991). Note that the luminescence intensity at the emission peak is increased more than

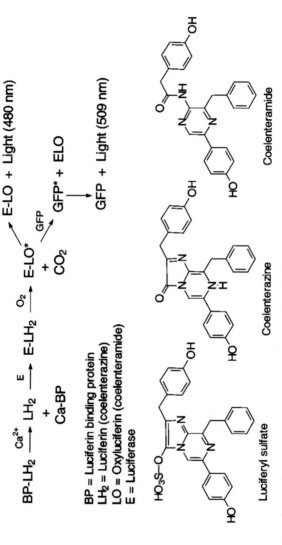

Fig. 4.6.2 A scheme showing the mechanism of the *Renilla* bioluminescence, and the chemical structures of coelenterazine derivatives involved.

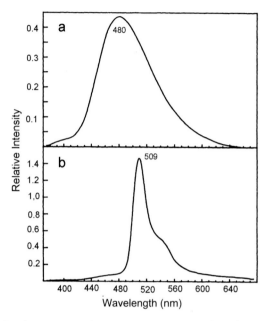

Fig. 4.6.3 Bioluminescence emission spectra measured with coelenterazine plus 1 μM *Renilla* luciferase in the absence (a) and presence (b) of 1 μM *Renilla* GFP. From Lorenz *et al.*, 1991.

threefold in the presence of GFP; this must be due to the fluorescence quantum yield of the GFP (0.8) that is significantly higher than that of the light emitter of coelenterazine (i.e. the amide anion of coelenteramide). See the next section for further information on GFP.

4.7 Green Fluorescent Protein (GFP)

Green fluorescent protein was first discovered from the jellyfish *Aequorea* (Shimomura *et al.*, 1962). Many luminous coelenterates emit green light due to the presence of a green fluorescent protein in their photogenic cells. In photogenic cells, the light energy produced by a photoprotein or a coelenterazine-luciferase system is transferred to the molecules of GFP, resulting in the emission of green light from the GFP molecules. The process involves a radiationless energy transfer mechanism, which has been discussed in Section 4.1.8. The bioluminescent organisms of Hydrozoa and Anthozoa frequently contain a GFP (Morin and Hastings, 1971b; Ward, 1979; Herring, 1983; Prasher, 1995). For example, hydrozoans *Aequorea*, *Obelia*, *Mitrocoma* (*Halistaura*) and *Phialidium*, and anthozoans the sea pansy *Renilla*, the sea pen *Ptilosarcus* and the sea cactus *Cavernularia*, all contain a GFP. Green fluorescent protein has not been found in Scyphozoa and Ctenophora.

General properties. Probably all GFPs are acidic proteins (pI 4.6–5.4) with molecular weights about 27,000. The chromophore structures of all GFPs are essentially identical, and are the same as that of *Aequorea* GFP (Shimomura, 1979). Most GFPs exist usually as dimer (Ward *et al.*, 1980). *Aequorea* GFP is monomer in dilute aqueous solutions, but its dimeric form predominates at concentrations over 5 mg/ml (Ward and Comier, 1979). Only two kinds of GFPs, those from *Aequorea aequorea* and *Renilla reniformis*, have been characterized in detail, of which *Aequorea* GFP is already described in Section 4.1.8. Although the Kekulé structures of the chromophores of *Aequorea* GFP and *Renilla* GFP are identical, *Renilla* GFP has a much greater extinction coefficient (Table 4.7.1), together with a strong tendency to dimerize and a high resistance to pH-induced conformational changes and denaturation (Ward, 1981; Ward and Bokman, 1982).

Table 4.7.1   A Comparison of *Aequorea* GFP and *Renilla* GFP (Ward, 1998, modified)

	Aequorea GFP	*Renilla* GFP
Monomer molecular weight	27,000	27,000
Isoelectric point (pI)	4.6–5.1	5.34
Fluorescence emission maximum	508–509 nm	509 nm
Fluorescence quantum yield	0.80	0.80
Molar absorption coefficient (ε) at peak wavelengths (as monomer)		
498 nm		133,000
475 nm	12,000–14,000	
397 nm	27,600	
280 nm	22,000	22,000

Fluorescence of GFP. The fluorescence emission maxima of all known GFPs are practically identical at 508–509 nm, except the GFPs of *Mitrocoma* and *Phialidium* which show the maxima at 497 nm and 498 nm respectively (Ward, 1998). The luminescence spectrum of an organism containing a GFP is usually indistinguishable from the fluorescence emission spectrum of the same organism, because the luminescence is in effect the fluorescence of GFP that is emitted in consequence of an energy transfer. There are considerable variations, however, in the wavelength of the fluorescence excitation maxima (and absorption maxima) of GFPs. Generally, GFPs show a fluorescence excitation peak in a range of 450–500 nm, in addition to the typical protein peak at 280 nm. As an exception, *Aequorea* GFP shows two peaks, at about 400 nm and 475 nm.

It is generally believed that the absorption (and fluorescence excitation) peak at about 400 nm is caused by the neutral form of the chromophore, 5-(*p*-hydroxybenzylidene)imidazolin-4-one, and the one in the 450–500 nm region by the phenol anion of the chromophore that can resonate with the quinoid form, as shown below (R^1 and R^2 represent peptide chains). However, the emission of light takes place always from the excited anionic form, even if the excitation is done with the neutral form chromophore. This must be due to the protein environment that facilitates the ionization of the phenol group of the chromophore. This is also consistent with the fact that the pK_a values of phenols in excited state are in an acidic range, between 3 and 5 (Becker, 1969), thus favoring anionic forms at neutral pH.

Neutral form   Anion

Recombinant GFPs. The cDNA of *Aequorea* GFP was first cloned by Prasher *et al.* (1992) and expressed in live organisms by Chalfie *et al.* (1994) and Inouye and Tsuji (1994a), as have already been mentioned (Section 4.1.8). Considering that the molecule of GFP contains a cyclic chromophore within its peptide chain, it is rather surprising that the cyclization and oxidation to form the chromophore occur spontaneously during expression. Green fluorescent protein can be expressed in various living systems as GFP itself, or as a tag of various proteins. GFP is now widely used as an important marker protein in biological and medical research. Various different recombinant GFPs with improved fluorescence characteristics have been prepared (Tsien, 1998; Chalfie and Kain, 1998). The discovery of the GFP-like proteins that fluoresce in orange (λ_{max} 538) and red (λ_{max} 583) in nonluminous anthozoan corals has further expanded the usefulness of these fluorescent proteins (Matz *et al.*, 1999). The research on GFPs is progressing rapidly.

4.8   The Ctenophores

The organisms of the phylum Ctenophora (comb jellies) are described here because of the fairly close relationship to coelenterates. Only about 100 species of organisms are known in this phylum. All are marine organisms and probably all are luminescent (Harvey, 1952; Herring, 1978a) except *Pleurobrachia* (Haddock and Case, 1995). The genus *Mnemiopsis* (Fig. 4.8.1) is very common along the north Atlantic coast. It is well known that *Mnemiopsis leidyi*, a North American species, was introduced to the Black Sea in the early 1980s in the ballast water of a ship, and has caused millions of dollars of damage to the anchovy fishery there by an explosive propagation (Shushkina *et al.*, 1990).

Ctenophores are characterized by the eight rows of comb-like plates that aid in swimming. They are among the brightest of luminous organisms. However, in some species, organisms that have been in the light for some time do not luminesce upon stimulation, and they become luminescent after they have been kept in the dark for

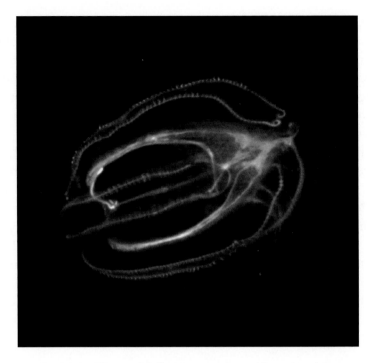

Fig. 4.8.1   The comb jelly *Mnemiopsis* sp. About 4 cm long.

about a half hour. The occurrence of a Ca^{2+}-sensitive photoprotein in *Mnemiopsis* was first reported by Hastings and Morin (1968; 1969a).

The photoproteins of *Mnemiopsis* sp. and *Beroe ovata*. We owe the biochemical knowledge on the luminescence of ctenophores largely to the pioneering efforts of Ward and Seliger (1974a,b). They extracted Ca^{2+}-sensitive photoproteins from 30,000 dark-adapted specimens of *Mnemiopsis* sp. (600 kg) and 1,000 dark-adapted *Beroe ovata* (41 kg) with a solution containing EDTA, and purified the photoprotein by ammonium sulfate fractionation and a series of chromatographic procedures. Due to the high photosensitivity of the photoproteins, all processes of purification were performed under dim red lighting at 0–4°C. The purified photoproteins from *Mnemiopsis* sp. and *Beroe ovata* were named mnemiopsin and berovin. Both photoproteins emitted light upon the addition of Ca^{2+}, regardless of the presence or absence of oxygen. The amount of purified photo-protein obtained from the specimens of *Mnemiopsis* was only about 2 mg. Mnemiopsin was further resolved into two functionally identical forms, mnemiopsin-1 and mnemiopsin-2, by DEAE-cellulose chromatography. Both mnemiopsin-1 and mnemiopsin-2 could be further resolved into 2 or 3 isoforms by high-resolution chromatography. Berovin was obtained as a single peak.

The molecular masses of mnemiopsin-1, mnemiopsin-2, and berovin are 24 kDa, 27.5 kDa and 25 kDa, respectively. The luminescence spectra of these three photoproteins in the presence of Ca^{2+} are identical (λ_{max} 485 nm), though the *in vivo* luminescence emission maxima are 488 nm with *Mnemiopsis* and 494 nm with *Beroe* (Fig. 4.8.2). Mnemiopsin-2 shows a broad absorption band near 435 nm, which disappears upon the addition of Ca^{2+}, accompanied by the appearance of a new absorption peak at 335 nm (Fig. 4.8.3). The light intensity of Ca^{2+}-triggered luminescence is strongly dependent on the pH of medium, showing sharp pH optima: 9.3 with mnemiopsin-1, 9.1 with mnemiopsin-2, and 8.5 with berovin (Fig. 4.8.4). The

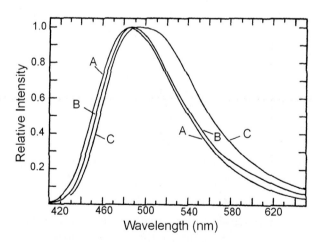

Fig. 4.8.2 Comparison of the *in vivo* and *in vitro* luminescence of the photoproteins of *Mnemiopsis* sp. and *Beroë ovata*: (**A**) *in vitro* luminescence spectra of mineopsin-1, -2, and berovin; (**B**) *in vivo* luminescence spectrum of *Mnemiopsis*, (**C**) *in vivo* luminescence spectrum of *Beroë*. From Ward and Seliger, 1974b, with permission from the American Chemical Society.

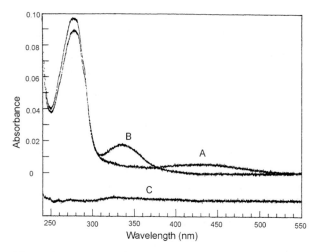

Fig. 4.8.3   Absorption spectra of pure mineopsin-2 (**A**), its luminescence product after the addition of Ca^{2+} (**B**), and the base line (**C**). Sample concentration 0.081 mg/ml. From Ward and Seliger, 1974b, with permission from the American Chemical Society.

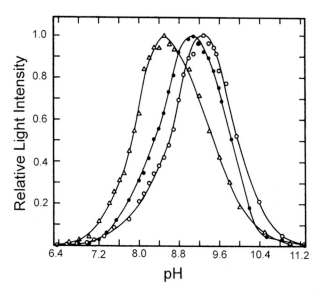

Fig. 4.8.4   Effect of pH on the light intensity of the Ca^{2+}-triggered luminescence of mnemiopsin-1 (○), mnemiopsin-2 (●), and berovin (△). From Ward and Seliger, 1974b, with permission from the American Chemical Society.

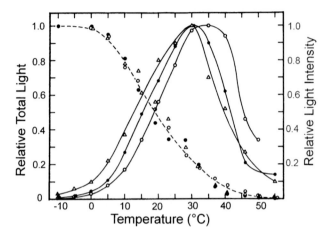

Fig. 4.8.5   Effects of temperature on the light intensity (solid lines) and total light (dashed line) of the Ca^{2+}-triggered luminescence of mnemiopsin-1 (o), mnemiopsin-2 (•), and berovin (△). From Ward and Seliger, 1974b, with permission from the American Chemical Society.

optimum temperature is found at about 30°C (Fig. 4.8.5). The quantum yield of mnemiopsin-2 is approximately 0.12.

Mnemiopsin is inactivated by the exposure to light over its entire absorption spectral range, and the inactivation cannot be reversed by keeping the inactivated material in the dark, which differs from the specimens of a live animal. The photoinactivation is accompanied by a partial loss of the 435 nm absorption band, which is probably due to the decomposition of the peroxidized coelenterazine in the protein.

Girsch and Hastings (1978) extracted and purified mnemiopsin from *Mnemiopsis leidyi* utilizing hollow fiber techniques. They studied the characteristics of the protein and the effects of various factors on the luminescence, such as pH, temperature, concentration of salts, and protein denaturants; they found that the key characteristics of the photoprotein they obtained are similar to those reported by Ward and Seliger (1974b). Anctil and Shimomura (1984) found that the photoinactivated mnemiopsin could be reactivated by incubation with coelenterazine in the dark at pH 9.0, in the presence of oxygen. The same authors also tried unsuccessfully to identify the coelenterazine derivatives existing in the photoinactivated mnemiopsin.

5

THE COELENTERAZINES

5.1 Discovery of Coelenterazine

In the early 1960s, two separate groups in the United States began the chemical studies of different bioluminescence systems. One was Dr. Milton J. Cormier's group at the University of Georgia studying the sea pansy *Renilla*, and the other was Dr. Frank H. Johnson's group at Princeton University studying the jellyfish *Aequorea*. The two groups did not even have the slightest idea that the light-emitting principles of their subjects were the same compound, coelenterazine. The Cormier group possibly isolated a trace amount of coelenterazine from *Renilla* as early as the mid 1960s, however without knowing its chemical identity (Hori and Cormier, 1965). The substance was later identified to be coelenterazine in 1977 by two groups (Inoue *et al.*, 1977a; Hori *et al.*, 1977).

On the other hand, the Johnson group isolated the Ca^{2+}-sensitive photoprotein aequorin from the jellyfish (Shimomura *et al.*, 1962), and, after considerable efforts, they isolated two compounds now called coelenteramine and coelenteramide from the photoprotein (Shimomura and Johnson, 1972, 1973c). Both the structures of coelenteramine and coelenteramide contained the skeleton of aminopyrazine that had been previously found in the structures of *Cypridina* etioluciferin and oxyluciferin (Kishi *et al.*, 1966a,b; McCapra and Chang, 1967; Stone, 1968). Based on this coincidence, they thought that the luminescence reactions of aequorin and *Cypridina* shared a common reaction mechanism, and deduced the chemical structure of a hypothetical luciferin "coelenterazine" that corresponds to the structure of *Cypridina* luciferin (Shimomura *et al.*, 1974b).

Soon after the hypothetical structure was published, coelenterazine was isolated as an actual substance from the liver of the luminous squid *Watasenia scintillans*, and it was chemically synthesized (Inoue *et al.*, 1975). The availability of synthetic coelenterazine led to the important discovery that the treatment of the luminescence product of aequorin with coelenterazine results in the regeneration of active aequorin (Shimomura and Johnson, 1975c), which consequently confirmed the presence of a coelenterazine moiety in the aequorin molecule. During the same period, it became increasingly evident that coelenterazine is involved as a luciferin in various bioluminescent organisms, such as the sea cactus *Cavernularia*, the sea pen *Ptilosarcus*, and the sea pansy *Renilla* (Shimomura and Johnson, 1975b).

The name of coelenterazine. We suggested calling the new luciferin "coelenterazine" (Shimomura and Johnson, 1975b), although Inoue *et al.* (1976) named it *Watasenia* preluciferin. The compound has been called by various other names, such as coelenterate luciferin, coelenterate-type luciferin, *Oplophorus* luciferin, and *Renilla* luciferin. However, these two-word chemical names are cumbersome to use and inconvenient when applying a chemical prefix or suffix. Currently, the name coelenterazine is widely and commonly used.

5.2   Occurrence of Coelenterazine

The occurrence of coelenterazine in marine bioluminescent organisms is extremely widespread. Coelenterazine functions as their light-emitting substance, usually as a luciferin or the functional group of a photoprotein. The compound has been found in a very wide variety of luminous organisms, including jellyfishes, hydroids, anthozoans, ctenophores, radiolarians, copepods, ostracods, squids, shrimps, and deep-sea fishes (Shimomura *et al.*, 1980; Campbell and Herring, 1990; Thomson *et al.*, 1995a). Those organisms are scattered among at least five phyla: Cnidaria, Ctenophora, Mollusca, Arthropoda, and Chordata. In addition, a bioluminescent arrow worm of the phylum Chaetognath has been found to contain coelenterazine (Haddock and Case, 1994). Coelenterazine has not been found in luminous organisms of the phylum Annelida or in terrestrial creatures.

Coelenterazine was also found in various nonluminous marine fishes and shrimps of plankton-feeding type, although their coelenterazine content levels are rather low (Shimomura, 1987). No coelenterazine has been found in the bottom-dwelling scavengers, such as lobsters and crabs. Surprisingly, a small amount of coelenterazine was found in the red beard sponge *Microciona prolifera*. Needless to say, those nonluminous organisms do not contain luciferase. Some data on the contents of coelenterazine and luciferase in bioluminescent and nonluminescent organisms are given in Table 5.1. The table shows a very wide range of variation in the contents of coelenterazine, which is due to the intrinsic nature of the specimens, including their sizes and ages, as well as the decomposition of coelenterazine before the assay. Thus, the amounts of coelenterazine shown should be considered as the minimum levels.

The existence of coelenterazine in various nonluminous organisms suggests that some of the coelenterazine-dependent luminous organisms might obtain coelenterazine from their food, either as the sole source of this substance or as a supplement to the coelenterazine biosynthesized in the body. In the case of the hydrozoan *Aequorea aequorea*, it was reported that the medusa is unable to produce its own coelenterazine and is dependent on a dietary supply of this compound for its capability of bioluminescence (Haddock *et al.*, 2001). The organisms that biosynthesize coelenterazine remain to be identified, but it seems to be a common opinion at present that at least copepods do make their own coelenterazine. According to Thomson *et al.* (1995), the shrimp *Systellaspis debilis* is capable of coelenterazine

Table 5.1   Contents of Coelenterazine and Luciferase Activity in Individual Specimen of Marine Luminous and Nonluminous Organisms. Luciferase Activities are Shown in the Parentheses in the Last Column

Organism Species	Body Weight (g)	Main Tissue Containing Coelenterazine[#]	Coelenterazine, nmol* (Luciferase, 10^{12} quanta/s)
LUMINOUS ORGANISMS			
Hydrozoan[†]			
Aequorea aequorea	50	margin of umbrella	1.3[a]
Halistaura cellularia	8	whole body	0.5[a]
Obelia geniculata		whole hydroid	3.3×10^{-4}[b]
Scyphozoan			
Periphylla periphylla	300	ovary	$1 \times 10^{-5}(10)$[‡c]
Anthozoan			
Cavernularia obesa	30	whole body	20.0 (66)[a]
Ptilosarcus gurneyi	60	whole body	20.0 (210)[a]
Renilla mülleri	6	whole body	2.0 (7.8)[a]
Ctenophore[†]			
Mnemiopsis leidyi	10	whole body	0.25[d]
Copepod			
Euaugaptilus magnus		whole body	0.033[b]
Pleuromamma xiphias		whole body	0.010[b]
Ostracod			
Conchoecia pseudodiscophora	0.002		0.00055[e]
Shrimp			
Oplophorus gracilirostris	2.3	hepatopancreas, stomach	5.0 (50)[f]
Oplophorus spinosus		hepatopancreas	1.3[b] 6.7[h]
Heterocarpus sp.	14	stomach	12.0 (200)[g]
Acanthephyra purpurea		hepatopancreas	0.85[b] 3.6[h]
Systellaspis debilis		hepatopancreas, stomach	24.6[b] 2.4[h]
Systellaspis lanceocaudata	21	stomach	1.5 (13)[f]
Sergestes lucens	0.4	hepatopancreas, stomach	0.01 (0.001)[f]
Sergestes prehensilis	1	stomach	0.0001 (2)[f]
Gnathophausia longispina	0.5	stomach	0.1 (0.7)[f]

(*Continued*)

Table 5.1   (*Continued*)

Organism Species	Body Weight (g)	Main Tissue Containing Coelenterazine[#]	Coelenterazine, nmol* (Luciferase, 10^{12} quanta/s)
Euphausiid[§]			
Meganyctiphanes norvegica	1.5	whole body	0.015
Euphausia pacifica	0.8	whole body	0.008
Squid			
Chiroteuthis imperator	150	liver	7.0 (0.13)[f]
Eucleoteuthis luminosa	80	liver	14.0 (0.07)[f]
Onychoteuthis borealijaponica	55	liver	3.8 (43)[f]
Watasenia scintillans	10	liver	120.0 (0)[f]
Symplectoteuthis luminosa	150	liver	7.2
Pterygioteuthis giardi		tentacles	46.0[b]
Pyroteuthis margaritifera		eyeball, tentacles, mantle photophores	2.6[b]
Sepiolina nipponensis	5	liver	0.005 (0.013)[f]
Fish			
Yarrella illustris	25	pyloric caeca	0.025 (0.02)[f]
Argyropelecus hemigymnus	3	stomach	0.05 (0.002)[f]
Neoscopelus microchir	30	liver	75.0 (0.003)[f]
Myctophum sp.		tail organ	0.1[b]
Diaphus elucens	32	liver, pyloric caeca	19.0 (0.07)[f]
Diaphus coeruleus	18	liver, pyloric caeca	12.0 (0.06)[f]
Lampadena sp.		tail organ	0.2[b]
Brittle star			
Amphiura filiformis	1	whole body	0.012 (1)
NONLUMINOUS ORGANISMS			
Herring *Clupea harengus*	170	liver, pyloric caeca	1.0[g]
Sand lance *Ammodytes*	1.5	liver	0.05[g]
Shrimp *Pandalus borealis*	13	hepatopancreas, stomach	0.12[g]

(*Continued*)

Table 5.1   (*Continued*)

Organism Species	Body Weight (g)	Main Tissue Containing Coelenterazine[#]	Coelenterazine, nmol* (Luciferase, 10^{12} quanta/s)
Shrimp *Pandalus platyuros*	5	hepatopancreas, stomach	0.16[g]
Shrimp *Pandalus danae*	5	hepatopancreas, stomach	0.015[g]
Shrimp *Heptacarpus sp.*	0.8	hepatopancreas	0.005[g]
Shrimp *Sicyonia ingentis*	15	hepatopancreas	0.018[g]
Red beard sponge *Microciona*		whole body	0.015/10 g[g]
Surface plankton mixture (243 μm net)			0.0007/10 g[g]

[#]Stomach was tested after removing the contents.
*1 nmol = 423 ng coelenterazine. Distribution of luciferase in the body may differ from that of coelenterazine.
[†]Coelenterazine exists as a photoprotein.
[‡]In one egg, about 1 mg.
[§]Coelenterazine does not contribute to bioluminescence in these organisms.
References: (a) Shimomura and Johnson, 1979b. (b) Campbell and Herring, 1990. (c) Shimomura *et al.*, 2001. (d) Anctil and Shimomura, 1984. (e) Oba *et al.*, 2004. (f) Shimomura *et al.*, 1980. (g) Shimomura, 1987. (h) Thomson *et al.*, 1995a.

biosynthesis, on the basis that the coelenterazine content of its eggs greatly increased during the course of development. However, the work does not take into account the possibility that an inactive form of coelenterazine, such as the enol-sulfate form or dehydrocoelenterazine, has been converted into active coelenterazine in the eggs; such a conversion would not be the true biosynthesis of coelenterazine.

A note on the assays of coelenterazine and luciferase activity. The methods for measuring coelenterazine and the corresponding luciferases are given in Appendix C5. Special attention must be paid to the fact that coelenterazine in aqueous buffer solutions spontaneously emits a low level of chemiluminescence in the absence of any luciferase, which is greatly enhanced by the presence of various substances, including egg yolk, BSA and various surfactants (especially, hexadecyltrimethylammonium bromide). Therefore, the utmost care must be taken in the detection and measurement of a low level of luciferase activity with coelenterazine, and the data must be carefully interpreted. On the other hand, in the detection of a trace amount of coelenterazine (pmol or nmol amount) with a luciferase, the loss of coelenterazine by auto-oxidation may become a serious problem.

5.3   Properties of Coelenterazine and its Derivatives

The unmodified form of coelenterazine, $C_{26}H_{21}O_3N_3$ (mass spec. M^+ 423), can be crystallized from methanol as orange-yellow prisms with a melting point of 175–178°C (dec). In methanol, coelenterazine is fluorescent in yellow, and its ultraviolet absorption spectrum shows a maximum at 435 nm (ε 9,800), as shown in Fig. 5.1.

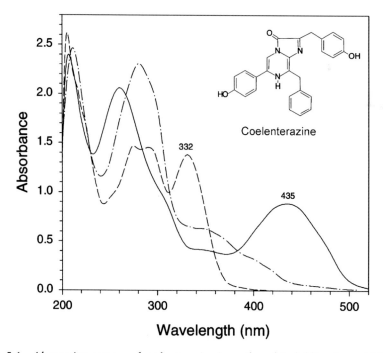

Fig. 5.1 Absorption spectra of coelenterazine in methanol (solid line, λ_{max} 435 nm, ε 9,800), with 10 mM HCl added (dash-dot line), and coelenteramide in methanol (dashed line, λ_{max} 332 nm, ε 15,000). Concentration: all 91.2 μM.

A large number of the analogues of coelenterazine have been synthe-sized (see Section 4.1.7). The absorption spectra of coelenterazine ana-logues are very similar to that of the original coelenterazine, except for e-coelenterazines. The properties of e-coelenterazines differ sig-nificantly from those of coelenterazines, particularly in their lumi-nescence properties. The absorption spectra of e-coelenterazine (λ_{max} 442 nm, ε 9,600) and e-coelenteramide (λ_{max} 347 nm, ε 19,500), both in methanol, are shown in Fig. 5.2.

The product of the luminescence reaction of coelenterazine, coe-lenteramide, is soluble in methanol, butanol, ethyl acetate and ether, and shows a strong blue fluorescence in these solvents, although this compound is only slightly fluorescent in aqueous solutions. Coelen-teramide shows an absorption maximum at 332–333 nm (ε 15,000) in methanol.

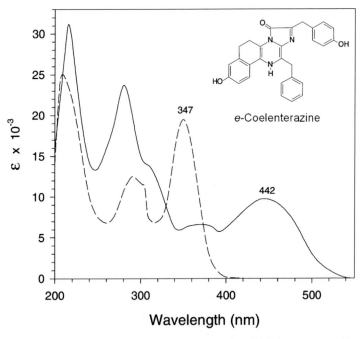

Fig. 5.2 Absorption spectra of e-coelenterazine (solid line, λ_{max} 442 nm, ε 9,600) and e-coelenteramide (dashed line, λ_{max} 347 nm, ε 19,500), both in methanol. Replotted from Teranishi and Goto, 1990.

Solubility and stability of coelenterazine. Coelenterazine is very poorly soluble in neutral aqueous buffer solutions, and the solutions are unstable in air. It can be easily dissolved in water in the presence of alkali, but the resulting solution is extremely unstable under aerobic conditions. Coelenterazine is soluble in methanol, and the solution is relatively stable. The stability is enhanced by the addition of a trace of HCl. A methanolic solution of coelenterazine can be stored for several days at $-20°C$, and a methanolic solution containing 1–2 mM HCl can be stored for several months at $-70°C$ under aerobic conditions without significant oxidation. In many other organic solvents, coelenterazine is less stable, and spontaneously auto-oxidized at significant rates. In dimethylformamide and DMSO, it is rapidly decomposed accompanied by the emission of chemiluminescence. e-Coelenterazines are generally less stable than coelenterazines.

The poor solubility of coelenterazine in neutral aqueous buffer solutions often hampers the use of this compound in biological applications. The simplest way to make an aqueous solution is the dilution of a methanolic 3 mM coelenterazine with a large volume of a desired aqueous buffer solution. If the use of alcoholic solvents is not permitted, dissolve coelenterazine in a small amount of water with the help of a trace amount of 1 M NaOH or NH_4OH, and then immediately dilute this solution with a desired aqueous buffer solution. However, because of the rapid oxidation of coelenterazine in alkaline solutions, it is recommended that the procedure be carried out under argon gas and as quickly as possible.

To circumvent these difficulties, a preparation of water-soluble coelenterazine has been developed (Teranishi and Shimomura, 1997a). The preparation contains coelenterazine and 50-times (by weight) of hydroxypropyl-β-cyclodextrin. To prepare this material, 0.1 ml of 3.0 mM coelenterazine in methanol and 0.2 ml of 45 mM solution of the cyclodextrin are mixed and dried under reduced pressure. The dried residue is extracted with 1.0 ml of 10 mM phosphate buffer, pH 7.0, containing 2 mM EDTA (if needed), and the extract (after centrifugation) is again dried under reduced pressure. With this preparation, an aqueous solution containing up to 3 mM coelenterazine can be made.

5.4   Chemi- and Bioluminescence Reactions of Coelenterazine

Coelenterazine emits "chemiluminescence" when dissolved in dimethyl sulfoxide (DMSO) or dimethylformamide (DMF) containing a trace amount of base. It also emits "bioluminescence" in aqueous media in the presence of a coelenterazine luciferase, such as *Renilla* luciferase or *Oplophorus* luciferase. In both cases, the luminescence reactions require molecular oxygen. The capability of coelenterazine to produce luminescence is attributed to the presence of the imidazopyrazinone structure in the molecule.

Thus, the 2-methyl derivative of the imidazopyrazinone (above) dissolved in DMSO spontaneously emits blue light (λ_{max} 450 nm) in the presence of air (Goto, 1968), like the 2-benzyl derivative (λ_{max} 475 nm), the 2-methyl-6(p-hydroxyphenyl) derivative (MCLA; λ_{max} 468 nm), and coelenterazine (λ_{max} 465 nm) under similar conditions (Fig. 5.3). The comparison of the luminescence spectra of these compounds shows that the 6-position substituent has little influence on the luminescence spectrum of coelenterazine derivatives, despite the apparent conjugation between the 6-phenyl ring and the imidazopyrazinone ring in the structures of MCLA and coelenterazine.

Based on the available knowledge on the chemiluminescence and bioluminescence reactions of various luciferins (firefly, *Cypridina*, *Oplophorus* and *Renilla*), the luminescence reaction of coelenterazine is considered to proceed as shown in Fig. 5.4 (p. 171). The reaction is initiated by the binding of O_2 at the 2-position of the coelenterazine molecule, giving a peroxide. The peroxide then forms a four-membered ring "dioxetanone," as in the case of the luminescence

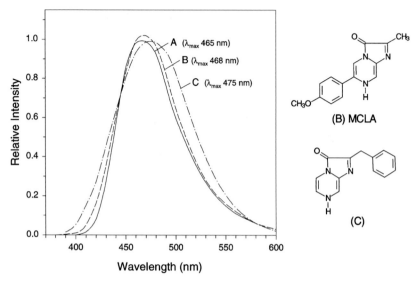

Fig. 5.3 Luminescence spectra of coelenterazine in DMSO (**A**), MCLA in DMSO (**B**), and compound C (2-methyl-3,7-dihydroimidazo[1,2-a]pyrazine-3-one) in DMSO plus 1 mM NaOH (**C**). Compound C was synthesized and the spectrum was measured by K. Teranishi in 2005.

reactions of firefly luciferin and *Cypridina* luciferin. The dioxetanone promptly decomposes as a result of the splitting of two bonds in a concerted fashion, producing CO_2 and the amide anion of coelenteramide in its excited state. The excited state of the amide anion of coelenteramide emits light when its energy level falls to the ground state, resulting in the emission of blue light (λ_{max} 450–470 nm).

Formation of the excited amide anion of coelenteramide as the light emitter in the luminescence reaction of coelenterazine was experimentally supported by the experiment of Hori *et al.* (1973a), in which 2-methyl analogue of coelenterazine was used as the model compound. The summary of their work is as follows: In the presence of oxygen, Ia and Ib in DMF emitted bright blue luminescence (λ_{max} 480 and 470 nm, respectively), and produced the reaction products IIa and IIb, respectively. The fluorescence emission of IIb (λ_{max} 470 nm) and that of the spent chemiluminescence reaction of Ib, both in DMF plus a base (potassium *t*-butoxide), were identical to the chemiluminescence emission of Ib in DMF. The fluorescence emission of IIa

(λ_{max} 530 nm) and that of the spent chemiluminescence reaction of Ia, in DMF plus base, were identical to the chemiluminescence emission of Ia (λ_{max} 530 nm) in the same solvent system but not to the chemiluminescence emission of Ib (λ_{max} 470 nm) under the same conditions. Thus, during chemiluminescence of Ia and Ib in DMF, the electronic excited states (i.e. light emitters) can be represented as the amide anions IIIa and IIIb, respectively.

Ia   R = OH
Ib   R = OCH$_3$

IIa   R = OH
IIb   R = OCH$_3$

IIIa   R = OH
IIIb   R = OCH$_3$

The decomposition of dioxetanone may involve the chemically initiated electron-exchange luminescence (CIEEL) mechanism (McCapra, 1977; Koo *et al.*, 1978). In the CIEEL mechanism, the singlet excited state amide anion is formed upon charge annihilation of the two radical species that are produced by the decomposition of dioxetanone. According to McCapra (1997), however, the mechanism has various shortfalls if it is applied to bioluminescence reactions. It should also be pointed out that the amide anion of coelenteramide can take various resonance structures involving the N-C-N-C-O linkage, even if it is not specifically mentioned.

The amide group of coelenteramide is an extremely weak acid; thus, it will be rapidly protonated in a neutral protic environment, changing into its neutral (unionized) form. If the rate of the protonation of the excited amide anion is sufficiently fast in comparison with the rate of its de-excitation, a part or most of the excited amide anion will be converted into the excited neutral species within the lifetime of the excited state of the amide anion, resulting in a light emission from the excited neutral coelenteramide (λ_{max} about 400 nm).

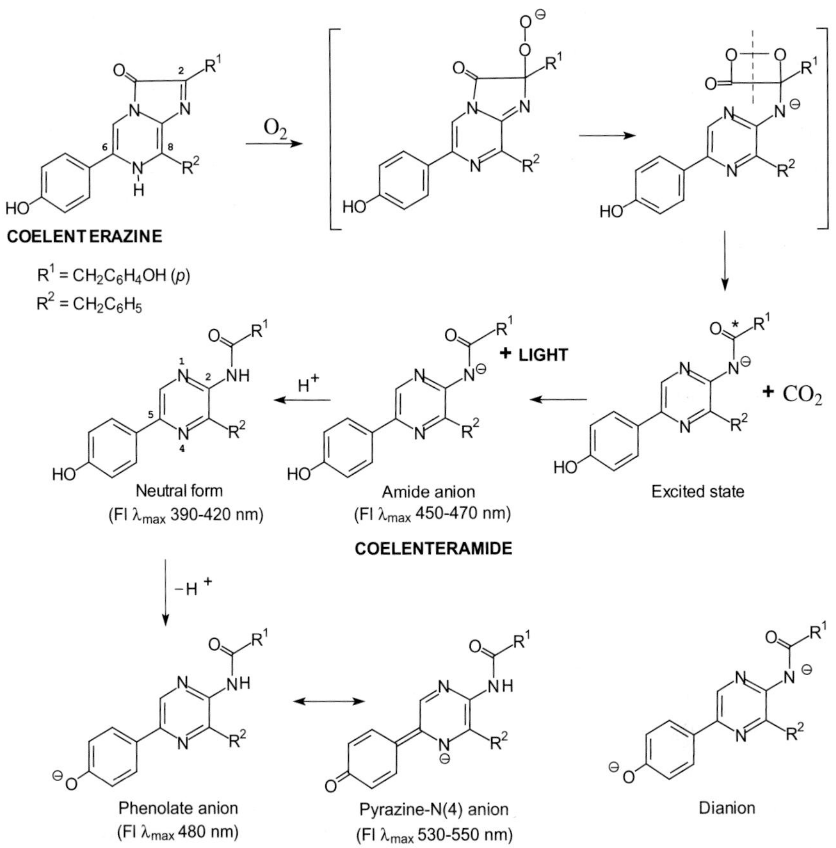

Fig. 5.4 Chemical mechanism of light emission in the bio- and chemiluminescence reactions of coelenterazine. The bottom row shows some of the fluorescence emitters of coelenteramide. The fluorescence characteristics of the dianion are unknown.

The rate of protonation may vary according to the structure of the light-emitter and the environment around the light emitter. In the case of chemiluminescence reactions in solutions, the hydrophobicity, permittivity (dielectric constant) and protogenic nature of the solvent are important environmental factors. In the case of bioluminescence involving a luciferase or photoprotein, the protein environment surrounding the light-emitter will be a crucial factor.

An example of light emission from the neutral coelenteramide can be seen in the Ca^{2+}-triggered luminescence of e-aequorin, a semisynthetic aequorin containing a moiety of e-coelenterazine. The luminescence spectrum of e-aequorin is bimodal (Fig. 4.1.15), with peaks at 405 nm (emission from the neutral form) and 465 nm (emission from the amide anion). The fluorescence emission spectrum of the spent solution after luminescence shows only one peak at 420 nm (emission from neutral form). It appears that the environment at the active site of e-aequorin is favorable for the protonation of the amide anion. The difference between the luminescence spectrum and the fluorescence spectrum of spent solution must reflect a significant change that takes place in the environment of the e-coelenteramide light-emitter after light is emitted. In the case of the native form of aequorin, the luminescence spectrum (λ_{max} 465 nm) and the fluorescence emission spectrum of the spent solution (λ_{max} 458 nm) are close, indicating that only a small change takes place in the environment of the coelenteramide light-emitter after light is emitted.

In the fluorescence of coelenteramide, three other light-emitting species have been identified in addition to the amide anion and the neutral form: phenolate anion, pyrazine-N(4) anion, and ion-pair state (Shimomura and Teranishi, 2000). The emission of fluorescence is influenced by various properties of the solvent, such as hydrophobicity, proton-donating nature, and most importantly, permittivity. In the presence of a base, coelenteramide is ionized and exists as phenolate anion (fluorescence emission λ_{max} roughly 480 nm, possibly with a low quantum yield) or pyrazine-N(4) anion (fluorescence emission λ_{max} 530–540 nm), depending on the solvent used. These are tautomers. In general, however, the occurrence of the latter is overwhelming and the former is seldom seen (it seems to occur in n-butanol; Shimomura and Teranishi, 2000). Incidentally, the fluorescence emission maximum at 530 nm of a coelenterazine analogue in DMF in the presence of potassium t-butoxide was incorrectly assigned to the dianion structure shown at lower right in Fig. 5.4 (Hori *et al.*, 1973a). The emission at 530 nm must be emitted from the pyrazine-N(4) anion of the same analogue; the fluorescence emission property of the dianion remains unknown.

5.5   Chemical Reactions of Coelenterazine

Various chemical reactions of coelenterazine are shown in Fig. 5.5. Coelenterazine (A) is oxidized with molecular oxygen in bioluminescence reactions, chemiluminescence reactions and also in spontaneous oxidation, giving coelenteramide (B) as the major product. Coelenteramide (B) is easily hydrolyzed by a dilute acid into coelenteramine (C). In the Ca^{2+}-triggered luminescence reaction of aequorin, the product consists of about 90% coelenteramide plus 10% coelenteramine (Shimomura and Inouye, 1999), in addition to apoaequorin, CO_2 and light. Considering that the bioluminescence reaction of *Cypridina* luciferin results in the formation of oxyluciferin (86%), etioluciferin (10%), and the α-keto acid $C_2H_5(CH_3)CHCOCOOH$ (10%) (see Fig. 3.1.8, p. 69; Shimomura and Johnson, 1971, 1973b), it is likely that aequorin luminescence reactions (also bioluminescence reactions of coelenterazine in general) produce the α-keto acid $HOC_6H_4CH_2COCOOH$ in an amount that corresponds to the amount of coelenteramine formed (about 10%). Although it is clear that the coelenteramine is not formed from coelenteramide, the mechanism involved in the formation of coelenteramine has not been elucidated. In the chemiluminescence reaction and spontaneous oxidation of coelenterazine, the ratio of coelenteramine to coelenteramide produced is greater than that in the bioluminescence reaction. In the case of *Cypridina* luciferin, the yields of etioluciferin and α-keto acid in chemiluminescence reaction have exceeded 50% under certain conditions (Shimomura and Johnson, 1971, 1973b).

Coelenterazine (A) is oxidized into dehydrocoelenterazine (D) by MnO_2 in a mixed solvent of ethanol and ether (Inoue *et al.*, 1977b). Dehydrocoelenterazine ($C_{26}H_{19}O_3N_3$) can be obtained as dark red crystals. It does not have the capability of chemiluminescence. The ultraviolet absorption spectrum (Fig. 5.6) shows its absorption maxima at 425 nm (ε 24,400) and 536 nm (ε 12,600) in ethanol. An addition of NaOH significantly increases the 536 nm peak at the expense of the 425 nm peak. Dehydrocoelenterazine can take a tautomeric structure of quinone type (not shown), in which the phenolic proton on the 2-substituent is shifted onto the N(7) of the imidazopyrazinone ring. Dehydrocoelenterazine can be readily reduced to

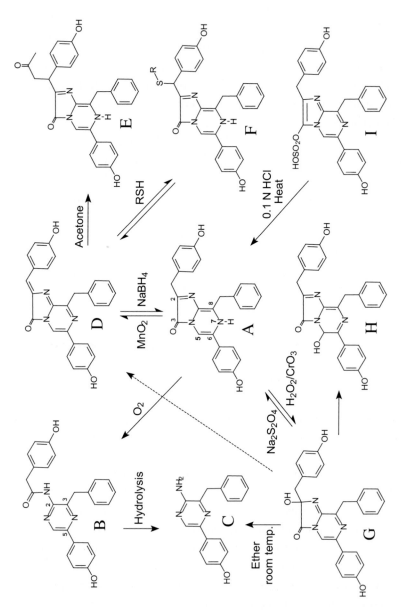

Fig. 5.5　Various chemical reactions of coelenterazine (shown at the center).

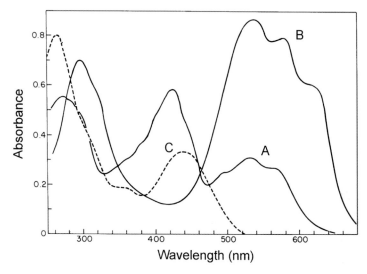

Fig. 5.6 Absorption spectra of dehydrocoelenterazine in ethanol with or without 0.01 M HCl (**A**); the same ethanolic solution plus 0.01 M NaOH (**B**); and coelenterazine in ethanol (**C**). Concentrations for A and B, approx. 23 μM; C, 33 μM. From Shimomura and Johnson, 1978.

coelenterazine by NaBH$_4$ (Inoue *et al.*, 1977b) and also by sulfhydryl compounds, such as 2-mercaptoethanol.

Dehydrocoelenterazine (D) forms an addition product (E) when treated with acetone in the presence of benzylamine (Takahashi and Isobe, 1993). Thiol compounds, such as DTT (dithiothreitol) and glutathione, also add to D, forming an equilibrium mixture containing F (Takahashi and Isobe, 1994). These adducts are chemiluminescent.

Coelenterazine (A) can be oxidized into 2-hydroxycoelenterazine (G) with H$_2$O$_2$ plus CrO$_3$ in ethanol containing phosphate buffer (Shimomura and Johnson, 1978). This compound is extremely unstable, and decomposes into coelenteramine when left standing in anhydrous ether at room temperature. It was originally obtained as "YC" (yellow compound) from aequorin by treating it with NaHSO$_3$, and it provided a strong support to the presence of coelenterazine-2-hydroperoxide in aequorin (Shimomura and Johnson, 1978). Teranishi *et al.* (1992), however, claimed the structure of YC to be 5-hydroxycoelenterazine (H), a stable compound; H is possibly formed from G by isomerization. The compound G can be converted into coelenterazine by reduction with $Na_2S_2O_4$, into coelenteramine by simply leaving it or by heating its solution, and into dehydrocoelenterazine (D) by treatment with H_2SO_4 in ether or with $HClO_4$ in dioxane (Shimomura and Johnson, 1978).

The enol-sulfate form (I), which is the precursor of the luciferin in the bioluminescence system of the sea pansy *Renilla* (Hori *et al.*, 1972), can be readily converted into coelenterazine by acid hydrolysis. The enol-sulfate (I), dehydrocoelenterazine (D) and the coelenterazine bound by the coelenterazine-binding proteins are important storage forms for preserving unstable coelenterazine in the bodies of luminous organisms. The disulfate form of coelenterazine (not shown in Fig. 5.5) is the luciferin in the firefly squid *Watasenia* (Section 6.3.1). An enol-ether form of coelenterazine bound with glucopyranosiduronic acid has been found in the liver of the myctophid fish *Diaphus elucens* (Inoue *et al.*, 1987).

5.6   Synthesis of Coelenterazines

Coelenterazine and its analogues have been synthesized using the reactions shown in Fig. 5.7 (Kishi *et al.*, 1972; Inoue *et al.*, 1975; Musicki, 1987; Shimomura *et al.*, 1989; Teranishi and Goto, 1990; Teranishi and Shimomura, 1997b). The condensation of coelenteramine with the α-ketoaldehyde to produce coelenterazine can also be done in one step by heating the two components in ethanol containing HCl, instead of the two steps shown in the figure. The scheme at the bottom is a new route of synthesizing coelenteramine by Pd-mediated cross coupling (Nakamura *et al.*, 1995; Wu *et al.*, 2000).

5.7   Coelenterazine Luciferases

There are many kinds of luminous organisms that utilize coelenterazine as their luciferin. These organisms possess luciferases to catalyze the luminescent oxidation of coelenterazine. Coelenterazine luciferases have been isolated from about 10 kinds of organisms, including the anthozoans *Renilla* and *Ptilosarcus*, the scyphozoan

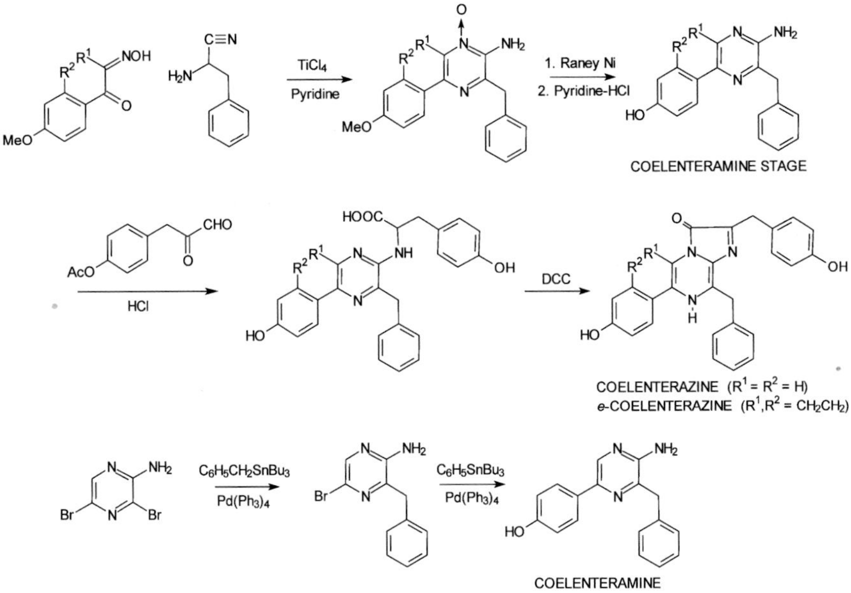

Fig. 5.7 Synthesis of coelenterazine. The bottom scheme shows a new method of synthesizing coelenteramine, an intermediate (Nakamura *et al.*, 1995; Wu *et al.*, 2000).

jellyfish *Periphylla*, three kinds of decapod shrimps, and two copepod species; some of them have been cloned. Some of the properties of these luciferases are shown in Fig. 5.8 and in part B of Appendix.

All of the luciferases cause the emission of a bluish light when they catalyze the oxidation of coelenterazine. However, there are some marked differences between the decapod shrimp luciferases and the cnidarian luciferases (Matthews *et al.*, 1977a,b). For example, the luminescence caused by the former (λ_{max} about 452 nm) is bluer than that caused by the latter (λ_{max} 470–480 nm), and the optimum pH of the former, about 8.5, is significantly higher than that of the latter (*Renilla*, 7.4; *Ptilosarcus*, 7.0). The optimum temperature of the decapod shrimp luciferases (35°C) is higher than those of *Ptilosarcus* (23°C) and *Renilla* (32°C).

The maximum specific activities of coelenterazine luciferases appear to be in a range of 10^{15}–10^{16} photons $\cdot$ s^{-1}mg^{-1}, but much

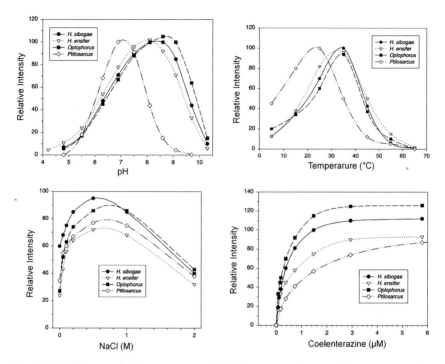

Fig. 5.8 Influence of pH, temperature, NaCl concentration, and the concentration of coelenterazine on the light intensity of luminescence reaction catalyzed by the luciferases of *Heterocarpus sibogae*, *Heterocarpus ensifer*, *Oplophorus gracilirostris*, and *Ptilosarcus gruneyi*. Buffer solutions used: 20 mM MOPS, pH 7.0, for *Ptilosarcus* luciferase; and 20 mM Tris-HCl, pH 8.5, for all other luciferases, all with 0.2 M NaCl, 0.05% BSA, and 0.3 μM coelenterazine, at 23°C, with appropriate modifications in each panel. Various pH values are set by acetate, MES, HEPES, TAPS, CHES, and CAPS buffers.

lower activities are frequently obtained. Ammonium sulfate precipitation often results in a significant decrease in the luciferase activity, possibly by causing an aggregation or unfavorable conformation of the luciferase molecules. It might be possible to restore the low activity to a range of 10^{15}–10^{16} photons $\cdot$ s^{-1}mg^{-1} by optimizing the molecular conformation of the luciferase, if a method of doing it could be found.

Luciferase activity on *e*-coelenterazine. In the presence of *Renilla* luciferase, the luminescence intensity of *e*-coelenterazine is more than 5 times higher than that of coelenterazine under the same conditions

(Inouye and Shimomura, 1997). With *Ptilosarcus* luciferase, the luminescence intensity of *e*-coelenterazine is also significantly higher than that of coelenterazine. With other coelenterazine luciferases, however, the luminescence intensity of *e*-coelenterazine is generally lower than that of coelenterazine; for example, the luminescence intensities of *e*-coelenterazine measured with the luciferases of the decapod shrimps, the jellyfish *Periphylla*, and the copepod *Pleuromamma*, were 50%, 4%, and 0.8%, respectively, in comparison with that of coelenterazine. Thus, the luminescence of coelenterazine catalyzed by *Pleuromamma* luciferase is suppressed by the addition of *e*-coelenterazine.

CHAPTER

6

LUMINOUS MOLLUSCA

In Mollusca, bioluminescence occurs in a great variety of organisms having distinctly different appearances, such as the classes Gastropoda (limpets, snails and sea hares), Bivalvia (clams), and Cephalopoda (squids and octopuses). All luminous molluscs currently known are marine organisms, except the New Zealand fresh water limpet *Latia neritoides* and the Malaysian land snail *Quantula* (*Dyakia*) *striata*. No information is yet available on the biochemical aspects of the *Quantula* luminescence.

Gastropoda. Luminous gastropods include *Latia neritoides* (discussed in Section 6.1) and *Planaxis labiosus*. The latter is a small marine snail (5–8 mm shell length) commonly found in shallow water at the stony beaches of the Hawaiian Islands and Hachijo Island in Japan. *Planaxis* emits light from its body only upon stimulation; it does not produce any luminous secretion (Haneda, 1958). According to my brief study, the snail is negative in the luciferin-luciferase reaction, and does not contain coelenterazine or *Cypridina* luciferin. When the crude extracts were tested with various chemicals, only sodium dithionite ($Na_2S_2O_4$) produced a significant level of light, suggesting that an active oxygen species, such as superoxide anion, might be involved in the luminescence reaction.

The order Nudibranchia (sea-slugs) contains several luminous genera (such as *Plocamopherus* and *Kalinga*) that show spectacular bioluminescence (Haneda, 1985). No information is available on their chemical aspects.

Bivalvia. The bivalve *Pholas* is historically important because the concept of "luciferin-luciferase reaction" was established with this clam (Dubois, 1887). It is the only bivalve that is well known and biochemically investigated. The details of the *Pholas* bioluminescence are given in Section 6.2.

Cephalopoda. The class Cephalopoda contains a large numbers of luminous species, mostly in the order Teuthoidea. Several of them are of great biochemical interest: for example, the firefly squid *Watasenia scintillans*, the purpleback flying squid *Symplectoteuthis oualaniensis*, the luminous flying squid *Symplectoteuthis* (*Eucleoteuthis*) *luminosa*, and *Onychoteuthis borealijaponicus* (*banksi*). Some of the luminous squids have their photophores and light organs distributed in a highly ornamental manner on their bodies (for example, *Watasenia scintillans* and *Chiroteuthis imperator*).

Some Myopsida and Sepioidea squids are known to harbor luminous bacteria to emit bioluminescence, for example, *Loligo*, *Uroteuthis*, *Doryteuthis*, *Semirossia* and *Sepiola* (Herring, 1978a). However, a large number of luminous squids utilize coelenterazine or its derivatives in their bioluminescence reactions, and they often contain a large amount of coelenterazine in their livers (Shimomura *et al.*, 1980). Although *Onychoteuthis banksi* has a luciferin-luciferase system utilizing coelenterazine as the luciferin (Shimomura *et al.*, 1980), some luminous squids appear to have a photoprotein-type luminescence system that involves coelenterazine. Details of the luminescence of *Watasenia scintillans*, *Symplectoteuthis oualaniensis* and *Symplectoteuthis* (*Eucleoteuthis*) *luminosa* are discussed in Section 6.3. According to Robinson *et al.* (2003), *Vampyroteuthis* contains a coelenterazine-luciferase luminescence system.

Only three kinds of octopus have been confirmed to be bioluminescent: *Japetella*, *Eledonella*, and *Stauroteuthis syrtensis* (Johnsen *et al.*, 1999). No information is available concerning the biochemistry of their luminescence.

6.1   The Limpet *Latia*

The New Zealand freshwater limpet *Latia neritoides* (Fig. 6.1.1) is the only known example of a freshwater luminous organism, with two possible exceptions: certain species of luminous bacteria and the larvae of certain species of fireflies. The limpet inhabits shallow clear streams in the North Island of New Zealand, clinging to stones and rocks. *Latia* has a small oval-shaped shell (6–8 mm long), and secretes a luminous mucus that emits a greenish glow around the body only when disturbed; the limpet does not show a spontaneous luminescence. The luminescence of *Latia* was first reported by Suter (1890) and further details including a positive luciferin-luciferase reaction were described by Bowden (1950). Both the luciferin and the luciferase have

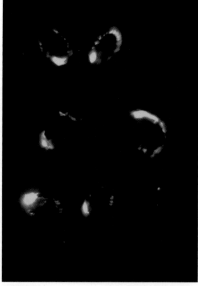

Fig. 6.1.1   The freshwater limpet *Latia neritoides* in day light (*left*) and in the dark (*right*).

been isolated and purified, and the structure of the luciferin has been determined (Shimomura and Johnson, 1968b). The bioluminescence reaction of *Latia* involves four components, luciferin, luciferase, a cofactor "purple protein" and molecular oxygen, to emit its bright green luminescence (Shimomura *et al.*, 1966b; Shimomura and Johnson, 1968c). The role of the purple protein is not completely clear.

Purification of *Latia* luciferin. The original purification method (Shimomura *et al.*, 1966b) was later simplified and improved (Shimomura and Johnson, 1968b). In the improved method, frozen limpets (30 g) preserved in dry ice are extracted with cold ethanol. The shells are crushed using a cold mortar and pestle, and the mixture is filtered. The filtrate is combined with 15 ml of *n*-butanol, and evaporated under reduced pressure to about 15 ml. The dark green solution (mostly butanol) is mixed with 15 ml of ethanol, and temporarily stored with dry ice. Several batches of crude extract prepared in this manner are combined and mixed with some water, and extracted with *n*-hexane. The hexane layer containing luciferin is washed with water, and poured onto a column of silicic acid (Mallinckrodt, 100 mesh, 4×30 cm). The active material adsorbed on the column is eluted with 25% benzene in *n*-hexane. The active eluate is evaporated to near dryness, and silicic acid chromatography is repeated with a column of a smaller size (2.5×20 cm; adsorption in *n*-hexane, elution with 8% ethyl acetate in *n*-hexane), giving a slightly yellowish luciferin solution. The solution is further purified by twice-repeated silicic acid chromatography (200–325 mesh, 1×15 cm; adsorption in *n*-hexane, elution with 18% benzene in *n*-hexane), yielding a colorless solution of luciferin. Purified *Latia* luciferin had a luminescence activity of 7.6×10^{15} photons/mg (Shimomura and Johnson, 1968b).

Purification of *Latia* luciferase and the purple protein. According to Shimomura and Johnson (1968c), frozen specimens of *Latia* (10 g) are vigorously shaken in 200 ml of cold 5 mM sodium phosphate buffer (pH 6.8) for 15 minutes. *Latia* luciferase is extracted into the buffer and the solution becomes turbid. Four batches of such turbid solutions are combined and centrifuged, and the clear supernatant is poured onto a column of DEAE cellulose (3.5×15 cm). The column is first washed with 5 mM sodium phosphate buffer containing 0.12 M NaCl (pH 6.8). Then luciferase is eluted with the same buffer containing 0.3 M NaCl. The active fractions containing both luciferase and the purple protein are combined, precipitated with ammonium sulfate, and centrifuged. The luciferase and the purple protein in the pellet are separated and purified by two steps of chromatography, first by gel filtration on a Sephadex G-200 column, followed by anion-exchange chromatography on a column of DEAE-cellulose (elution by a linear gradient of NaCl from 0.1 M to 0.5 M). In the gel filtration, luciferase is eluted before the purple protein. In the DEAE-cellulose chromatography, luciferase is eluted after the purple protein, and the two are completely separated.

Thirty-two years later, Kojima *et al.* (2000a) purified *Latia* luciferase by the following steps: ammonium sulfate fractionation, gel-filtration, affinity chromatography and Mono-Q anion-exchange FPLC.

Assay of luciferin and luciferase. To assay luciferin, 1 ml of 5 mM phosphate buffer, pH 6.8, containing crude luciferase (which contains the purple protein) is added to $5-10 \mu l$ of an ethanolic solution of luciferin, and the total light emitted is measured. To assay the activity of luciferase in crude material, 1 ml of 5 mM phosphate buffer, pH 6.8, containing a standard amount of luciferin is added to $5-10 \mu l$ of the sample, and the intensity of emitted light is measured. To assay purified luciferase, 1 ml of 5 mM phosphate buffer, pH 6.8, containing a standard amount of luciferin and a standard amount of purple protein is added to $5-10 \mu l$ of a sample, and the light intensity is measured (Shimomura and Johnson, 1968b,c).

Properties of *Latia* luciferin. *Latia* luciferin is a highly hydrophobic, fat-soluble compound, and volatile under vacuum. It is a colorless liquid, with an absorption maximum at 207 nm (ε approx. 13,700; Fig. 6.1.2). The chemical structure of *Latia* luciferin has been determined to be **1** ($C_{15}H_{24}O_2$), an enol formate of a terpene aldehyde **3** (Fig. 6.1.3; Shimomura and Johnson, 1968b). The enol formate group of *Latia* luciferin is unstable; the luciferin is spontaneously hydrolyzed

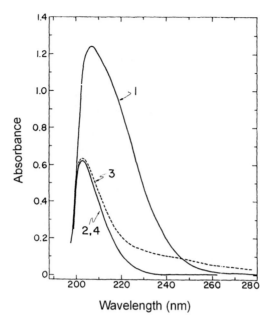

Fig. 6.1.2 Absorption spectra of *Latia* luciferin (**1**), the product of catalytic hydrogenation (**2**), the product of ammonolysis (**3**), and the product of hydrolysis (**4**), all in ethanol at the same molar concentration (89 μM). *Latia* luciferin has an ε value of 13,700 at 207 nm, and a bioluminescence activity of approximately 7.6×10^{15} photons/mg. From Shimomura and Johnson, 1968b, with permission from the American Chemical Society.

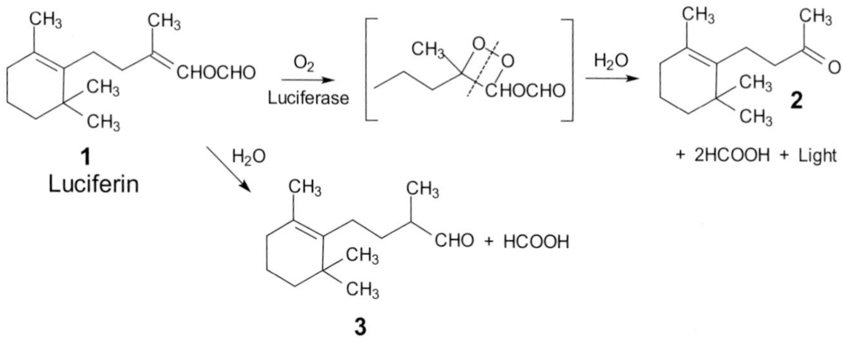

Fig. 6.1.3 Bioluminescence reaction of *Latia* and the hydrolysis of *Latia* luciferin.

in aqueous solvents, giving the aldehyde **3** and formic acid. The decomposition is very rapid in dilute alkaline solutions. Ammonolysis also converts luciferin **1** into aldehyde **3**. Compound **3** is not active in the luminescence reaction. The luciferin has been chemically synthesized (Fracheboud *et al.*, 1969; Nakatsubo *et al.*, 1970).

Latia luciferin emits a yellowish green light (λ_{max} 536 nm) in the presence of *Latia* luciferase, the purple protein, and molecular oxygen, decomposing itself into compound **2** (dihydro-β-ionone) and formic acid (Shimomura and Johnson, 1968c). Several analogues of *Latia* luciferin have been synthesized and their properties were investigated (Kojima *et al.*, 2000b). The results revealed that the trimethylcyclohexene ring in the luciferin structure is essential to the luminescence reaction. The study also showed that the substitution of the enol-formate group of luciferin with an enol-acetate group causes a 40% decrease in the activity of luciferin, and the substitution with an enol-ether group results in the complete loss of luciferin activity.

Properties of *Latia* luciferase and the purple protein. The absorption spectra of purified *Latia* luciferase and the purple protein are shown in Fig. 6.1.4. The sedimentation coefficient (s_{20}) of the

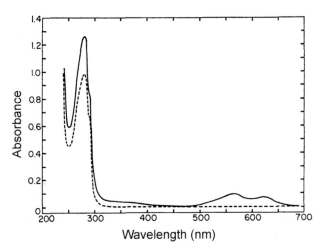

Fig. 6.1.4 Absorption spectra of *Latia* luciferase (dashed line) and the purple protein (solid line). From Shimomura and Johnson, 1968c, with permission from the American Chemical Society.

luciferase was 6.9×10^{-13}, and the molecular weight obtained by the sedimentation equilibrium method at 3°C was 173,000 (Shimomura and Johnson, 1968c), which agrees reasonably with the molecular mass 150 kDa obtained by size-exclusion chromatography by Kojima *et al.* (2000a). According to the latter authors, *Latia* luciferase is a glycoprotein consisting of inactive monomers of 31 kDa. The luciferase in a solution is satisfactorily stable for a few days at about 0°C in the presence of 0.1–1 mM EDTA, whereas 1 mM ascorbate and cysteine inactivates luciferase in 1–2 days (Shimomura and Johnson, 1968c).

The purple protein is reddish in solutions and purple in the ammonium sulfate precipitate. The molecular weight is 39,000 by the sedimentation equilibrium method. The purple protein is brightly red-fluorescent, but the fluorescence characteristics cannot be related to the luminescence of *Latia* (Fig. 6.1.5; Shimomura and Johnson, 1968c).

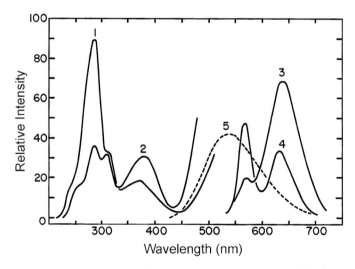

Fig. 6.1.5 Fluorescence spectra of the purple protein (1–4) and the luminescence spectrum measured with *Latia* luciferin, luciferase and the purple protein (5; λ_{max} 536 nm). Excitation spectra (1) and (2) were measured with emission at 630 nm and 565 nm, respectively. Emission spectra (3) and (4) were measured with excitation at 285 nm and 380 nm, respectively. From Shimomura and Johnson, 1968c, with permission from the American Chemical Society.

The role of purple protein. We reported over 30 years ago that the bright bioluminescence reaction of *Latia* requires the presence of luciferin, luciferase, molecular oxygen and a purple protein, though the function of the last named was not completely clear (Shimomura and Johnson, 1968c; Shimomura *et al.*, 1972). It seems appropriate to consider the purple protein as an activator or enhancer of the light-emitting reaction, rather than an essential requirement, for the following reasons: (1) the fluorescence of the purple protein is spectrally unrelated to the luminescence; (2) the purple protein can be replaced by ascorbate and DPNH in the luminescence reaction; (3) butanol extracts of acidified purple protein strongly intensify the luminescence; and (4) a weak, but significant, luminescence is emitted in the absence of the purple protein, as shown in Fig. 6.1.6. Item (4) is in agreement with the claim made by

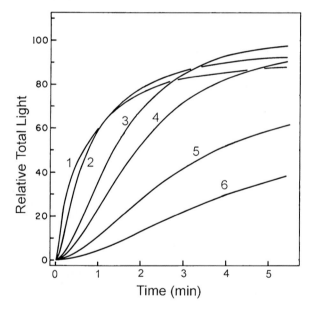

Fig. 6.1.6 Effect of the purple protein on the luminescence of *Latia* luciferin (0.16 μg) plus *Latia* luciferase ($A_{280,1\,cm}$ 1.2, 10 μl) in 5 ml of 5 mM sodium phosphate buffer, pH 6.8. The amounts of the purple protein solution ($A_{280,1\,cm}$ 0.6) used: 20 μl (curve 1), 5 μl (curve 2), 1 μl (curve 3), 0.5 μl (curve 4), 0.2 μl (curve 5), and none (curve 6). From Shimomura and Johnson, 1968c, with permission from the American Chemical Society.

Kojima *et al.* (2000a) that the purified luciferase exhibited a luciferin-luciferase reaction with *Latia* luciferin without any cofactor. Nevertheless, the purple protein is a conspicuous presence in the live organisms and it is highly likely that it enhances the bioluminescence of *Latia* in nature.

The luminescence reaction of *Latia*. The luciferin-luciferase luminescence of *Latia* is markedly intensified by the purple protein, ascorbate, and NADH, and more strongly by their combinations, and only briefly by Fe^{2+} (Shimomura and Johnson, 1968c; Shimomura *et al.*, 1972). Various characteristics of the luminescence reaction of *Latia* reported by Shimomura *et al.* (1966b) can be summarized as follows: The total light emission is optimum at pH 6.9, and the luminescence intensity is optimum at pH 6.4 (Fig. 6.1.7). The salts, such as NaCl and KCl, have little influence on luminescence at concentrations lower than 1 M. Temperature has a significant influence on the luminescence; the total light emission is doubled by lowering the temperature from 25°C to 8°C (Fig. 6.1.8). Cyanide (0.1–1 mM) and *o*-phenanthroline (0.1 mM) inactivate luciferase, suggesting the possible presence of Fe^{2+} in luciferase. The kinetics of the luminescence reaction is complicated,

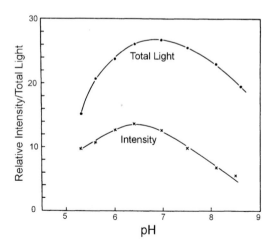

Fig. 6.1.7   Effect of pH on the initial light intensity and total light of *Latia* bioluminescence reaction in the presence of the purple protein, in 50 mM sodium phosphate buffer solutions having various pH values at 25°C (Shimomura *et al.*, 1966b).

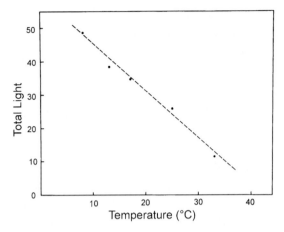

Fig. 6.1.8 Influence of temperature on the total light produced in the biolumi-nescence reaction of *Latia* in the presence of the purple protein, in 5 mM sodium phosphate buffer, pH 6.8 (Shimomura *et al.*, 1966b).

but the total light emitted is generally proportional to the amount of luciferin in the presence of a sufficient amount of luciferase, and the luminescence intensity is roughly proportional to the amount of luciferase in the presence of an excess amount of luciferin.

The quantum yield of *Latia* luciferin is surprisingly low. When an optimum concentration of purple protein was used together with luciferase, the quantum yield was 0.0030 at 25°C, and 0.0068 at 8°C (Shimomura and Johnson, 1968c). When 1 mM ascorbate and an opti-mum concentration of NADH (~0.25 mM) were added, the quantum yield was 0.009 at 25°C (Shimomura *et al.*, 1972).

The reaction scheme of *Latia* bioluminescence. Based on the structures of luciferin **1** (Ln) and the product of luminescence reaction **2** (OxLn), it was proposed that the luciferase-catalyzed luminescence reaction of *Latia* luciferin in the presence of the purple protein results in the formation of 2 moles of formic acid, as shown in the scheme A (Shimomura and Johnson, 1968c). However, when the luminescence reaction was carried out in a medium containing ascorbate and NADH (in addition to the purple protein) to increase the quantum yield, it was found that only one mole of formic acid was produced accompanied by the formation of CO_2 (Shimomura *et al.*, 1972); thus, the reaction scheme B was formulated in this case (XH_2 indicates a reducing agent such as ascorbate).

$$Ln\ (1) + O_2 + H_2O \xrightarrow{\text{Luciferase}} OxLn\ (2) + 2HCOOH + Light \quad (A)$$

$$Ln\ (1) + 2O_2 + XH_2 \xrightarrow{\text{Luciferase}} OxLn\ (2) + HCOOH + CO_2$$
$$+ X + H_2O + Light \qquad\qquad (B)$$

Because scheme B involves two oxygen molecules, it must be an overall result of two or more reactions. In scheme A, the reactants are satisfactorily equated with the products. Thus, it seems most likely that scheme A represents the essential part of the *Latia* luminescence reaction, and scheme B includes the reaction of scheme A. On the mechanism side, the light-emitting reaction probably involves a peroxide intermediate containing a four-membered dioxetane ring, as shown in Fig. 6.1.3, although the mechanism of the peroxide formation is unclear. In an effort to prove the dioxetane pathway, an ^{18}O-labeling experiment was performed. When the luminescence reaction was carried out with $^{18}O_2$ (instead of $^{16}O_2$ in air), the carbonyl group of the product 2 was labeled with ^{18}O atom; however, detection of ^{18}O in CO_2 or HCOOH was unsuccessful due to a flaw in the experimental method (Shimomura *et al.*, 1972). Thus, the involvement of the dioxetane intermediate has not been confirmed.

The light emitter in *Latia* luminescence. The purple protein is strongly fluorescent in red. Thus, at first glance, it would appear to be a most probable candidate for the light emitter or its precursor. However, this possibility was ruled out when we found that there is no way to relate the fluorescence of the purple protein to the bioluminescence spectrum. Thus, the luciferase must contain a chromophore that produces the light emitter.

Latia luciferase is colorless and normally nonfluorescent. However, the luciferase fluoresces visibly in alkaline solutions. The fluorescence is most prominent in a KCN solution, showing an emission spectrum that is very close to the bioluminescence spectrum and also to the fluorescence emission of a flavin (FAD) except for the 370 nm

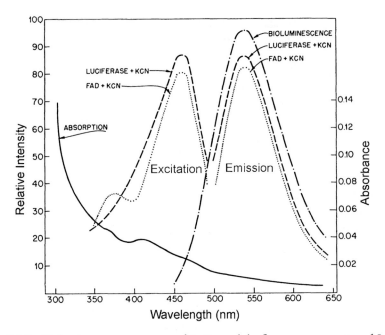

Fig. 6.1.9 Bioluminescence spectrum of *Latia*, and the fluorescence spectra of *Latia* luciferase ($A_{280,1\,cm}$ 1.1) and FAD (1.5 µg/ml), both in 0.1 M KCN. The fluorescence emission spectra were measured with excitation at 460 nm, and the excitation spectra were measured at 535 nm. The absorption spectrum of the luciferase ($A_{280,1\,cm}$ 1.1) included was measured in a phosphate buffer, pH 6.8. From Shimomura *et al.*, 1972.

region of its excitation spectrum (Fig. 6.1.9). Thus, it seems likely that the light emitter of *Latia* bioluminescence is a flavin or flavin-like group that exists in luciferase in a reduced nonfluorescent state or in a quenched state. Further study is needed to confirm the light emitter of *Latia* bioluminescence.

6.2   The Clam *Pholas dactylus*

The boring clam *Pholas dactylus* lives in holes bored into soft rocks. This species is distributed along the European coasts of the Atlantic Ocean and the Mediterranean Sea. Harvey (1952) mentioned that the clam has been known since early history, and that the generic name, *Pholas*, comes from the Greek word "pholas," meaning hidden

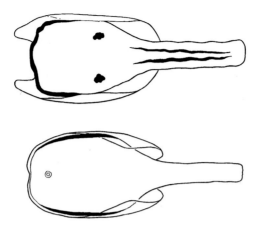

Fig. 6.2.1 Diagram showing the luminous areas of the clams *Pholas dactylus* (*top*; from Panceri, 1872) and *Rocellaria grandis* (*bottom*; from Haneda, 1939).

or lurking in a hole, and the specific name, *dactylus*, is also from the Greek word "daktylos," the forefinger, since the shape of the animal somewhat resembles a finger. The clam *Pholas dactylus* has five patches of luminous organs (Panceri, 1872; Fig. 6.2.1). The morphology and structure of the luminous organs of *Pholas* were studied in detail (Harvey, 1952; Herring, 1978). Though applications of *Pholas* bioluminescence have been commercially developed, the mechanism of the bioluminescence reaction has not been elucidated. The clam is consumed as a delicacy, and Michelson (1978) states that the species faces the threat of total extermination for several reasons, including human consumption. The only known luminous clam other than *Pholas* is *Rocellaria grandis*, discovered by Haneda (1939) in the Palau Islands (east of the Philippines).

On the basis of the luciferin-luciferase reaction discovered by Dubois (1887), Michelson, Henry and their associates studied the biochemistry of the *Pholas* bioluminescence for several years beginning in 1970 (Michelson, 1978). They isolated, purified, and characterized the luciferin and the luciferase, and published about a dozen papers in which the luciferin isolated was referred to as *Pholas* luciferin. Since the luciferin is clearly a protein, later authors called it "pholasin" following the traditional way of naming a photoprotein

(Roberts *et al.*, 1987). Therefore, "pholasin" is used in preference to "*Pholas* luciferin" in the following sections.

Purification of pholasin. The extraction of pholasin is carried out in the presence of diethyldithiocarbamate to inactivate *Pholas* luciferase, a copper protein. According to Henry *et al.* (1970) and Henry and Monny (1977), acetone powder (4 g) prepared from the luminous organs of *Pholas dactylus* is homogenized in 30 ml of 1 mM diethyldithiocarbamate, and centrifuged. The supernatant is made to contain 50 mM NaCl, and the pH is adjusted to 4.8 with 50 mM acetate buffer. The precipitate formed is removed by centrifugation. The supernatant is neutralized to pH 7.0, and fractionally precipitated with ammonium sulfate, taking a fraction between 50% and 80% saturation. The crude pholasin obtained is purified by gel filtration on a column of Sephadex G-100 using 10 mM phosphate buffer, pH 7.0, followed by anion-exchange chromatography on a column of DEAE-Sephadex prepared with 50 mM phosphate buffer, pH 7.0 (elution by NaCl gradient from 0.1 M to 0.8 M). The purified pholasin is dialyzed against 10 mM phosphate buffer, pH 7.0, then stored at 0°C.

Roberts *et al.* (1987) and Müller and Campbell (1990) used slightly different methods to extract and purify pholasin. They used 10 mM ascorbate, instead of diethyldithiocarbamate, to inhibit luciferase in the process of extraction and purification, which enabled them to obtain the purified preparations of both pholasin and the luciferase.

Assay of pholasin. Two different methods have been used. In the first method, light intensity or total light emission is measured when a standard solution of luciferase is added to a pholasin sample (Henry *et al.*, 1970). In the second method, total light emission is measured when 1 ml of a degassed solution of 0.3 mM $FeSO_4$ is injected into 2 ml of 0.15 M phosphate buffer, pH 7.0, containing a pholasin sample and 0.75 M NaCl (Michelson, 1978). The luminescence reaction is complete within 2 or 3 min.

Assay of *Pholas* luciferase. Two methods have been used. In the first method, light intensity is measured when a luciferase sample dissolved in 1 ml of 0.1 M phosphate buffer, pH 7.0, containing 0.5 M NaCl is injected into 2 ml of a standard pholasin solution made with the same buffer (Henry et al., 1975). The intensity is not strictly proportional to the luciferase concentration; therefore, a calibration is required (Henry et al., 1970). In the second method, the peroxidase activity of luciferase is measured with ascorbate and H_2O_2, because the luciferase is a true peroxidase (Henry et al., 1975). The assay mixture (1 ml) consists of a luciferase sample dissolved in 10 mM phosphate buffer, pH 8.5, containing 0.15 mM H_2O_2 and 0.1 mM ascorbate; H_2O_2 and ascorbate are added immediately before the assay. The rate of change in the A_{265} value is measured, and the activity is calculated by the initial rate of change. The rate of change in the absence of luciferase is used as the blank. One unit is defined as the amount of enzyme that induces a decrease of 0.1 A/min at 265 nm at 25°C. The assay is linear between 0.2 and 10 units.

Properties of pholasin. According to Henry and Monny (1977) and Michelson (1978), pholasin is a glycoprotein with a molecular weight of 34,600. Pholasin resembles the luciferase in amino acid composition, isoelectric point (pI < 3.5), and some other properties, but does not contain copper atoms. The ultraviolet absorption spectrum (Fig. 6.2.2) shows a bulge at about 325 nm, in addition to the protein peak at 280 nm. According to Müller and Campbell (1990), the functional group of pholasin is unrelated to coelenterazine or *Cypridina* luciferin, although their efforts to identify the chromophore were unsuccessful. They noted that a flavin-like moiety might be involved in the luminescence reaction based on the fluorescence of the extracts of pholasin.

Purification of *Pholas* luciferase (Michelson, 1978). Acetone powder of the light organs is extracted with 10 mM Tris-HCl buffer, pH 7.5, and the luciferase extracted is chromatographed on a column of DEAE Sephadex A-50 (elution by NaCl gradient from 0.1 M to 0.6 M). Two peaks of proteins are eluted, first luciferase, followed by a stable complex of luciferase and inactivated pholasin. The fractions of each peak are combined, and centrifuged in 40% cesium chloride

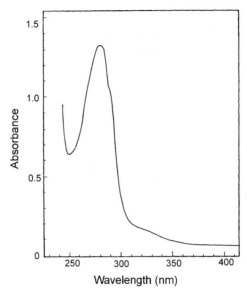

Fig. 6.2.2   Absorption spectrum of pholasin (0.85 mg/ml) in 10 mM acetate buffer, pH 4.8. From Henry *et al.*, 1973, with permission from Elsevier.

(w/v) at 50,000 rpm for 45 hr. The centrifuged solution is fractionated by sucking off with a peristaltic pump. Active fractions are pooled and further purified by gel filtration on a column of Sepharose 6B or Sephadex G-100.

Properties of *Pholas* luciferase.   The purified luciferase is a single, homogeneous protein according to ultracentrifugation and PAGE, and is relatively stable up to 45°C. It is an acidic glycoprotein (pI 3.5 or lower) having a molecular weight of 310,000, containing two copper atoms per molecule. Subunits with molecular weights 150,000 and 46,000 were found (Henry *et al.*, 1975; Henry and Monny, 1977). The luciferase is a peroxidase and oxidizes ascorbate in the presence of H_2O_2 (Henry *et al.*, 1975).

Luminescence reaction.   Pholasin undergoes an oxidative luminescence reaction in the presence of any of the following substances: *Pholas* luciferase, ferrous ions, H_2O_2, peroxidases, superoxide anions, hypochlorite and other oxidants. In all cases, molecular oxygen is required and pholasin is converted into oxypholasin in the reaction.

Thus, *Pholas* luciferase is clearly not a specific requirement for the luminescence of pholasin (Henry and Michelson, 1970; Henry *et al.*, 1973, 1975; Müller and Campbell, 1990). The luminescence reaction of pholasin with *Pholas* luciferase is optimum at pH 8–9, and at an ionic strength corresponding to about 0.5 M NaCl or KCl. The quantum yield of pholasin is 0.09 (Michelson, 1978). *Pholas* luciferase has two pholasin binding sites of equal affinity. The luminescence of pholasin elicited with *Pholas* luciferase has a maximum at 490 nm (Henry *et al.*, 1973, 1975), and that with Fe^{2+} shows a maximum at 484 nm (Fig. 6.2.3).

According to Reichl *et al.* (2000), the oxidation of pholasin by compound I or II of horseradish peroxidase induces an intense light emission, whereas native horseradish peroxidase shows only a small effect. The luminescence of pholasin by native myeloperoxidase (verdoperoxidase) is diminished by preincubation with catalase, which is interpreted as the result of the removal of a trace amount of naturally occurring H_2O_2 in the buffer (about 10^{-8} M) that forms compound I

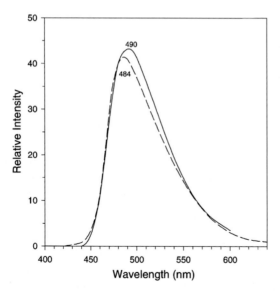

Fig. 6.2.3 Luminescence spectra of pholasin in the presence of horse radish peroxidase (solid line; from Henry *et al.*, 1973 with permission from Elsevier), and in the presence of $FeSO_4$ in 10 mM phosphate buffer, pH 7.0, containing 0.1 M NaCl (dashed line; Shimomura, 1980, unpublished).

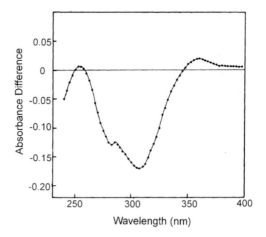

Fig. 6.2.4 Change in the absorption spectrum of pholasin (14.5 μM) caused by the luminescence reaction catalyzed by *Pholas* luciferase (1.1 μM). The curve shown is the differential spectrum between a cell containing the mixture of pholasin and *Pholas* luciferase (0.9 ml; in the sample light path) and two cells containing separate solutions of pholasin and the luciferase at the same concentrations (in the reference light path), all in 0.1 M Tris-HCl buffer, pH 8.5, containing 0.5 M NaCl. Four additions of ascorbate (3 μM) were made to the sample mixture to accelerate the reaction. The spectrum was recorded after 120 min with a correction for the base line. From Henry and Monny, 1977, with permission from the American Chemical Society.

and/or II. Indeed, the addition of H_2O_2 to a mixture of myeloperoxidase and pholasin gives an intense burst of light.

The luminescence reaction causes a change in the absorption spectrum of pholasin, indicating a structural change of the chromophore involved. The change is clearly seen in the differential spectrum between pholasin and the oxypholasin (Fig. 6.2.4), which reveals the disappearance of the absorption band of pholasin at around 307 nm (ε 11,800) accompanied by the appearance of a new band of oxypholasin at 360 nm (Henry and Monny, 1977). The spectral change of pholasin upon denaturation also might be of possible significance. When pholasin is treated with 5 M guanidine hydrochloride, the bulge of absorption spectrum at 325 nm disappears, accompanied by the appearance of fluorescence with an emission peak at 460 nm and a corresponding excitation peak at 365 nm (Fig. 6.2.5; Henry *et al.*, 1973, 1975).

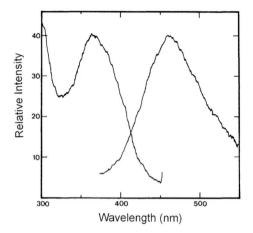

Wavelength (nm)

Fig. 6.2.5   Fluorescence spectra of pholasin after treatment with 5 M guanidine hydrochloride. *Left*, excitation spectrum measured at 460 nm; *right*, emission spectrum measured with excitation at 360 nm. From Henry *et al.*, 1973, with permission from Elsevier.

Recombinant apopholasin (Dunstan *et al.*, 2000). A full-length clone encoding apopholasin has been isolated from a cDNA library of the *Pholas dactylus* light organ, and has been expressed in cell extracts and insect cells. The protein contained 225 amino acid residues, including a signal sequence of 20 amino acids. The size of recombinant apoprotein expressed was 26 kDa, but the size expressed with the addition of microsomal membranes was 34 kDa, corresponding to the molecular mass of native pholasin. Both apoproteins, with and without signal sequence, were recognized by a polyclonal antibody for native pholasin. The incubation of an acidic methanol extract of *Pholas* light organs with recombinant apopholasin resulted in an ability of the protein to luminesce upon the addition of sodium hypochlorite, but the formation of the native photoprotein pholasin was not achieved.

6.3   Luminous Squids (Cephalopoda)

There are many kinds of bioluminescent squids. Some of them harbor luminous bacteria for their light emission (Harvey, 1952; Haneda, 1985), but all other luminous squids currently known utilize coelenterazine or its derivatives in their bioluminescence systems, and they often store a large amount of coelenterazine, its derivatives, and a luciferase in their livers. For example, *Watasenia scintillans* has large amounts of coelenterazine and dehydrocoelenterazine (Inoue *et al.*, 1975, 1977b), *Symplectoteuthis* (*Eucleoteuthis*) *luminosa* contains an extremely large amount of dehydrocoelenterazine (0.6 mg per specimen), and *Onychoteuthis borealijaponica* contains a strong luciferase activity (Shimomura *et al.*, 1980), in their respective livers.

The luciferin of *Watasenia scintillans* was reported to be coelenterazine disulfate, although its luciferase has not been isolated (Tsuji, 2002). The luminescence of *Symplectoteuthis oualaniensis* is caused by a photoprotein that can be regenerated with dehydrocoelenterazine (Takahashi and Isobe, 1993, 1994). The luminescence of the *Symplectoteuthis luminosa* photoprotein is strongly enhanced by the addition of H_2O_2 plus catalase (discussed later in detail). In the case of *Ommastrephes pteropus*, the homogenate of the light organs emits a flash of light by the injection of an oxidizing agent such as H_2O_2 (Girsch *et al.*, 1976).

6.3.1   *The Firefly Squid Watasenia scintillans*

The deep-sea squid *Watasenia scintillans*, "Hotaru-ika" in Japanese, is a seasonal delicacy, and commercially harvested in a large quantity (nearly 1,000 tons/year) with setnets in Toyama Bay, Japan. The season is from March to May, when enormous numbers of the squid come to the shallows to lay eggs. This species was first described as *Abraliopsis scintillans* by Watase (1905). The squid is small (mantle length 4–6 cm) and has five photophores around each eye, three black-pigmented photophores on the tip of each ventral arm, and numerous minute photophores covering the mantle and head. In the dark, the tiny photophores of the mantle and head shine blue, like stars in the sky, whereas the arm photophores radiate extremely intense light (Fig. 6.3.1).

Harvey (1917) noted that the fresh arm photophores of *Watasenia scintillans* do not show a luciferin-luciferase reaction, and Shoji (1919) using a gas chamber of purified hydrogen demonstrated that molecular oxygen is needed for the luminescence.

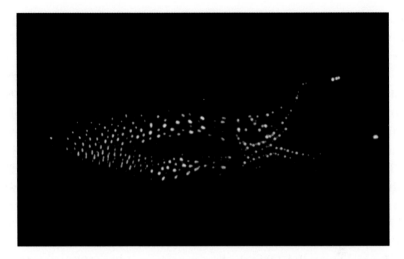

Fig. 6.3.1   The firefly squid *Watasenia scintillans*.

Goto *et al.* (1974) extracted the arm photophores, and isolated a compound that is highly fluorescent in blue. They determined the structure of this compound to be coelenteramide disulfate (structure **A** below), and named it "*Watasenia* oxyluciferin." Then, Inoue *et al.* (1975)

A

Watasenia oxyluciferin
Coelenteramide disulfate

B

Watasenia preluciferin
Coelenterazine

C

Watasenia luciferin
Coelenterazine disulfate

isolated a large amount of coelenterazine (**B**) from the livers of *W. scintillans*, 43 mg of crystals from 200 g of livers, and they named the compound "*Watasenia* preluciferin." It was the first time that coelenterazine was obtained by human hands as a pure substance, although the compound had been known to exist in the photoprotein aequorin (Shimomura *et al.*, 1974). A compound that fits the name "*Watasenia* luciferin," structure **C**, was soon isolated from the arm photophores of *W. scintillans* (Inoue *et al.*, 1976), and also from the photophores of the eyes and skin, and the livers (Inoue *et al.*, 1983). However, the luciferase that catalyzes the luminescent oxidation of the luciferin has not been isolated or identified.

The absorption spectra of W*atasenia* luciferin (coelenterazine disulfate) and *Watasenia* oxyluciferin (coelenteramide disulfate) are shown in Fig. 6.3.2. *Watasenia* luciferin in neutral aqueous solutions is auto-oxidized in air more rapidly than coelenterazine, and the compound emits a strong chemiluminescence in the presence of H_2O_2 (~ 10 mM) plus Fe^{2+} (~ 0.2 mM). *Watasenia* oxyluciferin is strongly fluorescent in aqueous solutions (λ_{max} 400 nm), 500 times stronger than the fluorescence of coelenteramide in aqueous media (Goto *et al.*, 1974).

Involvement of ATP in the luminescence. Tsuji (1985) found that homogenate of the light organs of *W. scintillans* emits light when ATP is added in the presence of Mg^{2+}. The luminescence reaction has a sharp pH optimum at 8.8 (Fig. 6.3.3), and the luminescence spectrum shows a peak at 470 nm (Fig. 6.3.4). The luminescence reaction

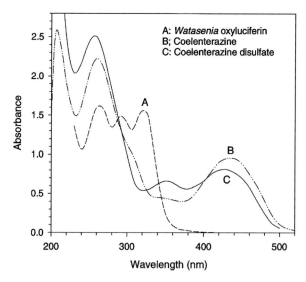

Fig. 6.3.2 Absorption spectra of *Watasenia* oxyluciferin (**A**), coelenterazine (**B**), and coelenterazine disulfate (**C**), all in methanol. Concentrations: 0.1 mM for B and C; undetermined for A.

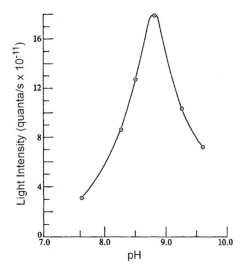

Fig. 6.3.3 Relationship between pH and the initial light intensity in the ATP-stimulated luminescence of the homogenate of *Watasenia* arm light organs in the presence of 1.5 mM ATP and 0.3 mM MgCl$_2$ (Tsuji, 1985).

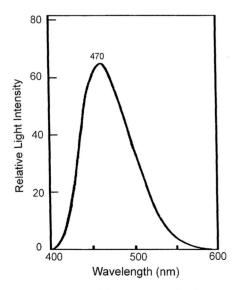

Fig. 6.3.4 Luminescence spectrum of the *Watasenia* bioluminescence reaction measured with a crude extract of light organs that contain particulate matters, in chilled 0.1 M Tris-HCl buffer, pH 8.26, containing 1.5 mM ATP. From Tsuji, 2002, with permission from Elsevier.

appears to be a luciferin-luciferase reaction that requires ATP, Mg^{2+} and molecular oxygen, in which the luciferin is coelenterazine disulfate (*Watasenia* luciferin) and the luciferase is in a form of insoluble membrane-bound protein.

Tsuji (2002) proposed a reaction scheme for *Watasenia* bioluminescence (Fig. 6.3.5) based on the results obtained with the crude extracts and particulate matters prepared from the light organs, although it would be extremely difficult to make a correct interpretation in the presence of numerous contaminants. The proposed mechanism somewhat resembles the mechanism of the firefly bioluminescence which involves the adenylation of luciferin. In the proposed mechanism, coelenterazine disulfate is adenylated on the active site of presumptive luciferase; however, the adenyl group is split off before the oxygenation of coelenterazine disulfate. Therefore, the adenylated form is not involved in the light-emitting reaction, making a distinct difference from the firefly bioluminescence reaction. Moreover, the role of the adenylation is unclear. It is difficult to believe that the adenylation is required just for a proper binding of coelenterazine disulfate to luciferase as suggested in the paper, because such a task is usually carried out by the enzyme protein. For elucidation of the mechanism, it would be essential to isolate and purify the luciferase.

6.3.2   The Purpleback Flying Squid Symplectoteuthis oualaniensis (Tobi-ika)

The squid *Symplectoteuthis oualaniensis* is widely distributed in the western Pacific and the Indian Ocean, and commercially fished in the waters of Okinawa in southern Japan from June to November. This squid has a large, oval patch of photophores on the dorsal surface of its mantle that produces an intense blue flash of light (mantle length 15–20 cm). The first biochemical study on the bioluminescence of this species has been reported by Tsuji and Leisman (1981). According to their study, the essential light-emitting components of this squid are membrane-bound. They homogenized and centrifuged the light organ in 50 mM Tris-HCl buffer, pH 7.2 or 7.6. The pellets obtained were suspended in the same buffer and centrifuged again, then resuspended

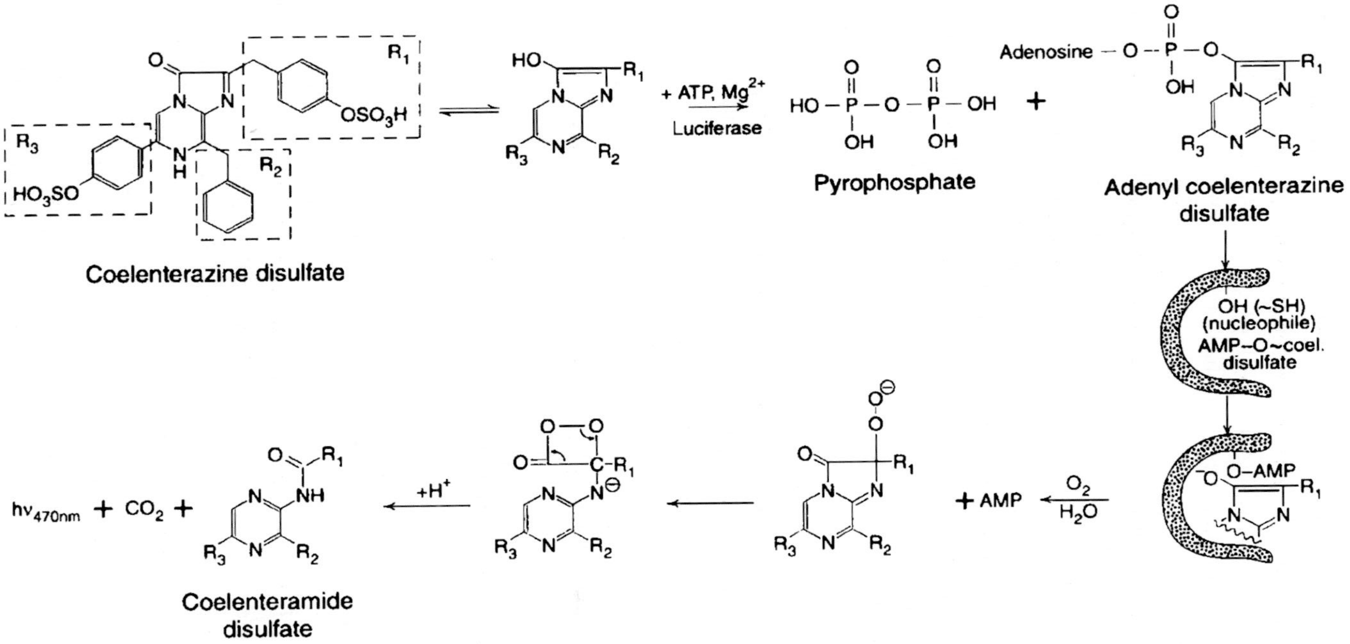

Fig. 6.3.5 A reaction scheme proposed by Tsuji (2002) for the *Watasenia* bioluminescence. The proposed mechanism involves the adenylation of luciferase-bound luciferin by ATP, like in the bioluminescence of fireflies. However, the AMP group is split off from luciferin before the oxygenation of luciferin, differing from the mechanism of the firefly bioluminescence. Thus the role of ATP in the *Watasenia* bioluminescence reaction remains unclear. Reproduced with permission from Elsevier.

in the same buffer and used in experiments. The suspension emitted light upon the addition of the salts of monovalent cations, such as K^+, Ru^+, Na^+, Cs^+, NH_4^+ and Li^+ (in the decreasing order of the effect). The presence of molecular oxygen was essential for light emission. Divalent ions, such as Ca^{2+} and Mg^{2+}, did not cause light emission. The optimum concentrations of KCl and NaCl were 0.6 M, and an optimum pH was found at 7.8 (Fig. 6.3.6). Analysis of the kinetics during the decay course of light intensity suggested that two light-emitting components are involved, one decaying faster than the other. The peak wavelength of the *S. oualaniensis* luminescence has been reported in various values: 456 nm (Tsuji and Leisman, 1981; Fig. 6.3.7), 470 nm (Takahashi and Isobe, 1993), and 480 nm (Takahashi and Isobe, 1994).

Involvement of dehydrocoelenterazine. Takahashi and Isobe (1993) reported that the luminescence of *Symplectoteuthis oualaniensis* is emitted from a photoprotein (later named symplectin) containing a dehydrocoelenterazine chromophore. According to the report, acetone powder prepared from light organs has the capability of light emission when 1 M KCl is added at room temperature. The extraction of the acetone powder with methanol gives a significant amount of dehydrocoelenterazine (structure **2** in Fig. 6.3.8) plus a trace amount of coelenterazine (**1**). However,

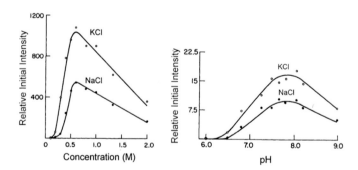

Fig. 6.3.6 Effects of salt concentration (*left* panel) and pH (*right* panel) on the initial light intensity emitted from the homogenate of the *Symplectoteuthis oualaniensis* light organ. The salt effect was tested in 50 mM Tris-HCl, pH 7.2, and the pH effect in the various buffers containing 0.5 M KCl or NaCl. From Tsuji and Leisman, 1981.

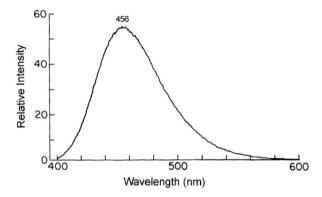

Fig. 6.3.7 Luminescence spectrum of a homogenate of the luminous organ of *Symplectoteuthis oualaniensis* in the presence of 0.5 M KCl (from Tsuji and Leisman, 1981). A homogenate suspension (1 ml) and 1 M KCl (1 ml), both made with 50 mM Tris-HCl, pH 7.6, containing 1 mM dithioerythritol, were mixed and the spectrum was measured 6 min after mixing. Note that the luminescence of the photoprotein symplectin isolated from the luminous organs showed a maximum at 470–480 nm (Takahashi and Isobe, 1993, 1994).

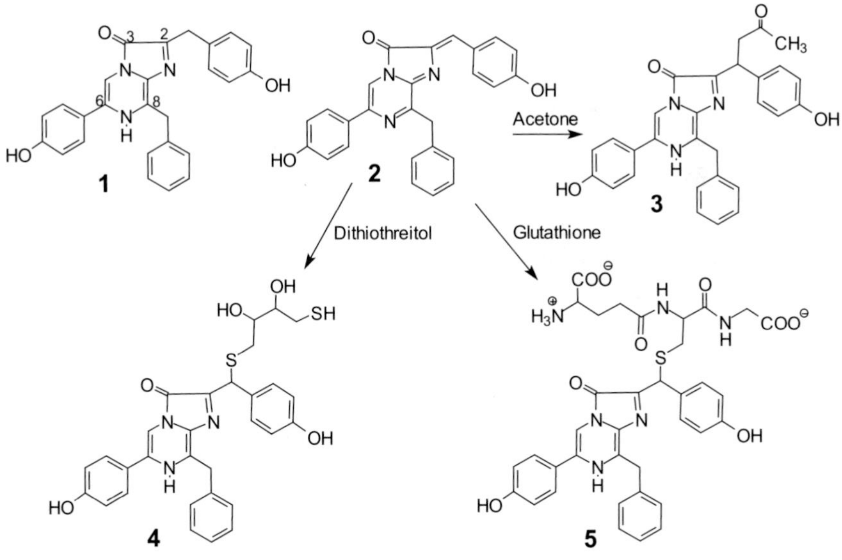

Fig. 6.3.8 Chemical structures of the compounds relevant to the luminescence reaction of symplectin. From Takahashi and Isobe, 1993.

when the acetone powder has been pretreated with 1 M KCl to cause luminescence, the extraction with methanol does not give any dehydrocoelenterazine, indicating that it has been consumed in the luminescence reaction. When the acetone powder (untreated with 1 M KCl) is extracted with a mixture of methanol plus acetone, compound **3** (an acetone adduct of **2**) is obtained, but this compound is not obtained when the acetone powder has been pretreated with 1 M KCl in advance. Compound **2** easily forms a stable adduct with acetone, thus compound **3** obtained above is considered to be an artifact due to the use of methanol-acetone mixture. The same authors also found that dithiothreitol and glutathione (reduced form) add to dehydrocoelenterazine, forming **4** and **5**, respectively, although the products cannot be isolated because the addition reactions are in equilibrium. Based on these results, the authors inferred that dehydrocoelenterazine (**2**) exists in the photoprotein of *S. oualaniensis* as an adduct similar to **4** and **5**.

Isolation of symplectin (Takahashi and Isobe, 1994). The photoprotein from *S. oualaniensis*, now named "symplectin", was partially purified at 4°C by the following method. A homogenate of the light organs made in a Tris-HCl buffer, pH 7.6, containing 0.25 M sucrose, 1 mM dithiothreitol and 1 mM EDTA (the basic buffer), was centrifuged, and the pellets obtained were suspended in the same buffer, then centrifuged again. Extraction of the pellets with the basic buffer containing 1 M KCl, followed by centrifugation, gave a crude extract of symplectin that was free of cell debris. The photoprotein in this crude extract was partially purified by size-exclusion chromatography on columns of Biogel A-0.5 m (BioRad) and Cellurofine GLC-300 m (Seikagaku Kogyo). Symplectin was eluted at the void volumes, indicating that the protein is probably in an aggregated form.

Luminescence activity was measured at room temperature by mixing a small part of the sample with the basic buffer containing 1 M KCl, pH 7.6. Light emission started upon warming the sample from 4°C to room temperature, and the light intensity gradually decreased with time, falling to about 10% of the initial value after about one and a half hours. At that point, an addition of dehydrocoelenterazine restored the luminescence intensity to about 50% of the initial value. The adducts 4 and 5 had similar effects in restoring the luminescence, but coelenterazine did not have such an effect. These properties may support that the photoprotein contains a dehydrocoelenterazine moiety in a form similar to 4 and 5.

The size of symplectin molecule. Fujii *et al.* (2002) used an improved method to extract and purify symplectin. They extracted the light organs first with 50 mM potassium phosphate buffer containing 0.25 M sucrose, 1 mM EDTA, 1 mM DTT, pH 7.6, and then re-extracted the residue with the same buffer containing 0.4 M KCl to remove soluble impurities. Symplectin in the residue was extracted with the same buffer containing 0.6 M KCl, but at pH 6.0. The 0.6 M KCl extract was purified by size-exclusion HPLC on a column of TSK G3000SW.

In the size-exclusion HPLC, symplectin was eluted as two oligomers (both >200 kDa) and a trace of 60 kDa monomer species. A limited tryptic digestion before HPLC increased the 60 kDa species at the expense of two high molecular weight species, and also resulted in the formation of 40 kDa and 16 kDa species. SDS-PAGE analysis of the two high molecular weight oligomers revealed that they consist mainly of the 60 kDa protein. The 60 kDa protein and the 40 kDa protein were fluorescent in the SDS-PAGE analysis, and both emitted light when warmed. The amino acid sequences of the 60 kDa, 40 kDa and 16 kDa species contained no similarity to known photoproteins but had a significant similarity to the carbon-nitrogen hydrolase domains found in mammalian biotinidase and vanin (pantetheinase).

Reconstitution of symplectin from aposymplectin. Isobe *et al.* (2002) tested the reconstitution of aposymplectin. Aposymplectin was prepared by the following method: a solution of extracted symplectin (1 ml) was mixed with 0.14 ml of 50 mM Tris buffer, pH 9.8, containing 0.6 M KCl and 1 mM DTT, and luminesced at 37°C for 30 minutes. Then ammonium sulfate (204 mg) was added to the mixture to precipitate aposymplectin. After leaving for 30 minutes at 0°C, the mixture was centrifuged, and the precipitate of aposymplectin was washed

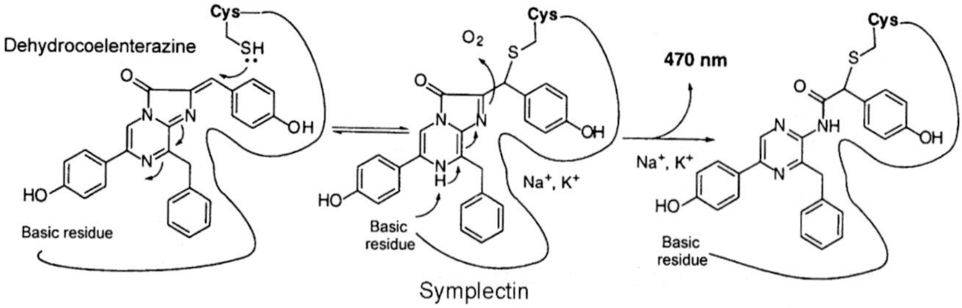

Fig. 6.3.9 An illustration showing the mechanism of the reconstitution and luminescence of symplectin. The binding of dehydrocoelenterazine with the SH group of a cysteine residue of aposymplectin (*left*) results in the formation of active symplectin (*center*). Symplectin is oxygenated at the C2 position, resulting in the formation of coelenteramide bound to aposymplectin (*right*), accompanied by the emission of light. From Isobe *et al.*, 2002, with permission from Elsevier.

twice with water (0.7 ml). Then the precipitate of aposymplectin was dissolved in 0.7 ml of 50 mM phosphate buffer, pH 6.0, containing 0.6 M KCl and 1 mM DTT. To reconstitute symplectin, the solution of aposymplectin (0.1 ml) and 2 μl of 2 mM dehydrocoelenterazine in DMSO were mixed and left standing for 20 minutes at room temperature. The luminescence of reconstituted symplectin can be initiated by adding four volumes of 50 mM Tris buffer, pH 9.8. The reactions involved are illustrated in Fig. 6.3.9.

The luminescence reaction of symplectin induced by warming is slow and not sufficiently bright to account for the bright luminescence of the live squid. The factors and conditions that cause the intense *in vivo* luminescence remain to be identified.

6.3.3 *The Luminous Flying Squid Symplectoteuthis luminosa (Suji-ika)*

The luminous squid *Symplectoteuthis* (*Eucleoteuthis*) *luminosa* (Fig. 6.3.10) is commonly fished along the Pacific coast of Ibaragi, northeast of Tokyo, from July to October. This species has a pair of long narrow strips of photophores on the ventral surface of the mantle and a pair of small patches near the eyes (called eye organs hereafter). The luminescence mechanism and chemical components involved in the light emission of this squid are presumed to be similar, if not identical, to those of *Symplectoteuthis oualaniensis* because they belong

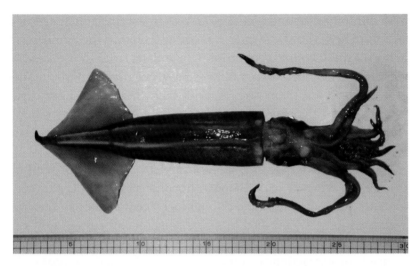

Fig. 6.3.10   The luminous flying squid *Symplectoteuthis* (*Eucleoteuthis*) *luminosa* (photo by Dr. Satoshi Inouye).

to the same genus. The following is a summary of a brief study made by the author using specimens sent from Japan by Dr. Satoshi Inouye. The specimens used had a body weight of about 150 g and mantle length of 18 cm.

The liver. In various luminous organisms, the livers and eggs often contain essential ingredients for bioluminescence, such as luciferin and luciferase. The liver of *S. luminosa* contains a large amount of dehydrocoelenterazine, which can be easily extracted with methanol. In our study, the liver (10 g) of a single specimen contained an unusually large amount of dehydrocoelenterazine (0.6 mg) plus a small amount of coelenterazine (3 µg). Dehydrocoelenterazine is readily reduced into coelenterazine with NaBH$_4$, and can be assayed using a coelenterazine luciferase, such as *Renilla* luciferase and *Oplophorus* luciferase.

Extraction and partial purification of photoprotein. The solubility and general luminescence characteristics of the *S. luminosa* photoprotein are similar to those reported for the *S. oualaniensis* photoprotein; the protein is soluble in buffer solutions containing 0.6–1.0 M salt but not in solutions containing 0.1–0.2 M salt, and the luminescence is pH-dependent. In the extraction of *S. oualaniensis*,

1 mM dithiothreitol and 1 mM EDTA were routinely included in all solutions. However, the former was found to retard the luminescence reaction of the *S. luminosa* photoprotein and the latter showed no clear advantage. Thus, these two additives have been omitted in our study of the *S. luminosa* photoprotein. Although the ventral light organs are much larger than the eye organs, the use of eye organs was preferred because of their significantly higher contents of non-aggregated photoprotein.

The eye light organs of *S. luminosa* were ground with sand in a mortar and extracted with 20 mM phosphate/1 M NaCl, pH 6.0, at 0°C. After removing insoluble cell debris by centrifugation, the crude extract was partially purified by size-exclusion chromatography on a column of Superdex 200 Prep (1 × 27.5 cm) using the same buffer as that used for extraction, resulting in two major activity peaks. The results are illustrated in Fig. 6.3.11. The fractions of elution volume 14–16.5 ml were stored at 5°C for use in various experiments.

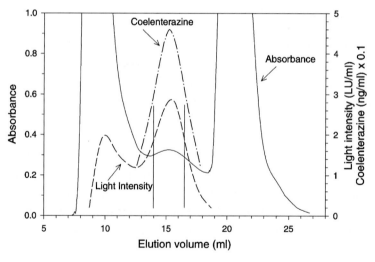

Fig. 6.3.11 Chromatography of an extract of the eye light organs of *Symplectoteuthis luminosa* on a column of Superdex 200 Prep (1×27.5 cm) in 20 mM phosphate buffer, pH 6.0, containing 0.6 M NaCl, at 0°C (monitored at 280 nm). Each fraction (0.5 ml) is measured for the initial intensity of H_2O_2/catalase-triggered luminescence and the content of dehydrocoelenterazine measured as coelenterazine after $NaBH_4$-reduction 1 LU = 6×10^8 photons.

Assay of photoprotein. The activity of the photoprotein was measured in 1 ml of 20 mM Tris-HCl buffer, pH 8.0, containing 0.6 M NaCl at room temperature. The intensity and total amount of light emitted were recorded. The luminescence intensity is markedly intensified by adding 5 µl of catalase solution (crystalline bovine liver catalase; 1.5 mg/ml) and 10 µl of 3% H_2O_2.

The addition of catalase and H_2O_2 increased the reaction rate (light intensity) of light emission more than 10 times. Thus, 90% of the total light was emitted in a period of 20–40 minutes at 20°C, compared to more than 3 hours required to emit 50% of the total light without the additions of catalase and H_2O_2. Considering that most of the H_2O_2 added is instantly decomposed by catalase and that catalase alone has no effect, the activation is probably due to a substance formed from catalase in the presence of H_2O_2. Because of the close resemblance between this photoprotein and symplectin, it would be reasonable to expect that the luminescence reaction of symplectin can also be enhanced by adding catalase and H_2O_2.

Characteristics of *Symplectoteuthis luminosa* photoprotein. In the size-exclusion chromatography noted above (Fig. 6.3.11), the luminescence activity was eluted as two peaks (at 10 ml and 15 ml), corresponding to the molecular masses of 400 kDa and 50 kDa, respectively. The former is probably an aggregate of the 50 kDa species, and the latter protein probably corresponds to the 60 kDa species of the *S. oualaniensis* photoprotein, symplectin. The colorless solution of the 50 kDa protein was fluorescent, with an emission maximum at 482 nm at pH 6.0 (excitation at 415 nm; nonfluorescent at pH 8.0), although the significance of this fluorescence is uncertain because the sample was not completely pure. Some data concerning the light emitting properties of the 50 kDa fraction are shown in Table 6.3.1. The data clearly show that the luminescence of *S. luminosa* photoprotein is enhanced by the addition of catalase and H_2O_2 (No. 4). The total light output is increased considerably by pretreatment of the sample with dehydrocoelenterazine (No. 5), whereas pretreatment with coelenterazine had no effect (not shown). When the sample had been pretreated with $NaBH_4$ to convert the dehydrocoelenterazine moiety

Table 6.3.1 Luminescence of a partially purified sample of *Symplectoteuthis luminosa* photoprotein (50 kDa fraction) measured with 20 μl of the photoprotein sample in 1 ml of 20 mM Tris-HCl buffer, pH 8.0, containing 0.6 M NaCl.[a] All at 20°C.

No.	Pretreatment of Sample and Treatment Time (min)	Reagents Added and Reaction Time (min)	Initial Intensity (LU/s)[b]	Total Light Emitted (LU)[b]
1	None	None (200)	0.01	47
2	None	H_2O_2	0.02	
3	None	Catalase	0.01	
4	None	Catalase/H_2O_2 (40)	0.22	99
5	Dehydro-coelenterazine (30)	Catalase/H_2O_2 (40)	0.28	122
6	NaBH$_4$ (20)	Catalase/H_2O_2 (40)	0.20	66
7	NaBH$_4$ (20), then dehydro-coelenterazine (30)	Catalase/H_2O_2 (40)	0.26	118
8	NaBH$_4$ (20)	*Oplophorus* luciferase (<3)	c	680[d]
9	Catalase/H_2O_2 (60), then NaBH$_4$ (20)	*Oplophorus* luciferase (<3)	c	5

[a]Reagents used: 5 μl of 1.5 mg/ml catalase solution (crystalline bovine liver); 10 μl of 3% H_2O_2 solution; 20 μl of 0.1 mM dehydrocoelenterazine solution in methanol; about 1 mg NaBH$_4$ powder; *Oplophorus* luciferase, amount not determined.
[b]1 LU = 6×10^8 photons.
[c]Light intensity is very high, and variable depending on the activity of *Oplophorus* luciferase used.
[d]About 25 before NaBH$_4$-treatment.

into free coelenterazine, the amount of coelenterazine detected was unexpectedly high (No. 8; total light 680 LU) even after taking into account the high quantum yield of the luminescence reaction of coelenterazine catalyzed by *Oplophorus* luciferase (0.3). The sample used probably contained a considerable amount of the dehydrocoelenterazine-binding protein already bound with dehydrocoelenterazine, in addition to the photoprotein. Further purification of the photoprotein is needed to characterize the bioluminescence reaction of *S. luminosa*.

***In vivo* luminescence of the *Symplectoteuthis* species.** The live specimens of both *S. oualaniensis* and *S. luminosa* emit bright flashes from their photophores upon nervous stimulation, differing from the photoproteins extracted from them that emit dim, long-lived luminescence when warmed from 4°C to room temperature. The luminescence of *S. luminosa* photoprotein, however, can be strongly enhanced by the addition of catalase and H_2O_2, as mentioned above. According to a recent communication from Dr. Masaki Kuse, Nagoya University, the luminescence of symplectin (*S. oualaniensis* photoprotein) can also be strongly intensified by the addition of H_2O_2 in the presence of catalase. Thus, it is of great interest to uncover the mechanism of the luminescence enhancement caused by catalase plus H_2O_2; the same or a closely related mechanism might be involved in the bright *in vivo* luminescence of *Symplectoteuthis*.

CHAPTER

7

ANNELIDA

In the phylum Annelida, bioluminescent species are found in two classes, Polychaeta (marine worms) and Oligochaeta (earthworms). Polychaeta contains a number of interesting luminous organisms in various families, such as chaetopteridae (*Chaetopterus*), polynoidae (*Harmothoë* and *Polynoë*), syllidae (*Odontosyllis* and *Pionosyllis*), and tomopteridae (*Tomopteris*). Oligochaeta contains many species of luminous earthworms in Megascolecidae, such as *Lampito*, *Diplocardia*, and *Octochaetus*.

7.1 The Tubeworm *Chaetopterus variopedatus*

This bizarre-looking worm lives in a U-shaped parchment-like tube up to 25 cm long (Fig. 7.1.1), which is usually buried in sand or mud, exposing only the openings of both ends. It is a filter feeder, and very widely distributed, sometimes forming dense colonies on the seabed. In fact, this species propagated so densely in the Los Angeles Harbor around 1965, Dr. Rimmon Fay was able to collect 10,000

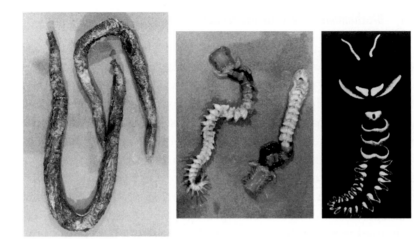

Fig. 7.1.1   *Left*: Two parchment-like tubes inhabited by *Chaetopterus variopedatus*. The tubes are usually buried in muddy sand, exposing only the openings at both ends. *Center*: The worms after removal of the casings. *Right*: Diagram showing luminous area (after Panceri, 1878); the third pair from the top corresponds to the 12th segment.

specimens in a matter of days. Recently, an explosive increase and "invasion" of this and related species have been reported at various places in the world, including Hawaii, New Zealand and Australia (Coles *et al.*, 1997).

The body of *Chaetopterus variopedatus* consists of three distinct parts, an anterior "head," a middle region with specialized feeding structures, and a longer, regularly segmented posterior part (Fig. 7.1.1). When disturbed, the posterior segments emit a flash of light that lasts a few seconds, and a steadily luminous slime is secreted from various parts of the body. According to Harvey (1952), the luminescence requires molecular oxygen, but the luciferin-luciferase reaction is negative in the luminescence of its slime. Harvey also stated: "It is hard to imagine what the purpose of light can be to an animal which remains hidden in a tube on the sea bottom in mud or sand well below the surface, and which never wanders about."

Since the early studies by Panceri (1878), the luminescence of this animal has been well studied in the areas of morphology, histology and physiology, but only briefly in biochemistry.

7.1.1   *Biochemistry of the Luminescence of Chaetopterus variopedatus*

Currently, available information on the biochemical aspects of the *Chaetopterus variopedatus* bioluminescence comes from the two papers published nearly 40 years ago by Shimomura and Johnson (1966; 1968d). In their study, they extracted and purified the light-emitting substance from the 12th segment (wing-like aliform notopodia) plus adjacent tissues of this worm. After purification, the light-emitting principle was obtained as a protein. The protein emitted light in the presence of H_2O_2 and Fe^{2+}, and the total amount of light emitted from this protein was proportional to the amount of the protein luminesced. Because the protein isolated did not fit the commonly used terms of "luciferin" or "luciferase", a new term "photoprotein" was proposed to categorize this kind of proteins (Shimomura and Johnson, 1966), as mentioned in the Introduction of this book.

Extraction of luminescence material. About 50 g of dry ice-frozen 12th segment material (from about 250 specimens) was homogenized with 250 ml of cold saturated ammonium sulfate solution containing 20 g of additional solid ammonium sulfate, a trace of sodium bicarbonate and some glass beads (6 mm diameter), in a Mason jar attached to an Omnimixer; it is important to use a Teflon stirrer blade (specially hand-made) in place of the usual stainless steel blade (to avoid contamination with ferrous ion). At this stage, the homogenate was able to emit light simply by diluting with water.

The homogenate was centrifuged at 30,000 g for 30 min at 0°C. The clear supernatant was discarded. The precipitate and floating material were stirred into 250 ml of 10 mM Tris-HCl buffer, pH 7.2, containing 0.2 mM oxine (8-hydroxyquinoline). The extremely viscous mixture obtained was allowed to stand for 1 hour in a refrigerator, during which time the viscosity decreased considerably. The mixture was centrifuged at 30,000 g for 20 min. The precipitate was discarded, and 100 g of ammonium sulfate was added to the supernatant, resulting in salting out of the photoprotein. After centrifugation, the precipitate was combined with similarly prepared precipitates from two or three additional batches. The combined precipitates were dissolved in 100 ml of 10 mM phosphate buffer, pH 7.0, containing 0.1 mM oxine, and dialyzed for 3 hr against the same buffer at 0°C.

Purification of photoprotein. The dialyzed photoprotein solution was centrifuged to remove precipitates, and then subjected to fractional precipitation by ammonium sulfate, taking a fraction precipitated between 30% and 50% saturation. The protein precipitate was dissolved in 50 ml of 10 mM sodium phosphate, pH 6.0, containing 0.1 mM oxine ("pH 6.0 buffer"), dialyzed against the same buffer, and the dialyzed solution was adsorbed on a column of DEAE-cellulose (2.5 × 13 cm) prepared with the pH 6.0 buffer. The elution was done by a stepwise increase of NaCl concentration. The photoprotein was eluted at 0.2–0.25 M NaCl and a cloudy substance (cofactor 1) was eluted at about 0.5 M NaCl. The photoprotein fraction was further purified on a column of Sephadex G-200 or Ultrogel AcA 34 (1.6 × 80 cm) using the pH 6.0 buffer that contained 0.5 M NaCl.

Crystallization. A solution of purified photoprotein was saturated with ammonium sulfate, and centrifuged. The precipitate was dissolved in a small amount of the pH 6.0 buffer, and ammonium sulfate was added to produce slight cloudiness. Crystals of photoprotein were formed in several hours (Fig. 7.1.2).

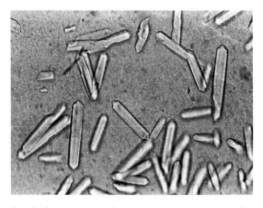

Fig. 7.1.2   Crystals of *Chaetopterus* photoprotein grown in a solution of ammonium sulfate. The width of the view field is approx. 300 µm.

Measuring the activity of photoprotein. The method of measuring the activity of this photoprotein had to be changed with the progress in purification. The initial extract in saturated ammonium sulfate was assayed by simply diluting a small sample amount with a large amount of water in the presence of molecular oxygen, measuring the resulting weak, long-lasting luminescence. After the dialysis step, it was discovered that the addition of H_2O_2 (or a peroxide) plus a trace of Fe^{2+}, causes brighter luminescence. The luminescence was short-lived, thus repeated additions of Fe^{2+} were needed for the assay of total light-emitting activity (Fig. 7.1.3). As the peroxide component, unidentified peroxides existing in old commercial dioxane and tetrahydrofuran

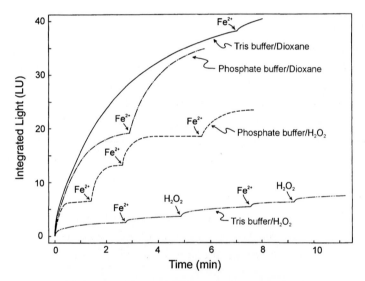

Fig. 7.1.3 Influence of the buffer and the type of peroxide on the luminescence reaction of *Chaetopterus* photoprotein. The reaction was initiated at zero time by the addition of a peroxide (old dioxane or H_2O_2) and $FeSO_4$ in each case, with successive additions of $FeSO_4$ or H_2O_2 at the time indicated with an arrow. In the experiments of the two upper curves, 10 µl of old dioxane and 1 µl of 10 mM $FeSO_4$ were added at zero time, followed by 1 µl of 10 mM $FeSO_4$ at each arrow. In the experiments of the two lower curves, 50 µl of 10 mM H_2O_2 and 20 µl of 10 mM $FeSO_4$ were added at zero time, followed by 50 µl of 10 mM H_2O_2 or 20 µl of 10 mM $FeSO_4$ at each arrow. All in 5 ml of 10 mM phosphate or Tris buffer, pH 7.2. From Shimomura and Johnson, 1966.

were significantly more effective than H_2O_2. During the purification of the protein by DEAE-cellulose chromatography, another activator (cofactor 1, noted above) was found and separated. Cofactor 1 was found to be a relatively heat-resistant protein. With further purification, the luminescence activity of photoprotein progressively decreased (in terms of luminescence intensity) when measured with optimum concentrations of Fe^{2+}, a peroxide and cofactor 1. The activity could be restored, however, by the addition of 0.1–0.2 mg/ml of an impure hyaluronidase preparation; with the purest photoprotein material, the activation of luminescence was more than 50-fold. A purer preparation of hyaluronidase gave less activation, indicating that the activating factor (cofactor 2) occurs as an impurity. Cofactor 2 is heat-stable, not dialyzable, resistant to acid and alkali, and soluble in 2:1 mixture of acetone and water. Perhaps it is a lipid, but its chemical identity remains unknown.

Thus, five factors — molecular oxygen, Fe^{2+}, a peroxide, cofactor 1, and cofactor 2 — are needed for bright luminescence of the purest preparation of *Chaetopterus* photoprotein. The presence of Fe^{2+} seems to be essential for the *in vitro* luminescence reaction to take place, despite the fact that the free form of Fe^{2+} is highly unlikely to occur in the living organisms. If Fe^{2+} is present in organisms, it must be in a form bound with a protein. The role of the peroxide is probably not directly related to the oxidation of luminophore, because the luminescence reaction requires molecular oxygen. The role of peroxide is probably to help the oxygenation of luminophore in conjunction with Fe^{2+}. The roles of cofactors 1 and 2 are probably entirely supportive, by protecting and extending the activities of Fe^{2+} and peroxide.

7.1.2 Properties of the Chaetopterus Photoprotein and its Luminescence Reaction

Spectral properties. The purified photoprotein is practically colorless, although its absorption spectrum (Fig. 7.1.4) shows a very slight absorption in the region of 330–380 nm in addition to the 280 nm protein peak. The solution of photoprotein is moderately blue fluorescent, with an emission maximum at 453–455 nm and an

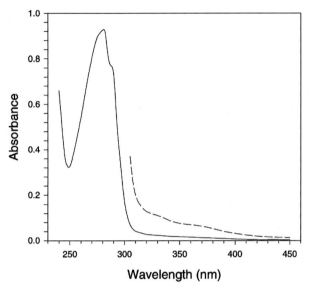

Fig. 7.1.4 Absorption spectrum of purified *Chaetopterus* photoprotein in 10 mM phosphate buffer, pH 6.0, containing 0.1 mM oxine and 0.2 M NaCl, measured against the buffer used. The dashed line indicates a 5-times expansion in absorbance.

excitation maximum at 375 nm (Fig. 7.1.5). The fluorescence does not change significantly after the luminescence reaction. The luminescence spectrum of purified photoprotein (λ_{max} 453–455 nm) matches exactly with the fluorescence emission spectrum, although the luminescence spectrum measured with the 12th segment is slightly red-shifted (λ_{max} 457–458 nm).

Molecular weight. The molecular weight of the photoprotein before crystallization (CPA) was 128,000 by the sedimentation method, and about 130,000 by gel filtration on Ultrogel AcA 34, whereas the molecular weight measured with the crystallized form (CPC) was 184,000 by the sedimentation method. It appears that the conditions of crystallization caused a change in the aggregation state. The comparison of the two molecular weights suggests the existence of a species with a molecular weight about 60,000, which might be the monomer. The initial luminescence intensity of CPC was about

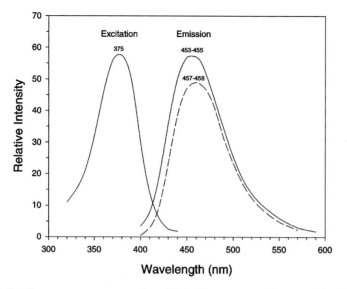

Fig. 7.1.5 Fluorescence spectra of purified *Chaetopterus* photoprotein (CPA) in 10 mM ammonium acetate, pH 6.7 (solid lines), and the bioluminescence spectrum of the luminous slime of *Chaetopterus* in 10 mM Tris-HCl, pH 7.2 (dashed line). Note that the luminescence spectrum of *Chaetopterus* photoprotein in 2 ml of 10 mM Tris-HCl, pH 7.2, containing 0.5 M NaCl, 5 μl of old dioxane and 2 μl of 10 mM FeSO₄ (λ_{max} 453–455 nm) matched exactly with the fluorescence emission spectrum of the photoprotein. No significant change was observed in the fluorescence spectrum after the luminescence reaction.

one half of that of CPA on the basis of the same weight, although the amounts of the total elicitable light were equal. Thus, under the same conditions, the luminescence of CPA is brighter than that of CPC, and the M_r 60,000 species, if available, might luminesce brighter than CPA does.

Luminescence reaction. In air, partially purified preparations of the photoprotein emit light in the presence of Fe^{2+} and a peroxide; however, highly purified preparations require two additional substances, cofactors 1 and 2, to emit the same intensity of light. Both cofactors can be substituted with a small amount of a crude extract of the photoprotein.

The luminescence is strongly affected by the peroxide selected and the type of buffer used (Fig. 7.1.3). The peroxides in aged dioxane and tetrahydrofuran are far more effective (long-lasting) than H_2O_2. Phosphate buffer is better than Tris buffer when H_2O_2 is used, whereas Tris buffer appears better than phosphate buffer when dioxane is used. The initial light intensity is strongly affected by the pH of the buffer (Fig. 7.1.6, left panel), showing a maximum at pH 7.7 with a sharp decrease at both sides (50% decreases at pH 6.5 and pH 8.3). The optimum temperature for the luminescence intensity is about 22°C (Fig. 7.1.6, right panel). The luminescence intensity is not significantly affected by the concentration of salt up to 1 M, when tested with NaCl.

To measure the total light emission of a photoprotein sample, it was necessary to add 1–5 µM Fe^{2+} several times due to the short effective life of Fe^{2+} under the conditions involved. The total amount of light measured in this manner was always proportional to the weight of photoprotein used, with both CPA and CPC (4.7×10^{13} photons/mg at 25°C; Shimomura and Johnson, 1968d). Thus, the quantum yield is estimated at roughly 0.01 for CPA, and 0.015 for CPC.

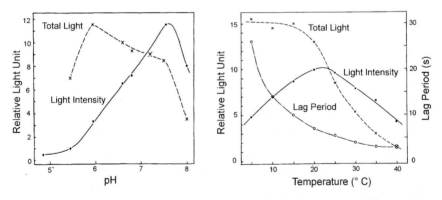

Fig. 7.1.6 Influence of pH and temperature on the luminescence of *Chaetopterus* photoprotein elicited by old dioxane and Fe^{2+} in 20 mM phosphate buffer. *Left* panel: the effect of pH in phosphate buffer solutions of various pH values, at 22°C. *Right* panel: the effect of temperature at pH 7.2. Luminescence was initiated by the injection of Fe^{2+}. The time lag of the light emission after the Fe^{2+} injection was also shown in the right panel. From Shimomura and Johnson, 1966.

No information is available on the chemical nature of the luminophore, although the photoprotein must contain a chromophore to emit luminescence and fluorescence. Acid treatment of the protein, followed by extraction with organic solvents, did not yield coelenteramide or coelenteramine, indicating that this luminescence system is unrelated to coelenterazine. A flavin (FAD) was found in partially purified preparations of photoprotein, but not in highly purified preparations.

7.2 The Bermuda Fireworm *Odontosyllis*

Luminous syllidae worms are widely distributed and many of them show striking luminescence. It is possible that Columbus saw the bioluminescence display of *Odontosyllis* in 1492 in the Caribbean on his first voyage to the new continent (Harvey, 1952). In the present day, the most well known example is the Bermuda fireworm *Odontosyllis enopla*. Like other species of *Odontosyllis*, the Bermuda fireworms are very small but they show a spectacular bioluminescence display that is correlated with the lunar cycle. The luminescence display takes place during a period of several days following a full moon. It begins about one hour after sunset and lasts only for a period of 10–15 min. It begins with the sudden appearance of a swarm of brilliantly luminescent females (about 2 cm long and 2 mm wide) at the surface of the water, each worm moving quickly in a tight circle. Within a few seconds, numerous males (about 1 cm long, also brightly luminescent) appear, and they dart toward the females from all directions, attracted to the light.

Collection of specimens. To collect specimens, a specially made net fitted with a bag of Dacron gauze, 30 cm in diameter and 40 cm deep, is used. In order to collect as many worms as possible in a limited period of time, worms are gathered with this net, keeping the already collected specimens at the bottom of the net. When a substantial amount of specimens is collected in the net, the contents are transferred to a bucket containing seawater. At the end of the collection, the specimens in the bucket are separated from the seawater by filtration on a piece of Dacron gauze, and the worms are frozen with dry ice.

The fireworm *Odontosyllis* is also common in the Turks and Caicos Islands and other places in the West Indies, in addition to Bermuda.

History in the study of the *Odontosyllis* bioluminescence. Harvey (1952) demonstrated the luciferin-luciferase reaction with *O. phosphorea* collected at Nanaimo, British Columbia, Canada, and with *O. enopla* from Bermuda. McElroy (1960) partially purified the luciferin, and found that the luminescence spectrum of the luciferin-luciferase reaction of *O. enopla* is identical to the fluorescence spectrum of the luciferin (λ_{max} 510 nm), and also that the luciferin is auto-oxidized by molecular oxygen without light emission. Further investigation on the bioluminescence of *Odontosyllis* has been made by Shimomura *et al.* (1963d, 1964) and Trainor (1979). Although the phenomenon is well known, the chemical structure of the luciferin and the mechanism of the luminescence reaction have not been elucidated.

Preparation of luciferase stock solution (Trainor, 1979). Frozen specimens of *O. enopla* (6 g) were ground thoroughly with a mortar and pestle in 50 ml of water. The brightly luminescing mixture was allowed to stand overnight at 2°C to complete the light emission, and centrifuged to remove cell debris. The supernatant was dialyzed twice in water (900 ml each) at 2°C; first for 3 hr, then for 21 hr. The solution was further dialyzed in 20 mM magnesium acetate (900 ml) at 2°C for 18 hr, and then centrifuged to remove any precipitate. The transparent supernatant was stored frozen for use as the stock luciferase solution.

Assay of luciferin. A buffer solution of 20–50 mM Tris-HCl (pH 7.2; 1.9 ml) containing 20 mM magnesium acetate and luciferase (typically 0.1 ml of the luciferase stock solution) is injected into a vial containing luciferin sample (1–10 μl) and 20 mM KCN (0.1 ml; made daily in 0.1 M Tris-HCl buffer, pH 7.2), and the resulting light emission is measured in terms of total light.

Extraction and purification of luciferin (Shimomura *et al.*, 1963d). Frozen specimens of *O. enopla* (100 g) were treated with one liter of boiling ethanol for 1 minute to destroy enzymes. The mixture was quickly cooled in an ice bath, and filtered on a Büchner funnel with a small amount of Filtercel. The filter cake was first extracted with 600 ml of boiling 20% methanol for 1 min; the mixture was cooled and filtered as before, saving the filtrate. The residue was extracted again with 400 ml of boiling 25% methanol, cooled, and filtered, saving the filtrate. The two filtrates were combined and evaporated to about 85 ml under reduced pressure, and the solution was mixed with 1.5 volumes of ethanol and left standing at 0°C for 6 hr. The precipitate formed was removed by centrifugation. The clear supernatant was evaporated to 50 ml under reduced pressure. The concentrated solution was mixed with 100 ml of ethanol and 700 ml of n-butanol, left standing in a refrigerator for 2 hr, and then centrifuged.

The reddish-brown precipitate containing luciferin was dissolved in 95 ml of 35% ethanol and mixed with 30 ml of absolute ethanol. The mixture was divided into 3 portions, and the luciferin in each portion was purified by chromatography on a column of DEAE-cellulose (2×4 cm) at 0°C. The sample was adsorbed at the top of the column as a sharp purple-colored band, with a yellow band immediately below. Upon elution with a mixture of 50 ml of ethanol, 60 ml of water and 8 g NaCl, the yellow band was eluted first, followed by a pink band, and finally the band of luciferin (colorless). The fractions containing 70 to 75% of the total activity were saved. Two other portions were chromatographed in the same manner, and the luciferin fractions from all three portions were combined, and further purified by twice-repeated chromatography on DEAE-cellulose, first on a column of 2×5 cm, followed by a column of 1×6.5 cm. Between each chromatography, the luciferin solution was desalted by adding ethanol and left standing at −30°C.

Trainor (1979) modified the above method: (1) In the initial extraction, luciferin was extracted with 50 mM acetate buffer (pH 4.75) at 95°C, instead of boiling 20% methanol, to increase the extraction yield. (2) In the DEAE-cellulose chromatography, the column, on which the luciferin sample had been adsorbed, was washed with the following solvents before the elution of luciferin: water, 10 mM HCl in methanol, methanol, and NaCl-saturated methanol. (3) To eliminate salts in purified luciferin, the solution was evaporated to dryness, and the luciferin in the residue was extracted with methanol. The methanol extract was chromatographed on a column of Sephadex LH-20 to separate luciferin from salts, using methanol as the solvent.

Properties of luciferin. The luciferin of *Odontosyllis* is a highly polar substance. It is soluble in water, methanol, and DMF, but practically insoluble in *n*-butanol, ethyl acetate and acetonitrile. The luciferin is strongly adsorbed onto DEAE-cellulose, even under acidic conditions, indicating that the molecule possesses a strong acidic functionality. Although the luciferin is unstable in the presence of air, it is quite stable in dilute methanol under argon at −20°C.

Odontosyllis luciferin is colorless and shows an absorption maximum at about 330 nm in aqueous ethanol (Figs. 7.2.1 and 7.2.2), and it undergoes various spectrum changes upon spontaneous oxidation

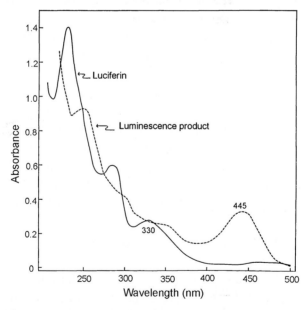

Fig. 7.2.1   Absorption spectra of *Odontosyllis* luciferin (solid line) and *Odontosyllis* oxyluciferin (dashed line), both in ethanol/water (5:6) containing 8% NaCl. To measure the latter curve, luciferin was first luminesced in the presence of luciferase, then luciferase was removed using a small column of DEAE cellulose. From Shimomura *et al.*, 1963d, with permission from John Wiley & Sons Ltd.

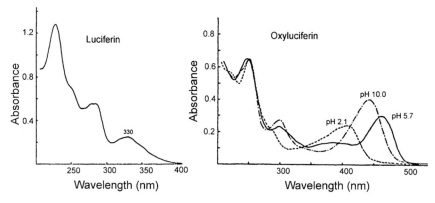

Fig. 7.2.2   Absorption spectra of *Odontosyllis* luciferin in methanol (*left* panel) and oxyluciferin in aqueous solution (*right* panel). From Trainor, 1979.

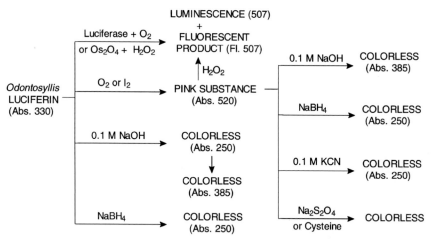

Fig. 7.2.3   Spectral changes of *Odontosyllis* luciferin caused by various reagents (Shimomura *et al.*, 1963d). The peak wavelengths (nm) of the absorption, luminescence and fluorescence spectra are shown in parentheses.

in air and by the addition of various reagents, as summarized in Fig. 7.2.3. In aqueous solutions, luciferin is rapidly auto-oxidized to a pink compound with an absorption maximum at 520 nm (Fig. 7.2.4), which can be turned colorless by treatment with NaOH, NaBH$_4$, KCN, or dithionite. The pink compound was also obtained by the oxidation of luciferin with I$_2$, Ce (IV), and horseradish peroxidase

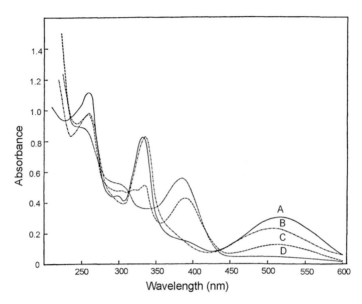

Fig. 7.2.4 Absorption spectra of the pink substance obtained from *Odontosyllis* luciferin by auto-oxidation in ethanol/water (5:6) containing 8% NaCl: A, in water (λ_{max} 520 nm); B, in 30 mM HCl; C, in 0.1 M NaOH; D, after heating C at 90°C for 5 min, then cooled. The concentrations of the substance are normalized. From Shimomura *et al.*, 1963d, with permission from John Wiley & Sons Ltd.

plus H_2O_2. In 50 mM NaOH, luciferin is gradually converted into a compound having an absorption maximum at 385 nm (Fig. 7.2.5).

Oxyluciferin. During the luminescence reaction catalyzed by luciferase, luciferin is converted into a fluorescent compound, oxyluciferin, accompanied by the emission of greenish-blue light that spectrally matches the fluorescence of oxyluciferin (Fig. 7.2.6). The absorption spectrum of oxyluciferin is shown in Figs. 7.2.1 and 7.2.2.

According to Trainor (1979), oxyluciferin can be prepared from luciferin with a good yield by heating luciferin in DMF at 70°C for 10 hr under a slightly elevated pressure of oxygen. The product is purified by TLC on silica gel, by developing the plate with acetone/water (14:1) and eluting the compound with water before the plate completely dries up. A further purification can be done by HPLC on a C18 column using pure water as the solvent. The pK_a values of oxyluciferin

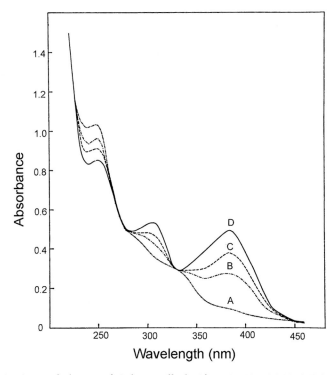

Fig. 7.2.5 Spectral change of *Odontosyllis* luciferin in 50 mM NaOH in air: **A**, at zero time; **B**, after 15 min; **C**, after 30 min; and **D**, after 80 min (λ_{max} 385 nm). From Shimomura *et al.*, 1963d, with permission from John Wiley & Sons Ltd.

estimated from the spectral change of 440–470 nm region were 3.65 and 6.77 in water, and 4.38 and 7.56 in 50% ethanol; the shifts in the pK_a values imply the presence of oxy-acid group(s).

In addition to the reactions of luciferin shown in Fig. 7.2.3, Trainor (1979) made an interesting discovery that arylsulfatase from *Pattela vulgata* (Sigma) converts luciferin into a violet compound (treatment conditions: pH 5.05, 37°C, 4 hr). The compound showed absorption peaks at 344 nm and 560 nm (Fig. 7.2.7), and the peaks shifted reversibly to 347 nm and 547 nm in acid, and to 364 nm and 720 nm in an alkaline solution. The violet compound was also obtained from the pink compound by treatment with arylsulfatase, or from luciferin by heating in 50% trifluoroacetic acid at 100°C for 1 hr, followed by an addition of oxygenated water. These results, together with the

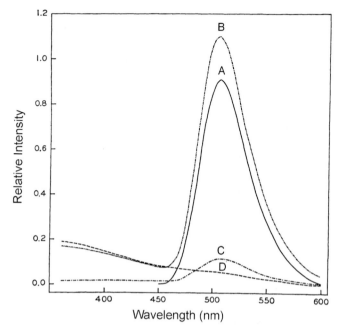

Fig. 7.2.6 (A) Luminescence spectrum of *Odontosyllis* luciferin-luciferase reaction in 30 mM magnesium acetate, pH 7.0 (λ_{max} 510 nm). (B) Fluorescence emission spectrum of the product of the luminescence reaction, 45 min after the start of the reaction. (C) Fluorescence emission spectrum of the luciferin used in (A). (D) Fluorescence emission spectrum of the luciferase used in (A). From Shimomura *et al.*, 1963d, with permission from John Wiley & Sons Ltd.

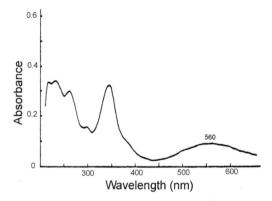

Fig. 7.2.7 Absorption spectrum of the violet compound produced by the treatment of *Odontosyllis* luciferin with arylsulfatase, in 50 mM acetate buffer, pH 5.05. From Trainor, 1979.

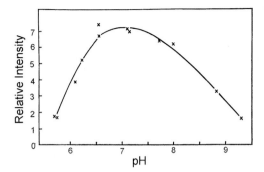

Fig. 7.2.8   Influence of pH on the luminescence intensity of the *Odontosyllis* luciferin-luciferase reaction at room temperature. From Shimomura *et al.*, 1963d, with permission from John Wiley & Sons Ltd.

ion-exchange behavior and other properties of luciferin, suggest that the luciferin molecule contains a functional group of phenolic sulfate.

Luciferase-catalyzed luminescence of luciferin. Odontosyllis luciferin emits light in the presence of Mg^{2+}, molecular oxygen and luciferase. The relationship between the luminescence intensity and the pH of the medium shows a broad optimum (Fig. 7.2.8). The luminescence reaction requires a divalent alkaline earth ion, of which Mg^{2+} is most effective (optimum concentration 30 mM). Monovalent cations such as Na^+, K^+, and NH_4^+ have little effect, and many heavy metal ions, such as Hg^{2+}, Cu^{2+}, Co^{2+} and Zn^{2+}, are generally inhibitory. The activity of crude preparations of luciferase progressively decreases by repeated dialysis and also by concentrating the solutions under reduced pressure. However, the decreased luciferase activity can be completely restored to the original activity by the addition of 1 mM HCN (added as KCN). The relationship between the concentration of HCN and the luciferase activity is shown in Fig. 7.2.9. Low concentrations of I_2 and $K_3Fe(CN)_6$ also enhance luminescence, but their effects are only transient.

Does HCN actually exist in *Odontosyllis enopla* as an activator of luminescence? Its existence appears likely based on the following two observations. (1) Using benzidine-copper acetate reagent (Feigl, 1960), a faint positive response of HCN was obtained from

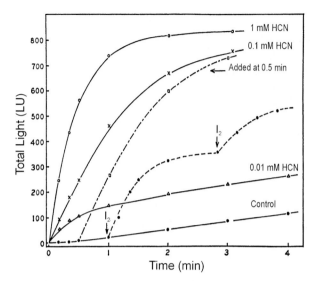

Fig. 7.2.9 Influence of cyanide and iodine on the *Odontosyllis* luciferin-luciferase luminescence reaction. Luciferin solution (0.1 ml) was first mixed with a HCN solution (0.1 ml), and then the mixture was injected into 8 ml of 20 mM magnesium acetate containing luciferase. The concentrations of HCN shown in the figure are the final concentrations. In the control experiment, HCN was omitted. In the experiment labeled "added at 0.5 min," 0.1 ml of HCN solution was injected to the control mixture 0.5 min after the start of the luminescence reaction to give a final concentration of 0.1 mM HCN. Arrows indicate the injection of a solution of I_2-3KI to the control mixture to give a final concentration of 0.1 mM I_2. From Shimomura *et al.*, 1963d, with permission from John Wiley & Sons Ltd.

the gas phase of a flask that contained 0.5–1 gram of the worms in 5–10 ml of diluted sulfuric acid. (2) A vacuum distillate obtained from a crude solution of luciferase, collected in a trap immersed in dry ice acetone bath, showed an activation of luminescence nearly equal to the effect of 0.01 mM HCN (Shimomura *et al.*, 1964). Probably HCN occurs as an adduct in the live organisms.

7.3   Luminous Earthworms (Oligochaeta)

In the class Oligochaeta, bioluminescent genera are found in three families: Enchytraeidae, Lumbricidae, and Megascolecidae (Harvey, 1952; Herring, 1978), and the majority of them are in the last family (*Diplocardia, Lampito, Microscolex, Octochaetus, Pontodrilus*, etc.).

They are distributed worldwide and all are terrestrial, except *Pontodrilus* that is found at the tidal line.

Overview of the earthworm luminescence. The yellowish-green luminescence of earthworms has been frequently described since the 17th century, and some of the important facts on earthworm luminescence have been known for a long time. Luminous earthworms produce luminous slime (coelomic fluid) generally from the mouth, anus and pores of the body wall when stimulated. The cells of coelomic fluid are large, up to 20 µm in diameter, and packed with many faintly greenish granules that contain light-producing apparatus. The granules differ from those of coelenterates in resistance to cytolysis, dissolving in salt solutions as rapidly as in distilled water (Gilchrist, 1919; Skowron, 1926). Harvey (1952) found that the luminescence of coelomic fluid requires molecular oxygen, but he had a negative result in the luciferin-luciferase reaction, in contrast to later findings.

Wampler and Jamieson (1980) studied 12 species of luminous earthworms belonging to six genera (*Diplocardia, Diplotrema, Fletcherodrilus, Octochaetus, Pontodrilus* and *Spenceriella*) from the United States, Australia and New Zealand, and found that all of the species exude luminous coelomic fluid from their dorsal pores, except *Pontodrilus bermudensis* that exudes the fluid from the mouth. All of them emit luminescence of broad emission spectra with the peaks ranging from 500 nm to over 570 nm.

Cormier *et al.* (1966) found for the first time that H_2O_2 stimulates the luminescence of earthworms. It was followed by the discovery by Bellisario *et al.* (1971) that a luciferin and a luciferase are involved in addition to H_2O_2 in the *in vitro* luminescence, opening the gate to the chemical study of earthworm luminescence. Thus, the luciferase of *Diplocardia longa* was isolated and characterized (Bellisario *et al.*, 1972; Rudie *et al.*, 1981) and the structure of the luciferin was determined (Ohtsuka *et al.*, 1976). The possibility of a flavin being involved in the luminescence reaction has long been suspected but not proved.

Assays of earthworm luciferin and luciferase (Bellisario *et al.*, 1972). The standard assay mixture contains luciferin, luciferase

(5–25 µg), and 4.4 µmoles of H_2O_2 in 1 ml of 60 mM potassium phosphate buffer (pH 7.5). The reaction is started by the injection of H_2O_2 into the mixture of other components, and the peak intensity of the light emission is measured. The luciferin is obtained in the process of the isolation of the luciferase (from the DEAE-cellulose step), or by a simple chemical synthesis (Mulkerrin and Wampler, 1978).

Extraction and purification of *Diplocardia* luciferase (Bellisario *et al.*, 1972). About 50 specimens of *Diplocardia longa* (widespread in southern Georgia; about 30 cm in length) were electrically stimulated in 250 ml of 0.1 M EDTA at 4°C to exude coelomic fluid. The suspension of coelomic cells obtained was centrifuged at 480 g for 5 min. The pellets from 200 worms were combined and an acetone powder was prepared. The acetone powder obtained (about 10 g) was stable at –80°C for at least one year.

Acetone powder (1.5 g) was extracted 4 times with 0.1 M sodium borate buffer (pH 7.6) at 4°C, homogenizing each time with a glass grinder equipped with a Teflon pestle. The first extraction was carried out with 80 ml of the buffer, followed by 3 times with 70 ml each of the buffer. Each homogenate was centrifuged at 16,000 g for 5 min. All supernatants were combined.

The pooled viscous supernatant from 4.5 g of acetone powder was poured onto a column of DEAE-cellulose (9 × 14 cm) prepared with the extraction buffer. The luciferin passed through the column without any retention. The luciferase bound was eluted with 0.1 M Na_2HPO_4, and the fractions having A_{280} values greater than 0.5 were saved. Ammonium sulfate (600 g/liter) was added to the combined fractions, and the mixture was stirred for 1 hr, then centrifuged at 28,000 g for 15 min. The pellets obtained were stirred for 10 min with 40 ml of 40% saturated ammonium sulfate (pH 7.5), and then centrifuged at 18,000 g for 15 min, discarding the supernatant. The pellet was extracted with 25 ml of 0.1 M potassium phosphate buffer (pH 7.5), and insoluble matter was removed by centrifugation (27,000 g for 20 min).

The yellow-brown supernatant was chromatographed on a column of Sephadex G-150 (4.9 × 82.5 cm) in 0.1 M potassium phosphate buffer (pH 7.5). The luciferase fractions eluted near the void volume of the column were combined, concentrated to about 10 ml with an Amicon ultrafiltration cell, and then dialyzed against 1 liter of 50 mM potassium phosphate (pH 7.0) for 48 hr with three changes of buffer. The luciferase was then adsorbed onto a column of hydroxylapatite (2.1 × 10 cm) prepared with 50 mM potassium phosphate (pH 7.0), and then eluted with 0.2 M potassium phosphate (pH 7.5), obtaining a single peak; the A_{280} peak coincided with the luciferase activity peak.

Improved method. An improved method that gives luciferase preparations of much higher specific activities was worked out by Rudie (1977) and described by Mulkerrin and Wampler (1978). In this procedure, the coelomic cells obtained from the worms by stimulation with a magneto generator were directly extracted with 0.1 M sodium borate, pH 7.5, containing 0.125 g/l sodium azide, 33 mg/l catalase (to remove H_2O_2), and 75 mg/l dithiothreitol. The extract was purified by column chromatography, first on DEAE-cellulose, then on CM (carboxymethyl)-cellulose, and finally on hydroxylapatite. The elution procedures used were as follows: DEAE-cellulose chromatography, with a linear increase of NaCl concentration from 0 to 0.2 M in 0.1 M borate buffer (pH 7.5); CM-cellulose chromatography, with a linear increase of NaCl concentration from 0 to 0.2 M in 10 mM phosphate buffer (pH 6.8); hydroxylapatite chromatography, with 0.2 M phosphate (pH 7.5).

Purification of luciferin (Rudie *et al.*, 1976). The luciferin fractions from the DEAE-cellulose chromatography of luciferase were combined and concentrated in a freeze-dryer. The concentrated solution was saturated with ammonium sulfate, and extracted with methyl acetate. The methyl acetate layer was dried with anhydrous sodium sulfate, concentrated to a small volume, then applied on a column of silica gel (2 × 18 cm). The luciferin adsorbed on the column was eluted with methyl acetate. Peak fractions of luciferin were combined, flash evaporated, and the residue was extracted with methanol. The methanol extract was concentrated (1 ml), then chromatographed on a column of Sephadex LH-20 (2 × 80 cm) using methanol as the solvent. The luciferin fractions eluted were combined and flash evaporated. The residue was further purified by two steps of column chromatography on silica gel, using chloroform containing 2.5% methanol (first step) and chloroform containing 10% ethanol (second step) as the solvents.

Structure of luciferin (Ohtsuka *et al.*, 1976). The luciferin of *Diplocardia longa* is a colorless liquid, and fairly stable at room temperature. It is soluble in polar organic solvents (methanol, ethanol, acetone, and methyl acetate) but insoluble in nonpolar solvents like hexane and carbon tetrachloride. Based on the chemical properties and spectroscopic data, the following chemical structure was assigned to the luciferin.

Properties of *Diplocardia* luciferase (Bellisario *et al.*, 1972). Purified *D. longa* luciferase is fairly unstable. It lost 20–60% of activity in 2 weeks when stored at 4°C in 0.1 M potassium phosphate, pH 7.5, whereas 90% of activity was maintained when stored at −80°C for 1 month. The molecular weight of luciferase is calculated to be 320,000 from the sedimentation equilibrium data obtained by short-colum Yphantis method, and 291,000 from the diffusion constant $(D_{20,w} = 2.25 \times 10^{-7}\ cm^2sec^{-1})$ and sedimentation coefficient (7.3 S). The calculated friction ratio (2.10) indicates a highly asymmetric nature of the molecule. The analysis of luciferase by SDS-PAGE showed three subunits, 71,000, 58,000, and 14,500, suggesting a minimum molecular weight of luciferase to be 143,500, which is approximately half of the molecular weight determined by sedimentation experiments. The ultraviolet absorption spectrum of luciferase (Fig. 7.3.1) is essentially that of a simple protein, with an absorption peak at 278 nm $(A_{278}/A_{260} = 1.8)$. The fluorescence is also that of a typical protein.

Luciferase is irreversibly inactivated by H_2O_2, and strongly inhibited by KCN, diethyldithiocarbamate and *o*-phenanthroline. Electron paramagnetic resonance spectroscopy (EPR) of luciferase showed a

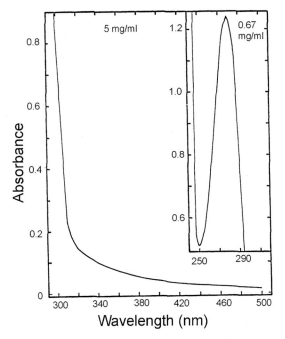

Fig. 7.3.1 Absorption spectra of the luciferase of the earthworm *Diplocardia* (5 mg/ml and 0.67 mg/ml) in 50 mM potassium phosphate, pH 7.5. From Bellisario *et al.*, 1972, with permission from the American Chemical Society.

large Cu^{2+} signal indicating a cupric content of 0.88 g-atom/mole of enzyme. Atomic absorption spectroscopy revealed the presence of 0.8 g-atom of copper, 0.28 g-atom of iron and 0.24 g-atom of zinc, respectively, per mole of luciferase. The removal of 90% of the copper in luciferase did not affect the luminescence activity, indicating the lack of correlation between the Cu^{2+} content and luciferase activity.

However, Rudie *et al.* (1981) reported the presence of firmly bound EPR-silent copper in their luciferase preparation that was 40 times more active than that obtained by Bellisario *et al.* (1972). Thus, copper may be a functional part of luciferase. According to Rudie *et al.* (1981), the luciferase contains carbohydrate (6%), lipid (2%), copper (up to 4 g-atom per mole), and an unusually high content of proline plus hydroxyproline (11% by weight).

Luminescence reaction (Bellisario *et al.*, 1972). Mixing the luciferin, luciferase and H_2O_2 results in an emission of light, regardless of the presence or absence of molecular oxygen. The *in vitro* luminescence with partially purified luciferin and luciferase (λ_{max} 503 nm) was spectrally similar to the *in vivo* luminescence from freshly exuded slime (λ_{max} 507 nm) (Fig. 7.3.2). However, Ohtsuka *et al.* (1976) reported that the emission maximum was found at 490 nm when a pure sample of luciferin was used.

The *in vitro* luminescence showed a broad optimum pH, from 7.0 to 8.5. The decay of the light intensity was the first order with a

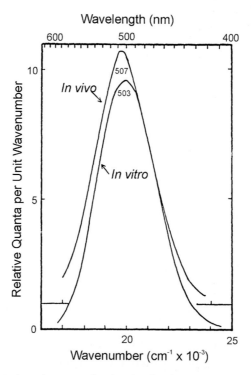

Fig. 7.3.2 Comparison between the *in vivo* luminescence spectrum of a freshly exuded slime of *Diplocardia longa* and the *in vitro* luminescence spectrum measured with partially purified preparations of *Diplocardia* luciferin and luciferase. Reproduced from Bellisario *et al.*, 1972, with permission from the American Chemical Society. Note that the *in vitro* emission maximum shifts to 490 nm when a sample of pure luciferin is used (Ohtsuka *et al.*, 1976).

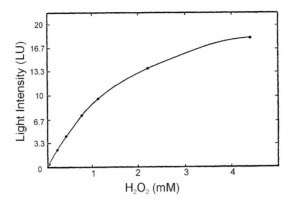

Fig. 7.3.3   Relationship between the concentration of H_2O_2 and the peak intensity of luminescence, when 0.2 ml of a H_2O_2 solution was injected into a mixture of 0.575 ml of 0.1 M potassium phosphate (pH 7.5), 0.025 ml of a solution of *Diplocardia* luciferin, and 0.2 ml of luciferase solution (0.12 mg). 1 LU $=$ 10^9 quanta/s. From Bellisario *et al.*, 1972, with permission from the American Chemical Society.

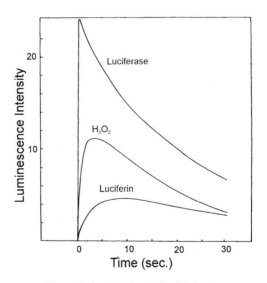

Fig. 7.3.4   Kinetic profiles of the *Diplocardia* bioluminescence reaction, when *Diplocardia* luciferase, H_2O_2, or *Diplocardia* luciferin was injected last. In each case, 0.1 ml of the last component was injected into 0.9 ml of the mixture of other components, to give the final concentrations: *Diplocardia* luciferase, 0.1 unit/ml; *Diplocardia* luciferin, 32 mM; and H_2O_2, 32 mM, in 0.1 M potassium phosphate buffer, pH 7.5. From Rudie *et al.*, 1981, with permission from the American Chemical Society.

half-time of about 12 sec. The half-time was independent of luciferase concentration, indicating that the luciferase does not turn over. The total light emission was proportional to the amount of luciferase in the presence of an excess amount of luciferin, a further indication that luciferase does not behave as a normal enzyme. The quantum yield with respect to luciferase was very low (0.002). The effect of the concentration of H_2O_2 in the luminescence reaction (Fig. 7.3.3) clearly shows a saturation phenomenon, and gives an apparent K_m value of about 2 mM for H_2O_2.

The light yield and kinetics of *in vitro* luminescence vary with the order of the addition of components, as shown in Fig. 7.3.4 (Rudie *et al.*, 1981). The light yield and initial rate are highest when luciferin and peroxide have been preincubated and the reaction is initiated by the injection of luciferase. Based on this and some other data, it was concluded that the true substrate of the *Diplocardia* luciferase is the H_2O_2 adduct of luciferin shown below (Rudie *et al.*, 1981).

When an excess amount of luciferin was preincubated with H_2O_2 before the addition of luciferase, the quantum yields of luciferase and the luciferin-H_2O_2 adduct were found to be 0.63 and 0.03, respectively (Rudie *et al.*, 1981). The aldehyde group of luciferin is probably converted into the corresponding acid in the luminescence reaction, although it has not been experimentally confirmed.

7.4   Polynoid Scaleworm *Harmothoë lunulata*

Bioluminescent polynoid scaleworms emit light only from the scales (elytra) that cover the dorsal surface of the worm in two imbricated rows, like shingles on a roof. When an animal is irritated, the scales become luminescent. If the stimulus is sufficiently strong, one or more of the scales becomes detached from the body, and the animal rapidly swims away leaving the glowing scale or scales behind. Drawings of *Harmothoë lunulata* and five other luminous scaleworms are shown in Fig. 7.4.1.

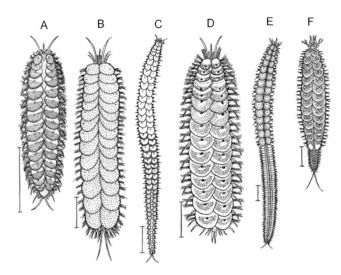

Fig. 7.4.1 Luminous species of polynoids. A, *Malmgrenia castanea*; B, *Gattyana cirrosa*; C, *Acholoë astericola*; D, *Harmothoë lunulata*; E, *Polynoë scolopendrina*; F, *Lagisca extenuata*. Scale 5 mm. Reproduced from Nicol, 1953, with permission from the Cambridge University Press.

The scales acquire a greenish fluorescence after luminescence, and this fluorescence is characteristic of a flavin (Lecuyer and Arrio, 1975). The luminescence reaction of polynoid worms requires molecular oxygen, but a luciferin-luciferase reaction has not been demonstrated (Harvey, 1952). The homogenate of scales emits light upon the addition of various reagents, such as sodium dithionite (Herrera *et al.*, 1974), ferrous ions (Lecuyer and Arrio, 1975), the xanthine-xanthine oxidase system (Nicolas, 1979), and Fenton's reagent (Nicolas, 1980). Nicolas *et al.* (1982) isolated and purified the light emitting substance of *Harmothoë lunulata*, obtaining a photoprotein. The photoprotein was named polynoidin.

Extraction and purification of polynoidin (Nicolas *et al.*, 1982). Specimens of *Harmothoë lunulata* were collected near Roscoff, Brittany, in France, and were anaesthetized with 0.1 mM tricaine in seawater. The scales were carefully removed with tweezers, and washed several times with seawater to remove tricaine. Then the scales were stored in liquid nitrogen until they were used.

About 300 frozen scales were ground with 5 ml of 50 mM phosphate buffer, pH 7.4, in a mortar, and then the muddy material was centrifuged. The green-fluorescent supernatant containing riboflavin was discarded. The pellet was extracted with 2 ml of 10 mM Tris buffer, pH 7.4, containing 0.6 M KCl and 1% Triton X-100, for 2 hr at 4°C with stirring. The mixture was centrifuged, and the pellet was extracted again using the same procedure. The two supernatants containing solubilized polynoidin were combined, and chromatographed twice on a column of Bio-Gel A-5m (1.5 × 80 cm; Bio-Rad) using 50 mM phosphate buffer, pH 7.4, containing 10 mM EGTA, 0.5 M NaCl and 0.5% Triton X-100, at 4°C. The fractions containing polynoidin were combined, desalted on a column of Sephadex G-75 (5 × 45 cm) equilibrated with 5 mM Tris buffer, pH 7.4, containing 5 mM EGTA, and then added to a column of DEAE-cellulose (1.5 × 5 cm) prepared with the same buffer. The photoprotein adsorbed on the column was eluted with 5 mM Tris buffer, pH 7.4, containing 5 mM EGTA, 1 M NaCl and 0.5% Triton X-100. The photoprotein was further purified by a repetition of the Bio-Gel chromatography.

Properties of polynoidin (Nicolas *et al.*, 1982). Polynoidin is a membrane protein insoluble in common buffer solutions not containing a surfactant. Triton X-100 is a suitable surfactant for solubilizing this photoprotein. The solubilized, purified polynoidin showed an absorption spectrum of a typical protein, and exhibited no special fluorescence both before and after luminescence. The luminescence spectrum of polynoidin initiated by Fenton's reagent (H_2O_2 plus Fe^{2+}) showed a maximum at 510 nm (Fig. 7.4.2), close to the 515 nm peak observed with stimulated scales (Nicol, 1957). The photoprotein lost 15% of its activity in 2 weeks at 4°C, whereas the loss was 95% at −20°C in one day and 60% at 20°C overnight, indicating that freezing is detrimental to the photoprotein activity. The molecular weight was estimated at roughly 500,000 based on gel filtration on a Bio-Gel A-5m column.

Luminescence reaction of polynoidin. According to Nicolas *et al.* (1982), purified polynoidin luminesces by the addition of sodium

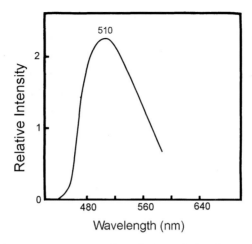

Fig. 7.4.2 Luminescence spectrum of a solution of the photoprotein polynoidin. Light emission was initiated by the additions of H_2O_2 (final conc. 3 mM) and Fe^{2+} (final conc. 0.1 mM). From Nicolas *et al.*, 1982, with permission form the American Society for Photobiology.

dithionite, xanthine-xanthine oxidase system and Fenton's reagent, like the homogenates of scales, suggesting a possible involvement of superoxide anion in the luminescence reaction. When Fenton's reagent was used, the luminescence was 30% brighter in phosphate buffer than in Tris buffer. Moreover, the luminescence intensity was significantly increased by including a complexing agent, such as EGTA, in both buffers. However, the injection of a polynoidin solution into the mixture of H_2O_2 and Fe^{2+} failed to produce light; Fe^{2+} must be added last to initiate light emission.

With both the xanthine-xanthine oxidase system and Fenton's reagent, the total light emitted was proportional to the amount of polynoidin used (Fig. 7.4.3, panel A), justifying categorizing polynoidin as a photoprotein. The initial light intensity was proportional to the amount of polynoidin used. The optimum concentration of H_2O_2 was found at 3 mM (panel B). The peroxide H_2O_2 could be replaced by the same concentration of various other peroxides, such as urea peroxide, *t*-butyl hydroperoxide, dimethylbenzyl hydroperoxide and ethyl hydroperoxide, but not by dicumyl peroxide. The most effective concentration of $FeSO_4$ was 0.1 mM (panel C). The addition of

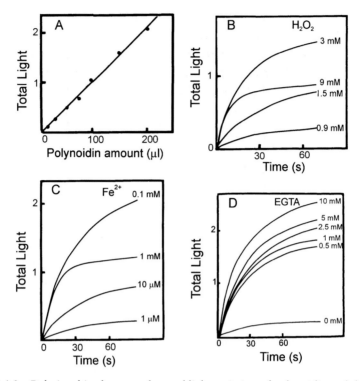

Fig. 7.4.3   Relationships between the total light emission of polynoidin and the concentrations of the photoprotein (**A**), H_2O_2 (**B**), Fe^{2+} (**C**), and EGTA (**D**). The basic buffer solution used was 50 mM phosphate buffer, pH 6.6, containing 3 mM H_2O_2, 0.1 mM $FeSO_4$ and 10 mM EGTA; the concentrations of H_2O_2, $FeSO_4$ and EGTA were varied in B, C and D, respectively. $FeSO_4$ was added last to initiate light emission. From Nicolas *et al.*, 1982, with permission from the American Society for Photobiology.

EGTA enhanced the luminescence up to 10 times at the concentration of 10 mM (panel D). DETAPAC (diethylenetriaminepentaacetic acid) was nearly as effective as EGTA, but EDTA (10 mM) diminished the light emission.

7.5   The Polychaete *Tomopteris*

Regarding the polychaete *Tomopteris*, the author has briefly examined its bioluminescence using a few specimens donated by Dr. Steven Haddock. The specimens did not show a luciferin-luciferase reaction, as noted by Harvey (1952). The luminescence could be elicited from the homogenate of a whole specimen with Fe^{2+} and sodium dithionite, suggesting that the luminescence is probably activated by superoxide anion, in resemblance to the luminescence systems of scaleworms, *Chaetopterus* and *Pholas*. The light-emitting substance appears to be a photoprotein tightly associated with small particles, and it could be extracted only partially with 50 mM Tris-HCl buffer, pH 7.8, containing 2% Triton X-100, 0.5 M KCl and 10 mM EDTA. The active substance in the particulate matter was surprisingly heat-resistant, withstanding a brief boiling.

CHAPTER

8

DINOFLAGELLATES AND OTHER PROTOZOA

The diverse group Protozoa consists of microscopic single-celled organisms. Of the various protozoan types, only Radiolaria and Dinoflagellata are known to contain luminous organisms (Harvey, 1952; Herring, 1978).

8.1 Radiolarians

Radiolarians are either solitary or colonial, and the colonial forms often grow to very large aggregates. Harvey (1926b) demonstrated that radiolarians *Thalassicolla* and *Collozoum* can luminesce in deoxygenated seawater containing platinized asbestos through which pure hydrogen has been passed for 45 minutes. The experiment indicated that these organisms are able to emit light even in the complete absence of oxygen, resembling hydromedusae, the scyphozoan *Pelagia*, and ctenophores. Campbell *et al.* (1981) isolated a Ca^{2+}-sensitive photoprotein from *Thalassicolla* sp., and named it thalassicollin. An unidentified solitary species and a colonial species (probably *Sphaerozoum* sp.) were also found to contain a Ca^{2+}-sensitive photoprotein (Campbell and Herring, 1990). Emission maximum of the Ca^{2+}-triggered luminescence of thalassicollin was found at 440 nm (Campbell and Herring, 1990) in close agreement with the *in vivo* luminescence of *Thalassicolla* (λ_{max} 450 nm; Herring, 1983; Latz *et al.*, 1987). Radiolarian photoproteins are of interest as they are the only known examples of Ca^{2+}-sensitive photoprotein other than the coelenterate (and ctenophore) photoproteins.

8.2 Dinoflagellates

Dinoflagellates are distributed worldwide and are almost ubiquitous in the surface water of the sea. The so-called phosphorescence of the sea is most commonly due to the bioluminescence of dinoflagellates (such as *Noctiluca*, *Gonyaulax* and *Pyrocystis*). The microscopic organisms of dinoflagellates emit sparkling luminescence in short flashes. For example, *Gonyaulax polyedra* is about 40 μm in diameter and emits flashes of light that last less than one-tenth of a second (Hastings and Sweeney, 1957). Dinoflagellates sometimes form colored patches called "red tide" on the surface of the sea due to the excessive growth (bloom) of the organisms, and these may emanate an unpleasant odor as well as toxins which can kill fishes.

Since the year 2001, the species name *Lingulodinium polyedrum* (Stein) Dodge 1989 is widely used in place of *Gonyaulax polyedra* Stein 1883. Considering that many important papers on dinoflagellate luminescence have been published using the old name, the name *Gonyaulax* is used in this book.

Overview of biochemical research. Although the luminescence of dinoflagellates has been studied since the middle of the 19th century by a number of scholars (Harvey, 1952), its chemical aspects were virtually unknown until J. Woodland Hastings took up the subject about 50 years ago. Biochemical research on dinoflagellate luminescence began with the discovery by Hastings and Sweeney (1957) that the light emission of the cell-free extracts of *Gonyaulax polyedra* requires three components: a heat-labile enzyme, a factor in the boiled extract, and a high concentration of salt, in addition to oxygen. The discovery suggested that dinoflagellate luminescence is essentially a luciferin-luciferase reaction.

Bode and Hastings (1963) isolated and characterized the luciferin and luciferase for the first time and demonstrated the G. *polyedra* luciferin-luciferase reaction in a particle-free solution. At about the same time, DeSa *et al.* (1963) discovered that a major fraction of the bioluminescence system of G. *polyedra* was associated with particles isolated at pH 8, which they named scintillons. Purified scintillons emitted a flash of light when the pH of the medium was lowered to 5.7. Scintillons were first believed to be crystal-like structures, but this was later shown to be incorrect (Fogel *et al.*, 1972). The scintillons were found to be organelles that contain luciferin, luciferase, luciferin-binding protein and other substances (Hastings, 1978). The luminescence spectra measured with luciferin plus luciferase, isolated scintillons, and living cells are all closely similar (Fig. 8.1; Hastings *et al.*, 1966). In the luciferin-luciferase reaction of dinoflagellates, the

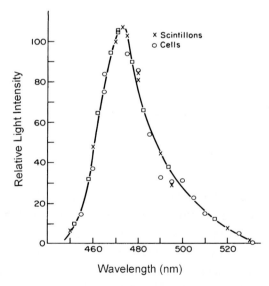

Fig. 8.1 Luminescence spectra of luciferin-luciferase reaction of the dinoflagellate *Gonyaulax polyedra* (*Lingulodinium polyedrum*) in a solution (solid line), isolated scintillons (×), and living *Gonyaulax* cells (○). From Hastings *et al.*, 1966.

luciferins and luciferases from various species cross-react in various combinations (Schmitter *et al.*, 1976).

The chemical structure of dinoflagellate luciferin was found to be a tetrapyrrole, possibly derived from chlorophyll (Nakamura *et al.*, 1989). The luciferase of *G. polyedra* was cloned (Bae and Hastings, 1994). Recently, the crystal structure of the third domain of the luciferase was determined (Schultz *et al.*, 2005).

8.2.1 *Cultivation and Harvesting of Dinoflagellates*

Cultures of *G. polyedra* (*L. polyedrum*) are grown at $20 \pm 5°C$ in a supplemented sea water medium (Hastings and Sweeney, 1957; Hastings and Dunlap, 1986), under cool-white fluorescent lighting of a 12-hr light/12-hr dark cycle. The cultures are inoculated at densities of 100 to 500 cells/ml. After 2–4 weeks, cells are harvested by vacuum filtration on a filter paper at cell densities of 7,000–15,000 cells/ml, yielding 0.3–0.7 g wet cells per liter of culture.

Pyrocystis lunula (clone T37) can be grown under light-dark cycles as well as under continuous illumination at $20 \pm 2°C$, in f/2 medium (Guillard and Ryther, 1962) with 0.5% soil extract instead of silicate (Guillard, 1974). The growth is somewhat slower and harvesting may be carried out about 40 days after inoculation, at a cell density of 15,000–20,000 cells/ml.

8.2.2 *Scintillons (Hastings and Dunlap, 1986)*

Extraction and purification of scintillons. Cells are homogenized in 0.1 M Tris-HCl, pH 8.5, containing 10 mM EDTA and 3 mM DTT. The cell debris are removed by low-speed centrifugation ($1,000\,g$), and the supernatant containing scintillons is centrifuged at high speed ($10,000\,g$). Scintillons in the pellet are resuspended in the same buffer, using 2 ml of the buffer per 10^7 cells extracted. Such preparations gradually lose their activities at 1°C (about 50% loss in 48 hr). The activity may be preserved by storing at −80°C in the presence of 20% glycerol.

The purification is done by sucrose density-gradient centrifugation (DeSa and Hastings, 1968; Hastings and Dunlap, 1986). Six sucrose solutions (8 ml each) with densities ranging from 1.29 to 1.14 g/cm^3 are layered and left standing for about 36 hr at 4°C to smooth the boundaries between the layers. The crude scintillon preparation is layered on the preformed gradient and centrifuged in a swinging bucket rotor for 2 hr at 3,200 rpm. Fractions are collected and measured for the absorbance, fluorescence and bioluminescence activity. The bands of mitochondria and chloroplasts appear at lower densities (1.18 g/cm^3); absorbance at 670 nm identifies chloroplasts. Fluorescence emission at 470 nm (excitation at 390 nm) shows luciferin, which coincides with the bioluminescence activity of scintillons. Further purification may be achieved by performing a second density-gradient centrifugation.

Assay of scintillon activity. A small volume (5–50 μl) of a scintillon sample is mixed with 1 ml of 50 mM Tris-HCl, pH 8.0, containing 10 mM EDTA and 1 mM DTT. To this mixture, 1 ml of 0.2 M sodium citrate, pH 5.2 (or 30 mM acetic acid) is injected and the light emission is measured.

8.2.3   *The Luciferase of Gonyaulax polyedra (Hastings and Dunlap, 1986)*

The active luciferase in the soluble luminescence system occurs in two molecular sizes, 130 kDa and 35 kDa. The 130 kDa luciferase is the native form and occurs in extracts made at pH 8, and if luciferase is extracted with a pH 6 buffer, 130 kDa luciferase is converted into 35 kDa luciferase by the action of a protease (Krieger and Hastings, 1968; Fogel and Hastings, 1971; Krieger *et al.*, 1974). The 130 kDa species is considered the naturally occurring form.

Extraction and purification of the 130 kDa luciferase. Dunlap (1979) and Dunlap and Hastings (1981) used the following method: the cells harvested from the cultures grown in light-dark cycles are extracted with 0.1 M Tris-HCl, pH 8.5, containing 10 mM EDTA and 3 mM DTT, and the cell debris (including scintillons) are removed by centrifugation (20 min at 27,000 g). The supernatant is fractionated by ammonium sulfate precipitation at pH 8.5, taking a fraction precipitated between 30 and 55% saturation. The precipitate is dissolved in 1 mM Tris-HCl, pH 8.5, containing 0.1 mM EDTA and 3 mM DTT.

The solution is dialyzed against the same buffer using a hollow fiber assembly, and then added onto a column of Affi-Gel Blue (50–100 mesh, 2×15 cm, Bio-Rad) prepared with the same buffer. The column is washed with the same buffer. Then luciferase is eluted with 50 mM Tris-HCl, pH 8.5, containing 5 mM EDTA, 3 mM DTT, and 0.5 M NaCl (Hastings and Dunlap, 1986, state that it may be preferable to omit the Affi-Gel step because of difficulties encountered).

The eluted luciferase is precipitated with ammonium sulfate. The precipitate is dissolved in 1 mM Tris-HCl, pH 8, containing 0.1 mM EDTA, 3 mM DTT and 0.1 M NaCl, and chromatographed on a column of Sephacryl S-300 (2.6×97 cm) using the same buffer. Luciferase is eluted in two peaks, corresponding to the molecular weights of about 420,000 (an aggregate) and 130,000, in a ratio of about 8:1. The fractions of these two peaks are pooled separately; the M_r 420,000 luciferase is concentrated by either ultrafiltration or precipitation with ammonium sulfate.

The concentrated luciferase solution is dialyzed overnight against 4 liters of 1 mM Tris-HCl buffer, pH 8.5, containing, 0.1 mM EDTA and 3 mM DTT. Then luciferase is further purified on a column of DEAE-BioGel A (1×25 cm, Bio-Rad) by elution with a linear increase of NaCl from 0 to 100 mM in the same buffer as that used in dialysis. The purified luciferase had a specific activity (based on initial maximum intensity) of approximately 8.5×10^{14} quanta $sec^{-1}ml^{-1}A_{280}^{-1}$.

The 35 kDa luciferase (Krieger _et al._, 1974). Harvested cells are extracted in 50 mM phosphate buffer, pH 6, containing 0.1 mM DTT. The luciferase in the extract is fractionally precipitated with ammonium sulfate, taking a fraction precipitated between 35 and 65% saturation. The precipitate is dissolved in 50 mM Tris-HCl, pH 8, containing 10 mM EDTA and 1 mM DTT, and this solution is chromatographed on a column of Sephadex G-100, using the same buffer (Fig. 8.2). Luciferase is eluted at the position corresponding to a molecular size of about 35 kDa. The luciferase activity can be measured at both pH 6.3 and pH 8. A luciferin-binding protein, which inhibits the luminescence reaction at pH 8 (but not at pH 6.3), is eluted at the position corresponding to approximately 120 kDa.

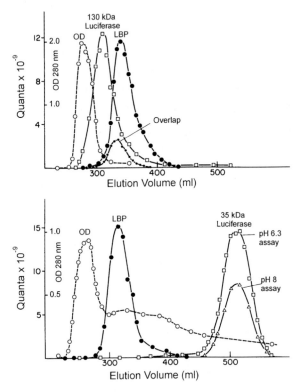

Fig. 8.2 Gel filtration on a column of Sephadex G-100 at pH 8 (both panels) of the crude extract of *Gonyaulax polyedra* cells prepared at pH 8 (*upper* panel) and prepared at pH 6 (*lower* panel). The activities of the 35 kDa and 130 kDa luciferases are measured by the addition of an excess of luciferin at pH 6.3 (□) or at pH 8 (△). The activity of the luciferin-bound LBP (luciferin-binding protein) in the upper panel is measured after the addition of an excess of 35 kDa luciferase at pH 6.3 (●). In the lower panel, the LBP activity can be obtained by the addition of an excess of luciferin at pH 8, followed by the removal of unbound luciferin with a small column of Sephadex G-25 before the luminescence assay of bound luciferin at pH 6.3 (see the Section 8.2.8). The "Overlap" in the upper panel is the light emission resulting from the mixing of an aliquot of the fractions with pH 6.3 buffer. From Fogel and Hastings, 1971, with permission from Elsevier.

Assay of the activities of luciferin and luciferase. Small volumes (5–50 μl) of luciferase and luciferin (Section 8.2.4) are mixed in 2 ml of 0.2 M citrate buffer, pH 6.3, or 0.2 M phosphate buffer, pH 8. The measurement is made in terms of the total light emission or the initial maximum light intensity that is reached within a few seconds. The

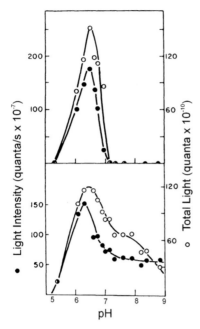

Fig. 8.3 Effect of pH on the bioluminescence of *Gonyaulax* luciferin in the presence of 130 kDa luciferase (*top*) and 35 kDa luciferase (*bottom*): light intensity at 1 min (●), and total light (○). From Hastings and Dunlap, 1986, with permission from Elsevier.

35-kDa luciferase can be assayed at either pH 6.3 or pH 8, but the 130-kDa luciferase can be assayed only at pH 6.3 (see Fig. 8.3).

Molecular characteristics of luciferase. A molecule of the luciferase of *G. polyedra* comprises three homologous domains (Li *et al.*, 1997; Li and Hastings, 1998). The full-length luciferase (135 kDa) and each of the individual domains are most active at pH 6.3, and they show very little activity at pH 8.0. Morishita *et al.* (2002) prepared a recombinant *Pyrocystis lunula* luciferase consisting of mainly the third domain. This recombinant enzyme catalyzed the light emission of luciferin (luminescence λ_{max} 474 nm) and the enzyme was active at pH 8.0. The recombinant enzyme of the third domain of *G. polyedra* luciferase was crystallized and its X-ray structure was determined (Schultz *et al.*, 2005). A β-barrel pocket putatively for substrate binding and catalysis was identified in the structure, and a mechanism for the regulation of luminescence activity by pH was presented.

8.2.4 Extraction and Purification of Dinoflagellate Luciferin

The luciferins of all species of dinoflagellate are believed to be identical or very similar to each other (Hastings and Bode, 1961; Hamman and Seliger, 1972). However, the light yield of bioluminescence per cell differs significantly by the species, and certain species, such as *Pyrocystis lunula*, contain much more luciferin than other species, such as *G. polyedra* (Seliger *et al.*, 1969; Swift and Meunier, 1976; Schmitter *et al.*, 1976). Based on the luciferin content, *P. lunula* was chosen as the source of luciferin.

Dunlap and Hastings (1981) and Nakamura *et al.* (1989) extracted and purified dinoflagellate luciferin by the following method: the cells of *P. lunula* are grown either in constant light or in light-dark cycles, and harvested by filtration on a filter paper. Cells are scraped off the filter paper and dispersed into a boiling extraction buffer (2 mM potassium phosphate, pH 8.5, containing 5 mM 2-mercaptoethanol), using 5 ml for the cells harvested from one liter of culture. The suspension is heated for 60 sec, then immediately chilled in ice. While cooling, the suspension is bubbled with argon gas. All subsequent operations, including chromatography, are performed in the cold and under argon atmosphere whenever possible. The pH of the chilled suspension is adjusted to 8.0–8.5 with 0.1 M NaOH. The white flocculent precipitate formed is removed by centrifugation at 27,000 g for 20 min at 0°C.

The reddish yellow solution is diluted with 4–5 volumes of cold water containing 5 mM 2-mercaptoethanol to reduce the conductivity to 0.7 mΩ^{-1} or less, and applied to a column of DEAE-cellulose (coarse grade; 5 × 15 cm) equilibrated with 2 mM potassium phosphate, pH 8.0, containing 5 mM 2-mercaptoethanol. The column is first washed with the cold equilibration buffer, then luciferin is eluted with a linear increase of potassium phosphate from 2 mM to 0.3 M, monitoring the effluent by fluorescence and the absorption at 390 nm. The rest of the purification method described below is adapted from the purification of the fluorescent compound F of euphausiid shrimp (Shimomura and Johnson, 1967).

The luciferin fractions are combined (400–700 ml) and rapidly concentrated to about 30 ml under reduced pressure at 30–40°C. The residue is diluted with 50 ml of ethanol, and the mixture is extracted with 90 ml of n-butanol. The strongly fluorescent butanol extract is loaded onto a column of basic alumina (Woelm basic alumina, grade 1, 1.6×9 cm) prepared with cold n-butanol, applying a slight positive pressure of argon to accelerate the flow rate. The luciferin adsorbed on the column is washed with 50 ml of cold n-butanol, and then eluted with 50% ethanol containing 0.6% ammonium hydroxide. The fluorescent fractions with an absorption peak at 390 nm are combined and concentrated under reduced pressure to 5–10 ml. The concentrate is diluted with one or two volumes of 50% ethanol containing a trace of Tris base. The highly fluorescent solution is loaded onto a column of DEAE-cellulose (2.5×8 cm) equilibrated with 10 mM sodium arsenate, pH 8, made with 50% ethanol. The column is washed with 100 ml of the arsenate buffer, and luciferin is eluted with 300 ml of the same buffer containing 0.1 M NaCl. To accelerate the chromatography process, a positive pressure of argon gas is used.

Purification by HPLC can be carried out as an alternative or in addition to the above DEAE-cellulose chromatography. The fluorescent fractions eluted from the alumina column are combined and concentrated to near dryness after addition of a trace amount of Tris base. The residue is extracted with 15 ml of ethanol. A small amount of insoluble matter (salts) is removed by centrifugation, and then the supernatant is concentrated to dryness. The residue is dissolved in 2–3 ml of aqueous 30% acetonitrile and divided into several portions. Then, each portion is purified by HPLC on a TSK DEAE-5PW column (0.75×7.5 cm) using 40% acetonitrile containing 85 mM NaCl and 3 mM $NaHCO_3$, monitoring at 388 nm. The luciferin fractions obtained from all HPLC runs are combined (total about 8.8 ml) and concentrated until salts begin to crystallize. The residual solution is diluted with approximately 10 ml of ethanol, and the precipitated salts are removed by centrifugation. The supernatant is concentrated to dryness, to give purified luciferin with a purity of 60–70% . When a small amount of 2-mercaptoethanol is added to the luciferin solution prior to the desalting process, the purity of the final product is increased to over 90%, which corresponds to an yield of approximately 0.003% on the basis of the weight of wet cells (3–4 mg of luciferin from 121 g wet cells harvested from 950 liters of culture).

8.2.5   Properties of Dinoflagellate Luciferin

The solution of purified dinoflagellate luciferin is yellow, showing absorption maxima at 245 and 390 nm in an aqueous solution and at 241 and 388 nm in 40% acetonitrile containing 85 mM NaCl and 3 mM $NaHCO_3$ (Fig. 8.4). The compound is strongly fluorescent in blue (excitation maximum at 390 nm, emission maximum at 474 nm; Fig. 8.5). The properties of this luciferin are nearly identical with those of the compound F of euphausiid shrimps (Section 3.2). The luciferin is rapidly oxidized in the presence of a trace of oxygen, and also inactivated by a weak acid, even by an acidity of pH 4 or the acidity

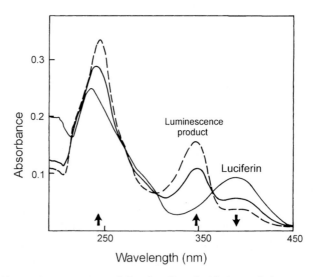

Fig. 8.4   Absorption spectrum of dinoflagellate luciferin, and the spectral changes caused by luminescence reaction after the addition of luciferase, in 0.2 M phosphate buffer, pH 6.3, containing 0.1 mM EDTA and BSA (0.1 mg/ml) (Nakamura et al., 1989). Reproduced from Hastings, 1989, with permission from the American Chemical Society and John Wiley & Sons Ltd.

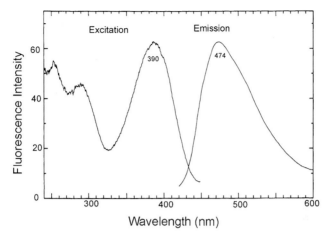

Fig. 8.5 Corrected fluorescence spectra of partially purified dinoflagellate luciferin obtained from *Dissodinium*. From Njus, 1975.

produced by dry ice; thus, purified luciferin cannot be stored with dry ice, although luciferin in the cells before extraction is unaffected by dry ice.

When the luciferin is allowed to auto-oxidize in air, simultaneous changes occur in the absorption spectrum, fluorescence spectrum and luminescence activity, eventually resulting in the formation of a blue compound that does not fluoresce or luminesce (Fig. 8.6). The luciferin is also oxidized by $K_3Fe(CN)_6$, resulting in the formation of a blue product that is closely similar, if not identical, to the blue compound produced by air oxidation just mentioned. The $K_3Fe(CN)_6$ oxidation is stoichiometric, like in the case of the fluorescent compound F of euphausiid. By titrating the luciferin with $K_3Fe(CN)_6$, the molecular extinction coefficient (ε value) for the 390 nm absorption peak of luciferin was estimated at 2.8×10^4, assuming that the reaction involved is a two-electron oxidation (Dunlap and Hastings, 1981). In the case of the fluorescent compound F, it has been confirmed that the reaction involved is a one-electron oxidation and that the ε value for 390 nm peak is 1.54×10^4 (Shimomura, 1995a). Based on the close resemblance between dinoflagellate luciferin and the compound F, it seems almost certain that the ferricyanide oxidation of dinoflagellate

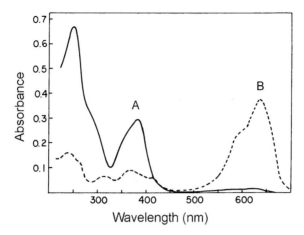

Fig. 8.6 Absorption spectra of slightly autoxidized dinoflagellate luciferin in 50% ethanol (**A**), and the blue oxidation product in methanol (**B**). From Dunlap *et al.*, 1981, with permission from the Federation of the European Biochemical Societies.

luciferin involves a one-electron oxidation, and thus, the ε value of the 390 nm peak should be approximately 1.4×10^4, rather than 2.8×10^4.

8.2.6   Chemical Structures of Dinoflagellate Luciferin and its Oxidation Products (Nakamura et al., 1989)

Mild chromic acid oxidation of luciferin ($CrO_3/KHSO_4/H_2O$, room temperature) yielded 3-methyl-4-vinylmaleimide (**1**, Fig. 8.7), 3-methyl-4-ethylmaleimide (**2**), and an aldehyde (**3**), whereas vigorous chromic acid oxidation ($CrO_3/2N\ H_2SO_4$, 90°C) gave hematinic acid (**4**) (Dunlap *et al.*, 1981). These results closely resemble the results of the chromic acid oxidation of the fluorescent compound F of euphausiid (p. 76), indicating a structural similarity between dinoflagellate luciferin and the compound F.

When a methanolic solution of luciferin was left at –20°C in the presence of air, most of the luciferin was oxidized in three days, based on its ^{1}H-NMR spectrum. The air oxidation product was purified by HPLC on a TSK DEAE-5PW column using 35% acetonitrile containing 85 mM NaCl and 3 mM NaHCO$_3$. The purified product in the HPLC eluent showed absorption maxima at 237 nm

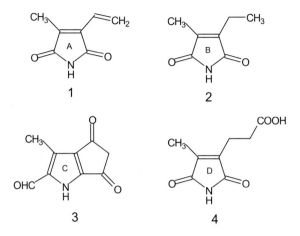

Fig. 8.7   The chromic acid oxidation products of dinoflagellate luciferin.

and 300 nm (shoulder). The FAB mass spectrum showed m/z 627 [$(M + 2H - Na)^+$], 649 [$(M + H)^+$] and 671 [$(M + Na)^+$]; and the exact masses were 649.2620 ($C_{33}H_{39}O_7N_4Na_2$, calcd 649.2614) and 671.2484 ($C_{33}H_{38}O_7N_4Na_3$, calcd 671.2433). Based on the ^{1}H NMR data, ^{13}C NMR data, mass spectral data, and the structure of the oxidation product of compound F, the structure **7** (Fig. 8.8) was assigned to the air oxidation product of luciferin.

The luciferin produces a blue oxidation product during its purification process. In the DEAE chromatography of luciferin, this blue compound is eluted before the fractions of luciferin. The fractions of the blue compound were combined and purified by HPLC on a column of Hamilton PRP-1 (7 × 300 mm) using methanol-water (8:2) containing 0.1% ammonium acetate. The purified blue compound showed absorption peaks at 234, 254, 315, 370, 410, 590 (shoulder) and 633 nm. High-resolution FAB mass spectrometry of this compound indicated a molecular formula of $C_{33}H_{36}O_6N_4Na_2$ [m/z 609.2672 $(M - Na + 2H)^+$, and m/z 631.2524 $(M + H)^+$]. These data, together with the ^{1}H NMR spectral data, indicated the structure of the blue compound to be **8**.

The luciferin showed an absorption spectrum identical with that of the fluorescent substance F of euphausiid (λ_{max} 388 nm), and gave

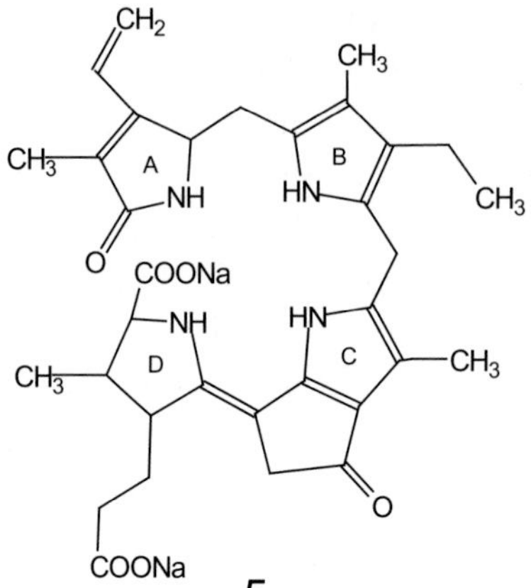

Fig. 8.8 The chemical structures of dinoflagellate luciferin (**5**), the product of luminescence reaction catalyzed by luciferase (**6**), air-oxidation product formed at −20°C (**7**), and the blue oxidation product (**8**). Note structural resemblance between these compounds and chlorophylls.

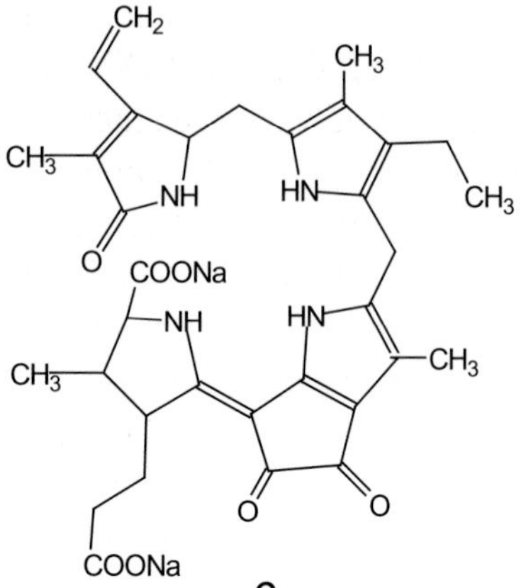

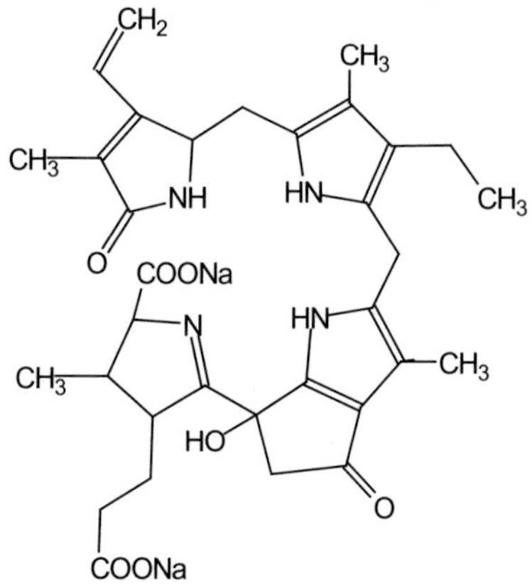

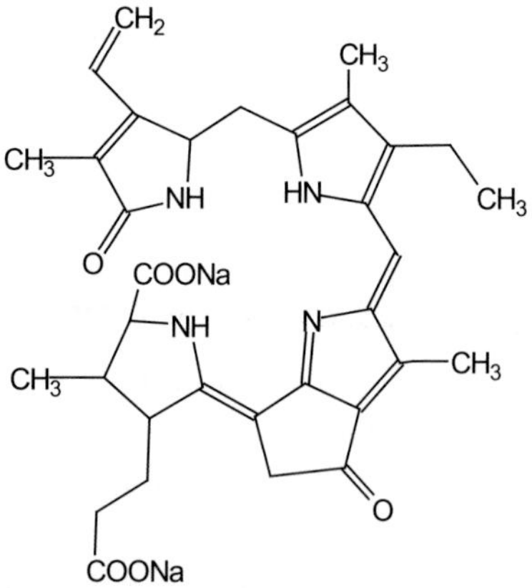

a FAB mass spectrum with molecular ions at m/z 633 [(M + H)$^+$] and 655 [(M + Na)$^+$]. Based on the comparison of the ^{1}H NMR, ^{13}C NMR and mass spectral data of luciferin with those of the compound F, the structure of dinoflagellate luciferin was formulated as **5**.

The product of the luminescence reaction of luciferin was obtained by luminescing a dilute solution of luciferin (A_{388} 0.1 or less) with luciferase at room temperature in 0.2 M phosphate buffer, pH 6.3, containing 0.1 mM EDTA and 0.1 mg/ml of BSA. After the completion of luminescence reaction, the solution (about 1 liter) was concentrated to 150 ml. The concentrated solution was mixed with 40 ml of ethanol, and extracted with 150 ml of n-butanol. The butanol layer was concentrated, the residue was extracted with 40 ml of ethanol, and the ethanol solution was concentrated again. The residual crude material was dissolved in 6 ml of 20% acetonitrile and purified by HPLC on a column of TSK DEAE-5PW, using 30% acetonitrile containing 50 mM NaCl and 3 mM $NaHCO_3$. The purified enzymatic oxidation product was nonfluorescent, with absorption peaks at 243, 262 and 348 nm. FAB mass spectrum showed m/z 647 $[(M + H)^+]$ and 669 $[(M + Na)^+]$, and the molecular formula obtained by high-resolution mass measurement was $C_{33}H_{36}O_7N_4Na_2$. Structure 6 was assigned for the enzymatic oxidation product.

8.2.7   Chemical Mechanism of Dinoflagellate Bioluminescence

The oxidation mechanism of dinoflagellate luciferin has been studied with model systems (Stojanovic and Kishi, 1994; Stojanovic, 1995). According to the results, the bioluminescence reaction of dinoflagellate luciferin might proceed as shown in the scheme in Fig. 8.9, which involves a series of intermediates (A, B, C and D) to give the final oxidation product 6. In the presence of luciferase, luciferin might yield a radical cation and superoxide radical anion (A), and the latter can deprotonate the radical cation at C.13[2] to form B, and then a radical recombination can yield the excited state of C. The intermediate C can be the light emitter itself or it could transfer energy to an unreacted molecule of luciferin 5. It is also possible that luciferin is directly oxygenated into C without the involvement of a radical recombination; then C might give a rearrangement product D in its excited state through the CIEEL mechanism. The chromophore structures of intermediates C and D are essentially the same as that of luciferin, explaining why the bioluminescence spectrum is very close to the fluorescence emission spectrum of luciferin. According to this mechanism,

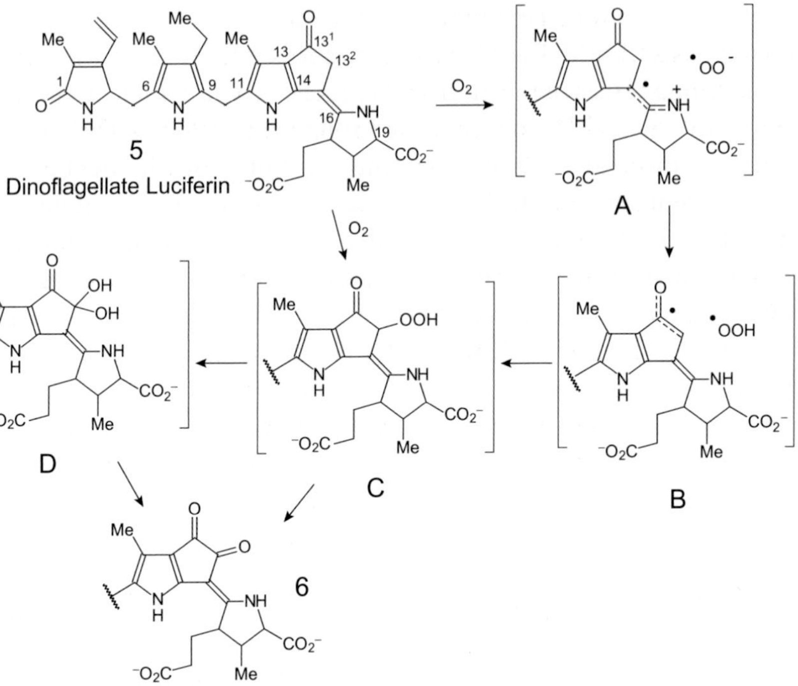

Fig. 8.9 Possible mechanisms of the bioluminescence reaction of dinoflagellate luciferin, based on the results of the model study (Stojanovic and Kishi, 1994b; Stojanovic, 1995). The luciferin might react with molecular oxygen to form the luciferin radical cation and superoxide radical anion (**A**), and the latter deprotonates the radical cation at C.13^2 to form (**B**). The collapse of the radical pair might yield the excited state of the peroxide (**C**). Alternatively, luciferin might be directly oxygenated to give **C**, and **C** rearranges to give the excited state of the hydrate (**D**) by the CIEEL mechanism. Both **C** and **D** can be the light emitter.

an important role of luciferase must be to facilitate deprotonation of the C.13^2 proton(s).

8.2.8   Luciferin Binding Protein of Dinoflagellates

Luciferin binding protein (LBP) binds luciferin at pH 8 but not at pH 6 (Fogel and Hastings, 1971); thus, LBP inhibits the luciferin-luciferase reaction at pH 8 but not at pH 6. Luciferin bound to LBP is stable, differing from the free form of luciferin that is extremely unstable. The molecular size of the *Gonyaulax* LBP was considered about 120 kDa for a long time, but it has been revised to about 75 kDa on the basis of the molecular cloning of this protein (Lee *et al.*, 1993).

Purified LBP is obtained from the crude LBP separated in the gel filtration of the 35 kDa luciferase on Sephadex G-100 (see Fig. 8.2). The fractions of crude LBP are combined and the protein is precipitated with ammonium sulfate (75% saturation). The precipitate is dissolved in a small volume of 10 mM Tris-HCl/5 mM 2-mercaptoethanol, pH 8, and a small amount of luciferin is added as a tracer. Then, the crude LBP is purified on a column of Sephadex G-200 (Hastings and Dunlap, 1986). The fractions of LBP are identified by luminescence produced by the addition of luciferase at pH 6.3; the luminescence due to the tracer luciferin is proportional to the amount of LBP in each fraction.

To assay the amount of LBP, first an excess amount of luciferin is added to the sample at pH 8 to saturate the binding site of LBP, and then the excess luciferin is removed by gel filtration using a small column of Sephadex G-25 (about 1 ml volume) also at pH 8. The luciferin-bound LBP is eluted at the void volume. To measure the amount of LBP, the following assay buffer is added to a small portion of the eluate: 0.2 M phosphate, pH 6.3, containing 0.25 mM EDTA, 0.1 mg/ml of BSA, and luciferase (Morse and Mittag, 2000). The total light obtained represents a relative amount of LBP; the absolute amount (the weight or the number of molecules) cannot be obtained because the quantum yield of the luminescence reaction is not known.

9

LUMINOUS FUNGI

9.1 An Overview on Fungal Bioluminescence

Luminous fungi are commonly seen on land, like the luminous bacteria in the sea. In woods, certain kinds of mushroom and decaying wood emit mysterious ghostly light at night. This phenomenon was known since ancient times and described by Aristotle (384–322 B.C.) and Pliny (A.D. 23–79), and the light was often called "shining wood" or "fox-fire." It is well known that Robert Boyle, in 1667, used his air pump to demonstrate the necessity of air in the light production of shining wood (Boyle, 1668). One of the most important early discoveries concerning fungal luminescence was made in the early 19th century through the careful observations of Derschau and others (Harvey, 1952, 1957); they found that the light of luminous wood came from the fungal mycelium in the wood.

A great variety of luminous fungi have been discovered in the past two centuries. Most of them belong to the class Basidiomycetes; the genus *Xylaria* of the class Ascomycetes might be an exception. Most species of luminescent fungi produce light during their mycelial stages when they are cultured on a suitable media and are vigorously growing. During the stage of fruiting bodies, some species are luminescent but many are entirely non-luminescent. The colors of light produced from all known species of luminescent fungi are similar, with their emission maxima close to 530 nm (Van der Burg, 1943; Endo *et al.*, 1970; Lavelle *et al.*, 1972), suggesting the involvement of a common light-emitter in all luminous fungi.

According to Wassink (1978), about 40 species of luminous fungi are known in nine genera (Table 9.1). Some species have multiple synonyms, which often cause confusions in their identification.

In addition to the above luminous species, Wassink listed 17 species as uncertain luminous species based on inconsistency in observations. This inconsistency might be due to the fact that the

Table 9.1   List of Luminous Fungi (Wassink, 1978)

Genus	Species
Armillaria	*A. mellea* ("Honey mushroom"; syn. *Armillariella mellea*), *A. fuscipes*
Pleurotus (syn. *Omphalotus*)	*P. olearius* ("Jack O'Lantern"; several synonyms, including *Clitocybe illudens* and *Omphalotus olearius*), *P. japonicus* (syn. *Lampteromyces japonicus*), *P. noctilucens*
Panellus (syn. *Panus*)	*P. stipticus* (syn. *P. stypticus*)
Mycena	*M. polygramma, M. tintinnabulus, M. galopus, M. epipterygia, M. sanguinolenta, M. dilatata, M. stylobates, M. zephira* (several synonyms), *M. parabolica, M. galericulata, M. avenacea, M. illuminans, M. chlorophos, M. lux-coeli, M. noctilucens, M. pruinosa-viscida, M. sublucens, M. rorida, M. manipularis, M. pseudostylobates, M. daisyogunensis, M. photogena, M. microillumina, M. yapensis, M. citrinella*
Omphalia	*O. flavida* (syn. *Mycena citricolor*)
Polyporus	*P. rhipidium*
Dictyopanus	*D. pusillus* (syn. *Polyporus rhipidium*), *D. luminescens; D. gloeocystidiatus, D. foliicolus*
Marasmius	*M. phosphorus*
Locellina	*L. noctilucens, L. illuminans*

luminosity of mycelial culture is highly dependent on the composition and conditions of the culture medium. A fungal culture that is brightly luminescent can become non-luminescent when cultured under slightly different conditions. Moreover, certain species, such as *Panellus stipticus* and *Mycena polygramma*, are known to exist in two distinct forms, a luminous form and a non-luminous variety (Wassink, 1948), making the matter more complicated. In the case of *P. stipticus*, it was believed for a long time that the North American form is luminescent, and the European variety is non-luminescent. However, it is now apparent that the non-luminescent form exists also in the North America.

Luminous fungi emit light only during certain stages of their life cycles. They might emit light from the fruiting body, or from the mycelium, or from both, depending on the species. Generally, the luminescence is much brighter with a young fruiting body and new fast-growing mycelium than with a mature fruiting body or older mycelium, although their luminescence intensities vary widely depending on the species and environmental conditions.

The luminescence of fungi is definitely not bright in comparison with most other luminous organisms. However, the luminescence of fungi continues for several days, day and night, like that of the luminous bacteria. For example a fruiting body of *P. stipticus* (Fig. 9.1) weighing 0.2 g typically emits $1-2 \times 10^{10}$ photons/s of light continuously for many days (Shimomura, 1989, 1992). Thus, the total amount of light emitted from a small luminous mushroom during its lifetime is comparable to, or more than, that emitted from a specimen of the most brightly luminescent organisms, such as the fireflies and the ostracod *Cypridina*.

9.2   Early Studies on the Biochemistry of Luminous Fungi

Efforts by various investigators to demonstrate a luciferin-luciferase reaction with luminous fungi were uniformly unsuccessful until Airth and McElroy (1959) obtained a positive result. Airth's group studied further details of the luciferin-luciferase reaction using the luciferin preparation obtained from *Armillaria mellea* and the luciferase preparation obtained from a species identified by them

Fig. 9.1 Fruiting bodies of the luminous fungus *Panellus stipticus* grown on a oak log, in daylight (*top*) and in the dark (*bottom*).

as *Collybia velutipes* (incorrect identification according to Wassink, 1978), and they formulated the following schemes to describe the luciferin-luciferase reaction of fungi (Airth and Foerster, 1962, 1964; not equated).

$$Ln + NADH + H^+ \xrightarrow{\text{Soluble enzyme}} LnH_2 + NAD^+$$

$$LnH_2 + O_2 \xrightarrow{\text{Particulate luciferase}} Ln + H_2O + Light$$

The preparations of luciferin (Ln, an electron acceptor) and "soluble enzyme" used were crude or only partially purified. The luciferase was an insoluble particulate material, possibly composed of many substances having various functions. Moreover, the luciferin-luciferase reaction was negative when both luciferin and luciferase were prepared from certain species of luminous fungus. It appears that the light production reported was the result of a complex mechanism involving unknown substances in the test mixture, and probably the crucial step of the light-emitting reaction is not represented by the above schemes.

Kuwabara and Wassink (1966) extracted 15 kg of the mycelium of *Omphalia flavida* (a Puerto Rican coffee-leaf fungus), and isolated a luciferin in a crystalline form. The luciferin was chemiluminescent with H_2O_2, and produced light using Airth's luciferase (λ_{max} 524 nm). No information was reported concerning the chemical nature of this luciferin.

Endo *et al.* (1970) isolated from the fruiting bodies of *Lampteromyces japonicus* (*Pleurotus japonicus*; commonly called Tsukiyotake in Japanese) four compounds, including illudin S, a toxic compound with UV absorption characteristics similar to the crystalline luciferin obtained by Kuwabara and Wassink, and ergosta-4,6,8(14),22-en-3-one that showed a fluorescence emission peak (530 nm) nearly identical with the bioluminescence emission peak of various fungi. From the same mushroom, Isobe *et al.* (1987, 1988) isolated riboflavin and lampteroflavin, of which both showed a fluorescence emission spectrum (λ_{max} 524 nm) almost identical with the bioluminescence of the mushroom. However, O'Kane *et al.* (1994) ruled out riboflavin as the light emitter of fungal bioluminescence on the basis of their spectral studies.

Considering that a luminous fungus may contain a number of fluorescent compounds and that the emission spectra of some of them might match that of luminous fungi, a spectral coincidence is only a marginal suggestion that an isolated compound might be the light emitter in fungal bioluminescence; it is important to show that the compound is actually involved in the light-emitting reaction.

9.3   Role of Superoxide in Fungal Luminescence

As already noted, the luciferin isolated from *Omphalia flavida* by Kuwabara and Wassink was chemiluminescent in the presence of H_2O_2, with or without Fe^{2+}. According to our studies, the five species of luminous fungus we investigated (*Panellus stipticus*, *Armillaria mellea*, *Mycena citricolor*, *Pleurotus japonicus*, and *Omphalotus olearius*) all contained chemiluminescent substances that are capable of emitting light in the presence of H_2O_2 and Fe^{2+}. In this chapter, such chemiluminescent substances are assumed to be the fungal luciferins that actually function as the luciferins in these fungi. With regard to luciferase, however, all our efforts were unsuccessful in demonstrating the existence of any protein (enzyme) that catalyzes the light emission of luciferin in any of these species. It does not prove but may imply the absence of luciferase in bioluminescent fungi. The requirement for H_2O_2 plus Fe^{2+} for light emission might suggest a possibility that an active oxygen species, such as superoxide anion, is involved in the luminescence reaction of luminous fungi.

Superoxide anion is an elusive substance with a short life, and is difficult to detect or measure. It is generally considered toxic to living cells. However, superoxide anion appears to be an essential requirement in various bioluminescence systems, including the tubeworm *Chaetopterus* (Shimomura and Johnson, 1966), the scaleworm *Harmothoë* (Nicolas *et al.*, 1982), and the clam *Pholas* (Michelson, 1978). The mechanism of the superoxide action might involve a one-electron oxidation of the luciferin molecule, which results in the formation of a luciferin free radical. The free radical would be immediately oxygenated by molecular oxygen into a peroxide (Shimomura, 1993), which is probably followed by the decomposition of the peroxidized luciferin that provides the energy for the light emission.

We have investigated the possible involvement of superoxide anion with six species of fungi, by analyzing the interrelationships between the light emission of the fungus (both fruiting body and growing mycelium) and the contents of luciferin, its precursor, superoxide dismutase (SOD), and catalase (Shimomura, 1992). The study resulted in several important findings as summarized below (refer to Fig. 9.2 and Table 9.2).

Table 9.2   Contents of Luciferin, the Precursor of Luciferin, and SOD, in 1 g of Fruiting Body or Mycelium of *Panellus stipticus, Armillaria mellea, Pleurotus japonicus, Omphalotus olearius, Mycena lux-coeli,* and *Mycena citricolor* (data taken from Shimomura, 1992)

Fungus Material[a]	Luminescence[b] $(10^{10}$ quanta/s)	Luciferin[c] $(10^{10}$ quanta)	Precursor[d] $(10^{10}$ quanta)	SOD[e] (units)
P. stipticus (frt) luminous form	9	200	15,000	58
P. stipticus (frt) non-luminous form	0.002	220	18,000	58
P. stipticus (myc) luminous form	0.4/8	70	70	160
A. mellea (myc) 2 min soaked in water	6/16	100	—	150
A. mellea (myc) 5 hr soaked in water	25/16	84	—	53
P. japonicus (frt)	>9	200	—	24
P. japonicus (myc)	0.8/21	70	—	160
O. olearius (myc)	0.8/16	60	—	125
M. lux-coeli (frt)	>9	850	650	120
M. citricolor (myc)	2/6	600	—	170

[a] Abbreviations: frt, fruiting body; myc, mycelium.
[b] Spontaneous light emission from fresh material. For mycelium, the two data shown are: "scraped mycelium"/"mycelium on agar before scraping."
[c] Chemiluminescence activity of the extract in the presence of Fe^{2+} and H_2O_2.
[d] The extract was activated with methylamine, and then chemiluminescence activity was measured. A minus sign $(-)$ indicates an unsuccessful quantitation due to various causes.
[e] Assayed by the luminol chemiluminescence method (Puget and Michelson, 1974). One unit corresponds to 0.35 unit of the cytochrome c method (McCord and Fridvich, 1969).

1. Light-emitting period. It is clearly shown that a luminous fungus emits light only during a certain period(s) of its life cycle, and is practically non-luminous before and after that period (Fig. 9.2).

2. Low luciferin content. Luminous fungus in various growth stages contains relatively small amounts of luciferin that can sustain the luminescence for a period of only 3–100 s, in resemblance to the situation found in luminous bacteria. In luminous bacteria,

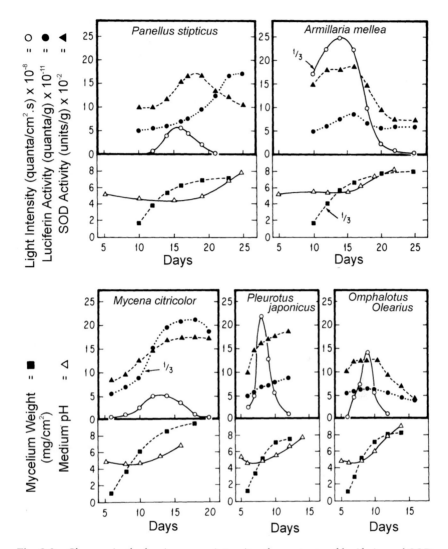

Fig. 9.2 Changes in the luminescence intensity, the contents of luciferin and SOD, and medium pH during the mycelial growth of five kinds of bioluminescent fungi: *Panellus stipticus*, *Armillaria mellea*, *Mycena citricolor*, *Pleurotus japonicus*, and *Omphelotus olearius*. The ordinate readings of the curves marked "1/3" should be multiplied by 3. For the SOD activity, see Table 9.2, footnote *e*. From Shimomura, 1992, with permission from Oxford University Press.

the bacterial cells contain a very small amount of fatty aldehydes (the luciferin) that can support the luminescence for about one second, and the steady state of bacterial luminescence is maintained by the continuous biosynthesis of the aldehydes (Shimomura *et al.*, 1974a).

3. *Generation of superoxide anion in luminous fungi (Shimomura, 1992)*. During mycelial growth, both luciferin and SOD start to appear at a very early stage of the growth, and the change of the SOD content is roughly parallel to that of the luciferin content, except in the case of *P. stipticus*. Considering that SOD is produced to remove toxic O_2^- in cells, the peak of the O_2^- formation probably occurs a little before the peak of SOD content, when the mycelial growth is most vigorous, metabolism is high, and the luminescence of mycelium is high, supporting the hypothesis that the luminescence of luciferin is caused by O_2^-.

4. *Difference between fruiting body and mycelium*. The luminescence intensities of both fruiting bodies and mycelia are in a relatively narrow range ($6-21 \times 10^{10}$ quanta $\cdot$ s^{-1}g^{-1}), despite the clear difference that the fruiting bodies are generally rich in luciferin and the mycelia are rich in SOD (Table 9.2).

5. *Comparison of luminescent strain and non-luminescent strain of P. stipticus*. The *P. stipticus* fruiting bodies of the bioluminescent strain and the so-called non-bioluminescent strain are identical in their appearance, except that the former get a more brownish color than the latter upon aging for 2–3 weeks. Although the contents of luciferin, the precursor and SOD are essentially equal in both strains, the light emission from the luminescent strain is 4,500 times brighter than that of the non-luminescent strain. The luminescence of the non-luminescent strain of *P. stipticus* (2×10^7 quanta $\cdot$ s^{-1}g^{-1}) is close to that of the fresh fruiting bodies of various ordinary non-bioluminescent fungi (10^5-10^6 quanta $\cdot$ s^{-1}g^{-1}).

6. *Regulation of light intensity in fungal luminescence*. If the luminescence reaction was caused by O_2^-, the luminescence intensity of the reaction might be controlled by SOD. The contents of SOD in various luminous fungi are generally very high, suggesting high levels of O_2^- generation in the tissue. If the SOD were distributed uniformly in the tissue, the high concentration of SOD would instantly decompose any

O_2^- generated; therefore, no light emission would be expected. For the occurrence of luminescence, the SOD activity in the tissue must be distributed unevenly; very low levels of SOD activity should be distributed at the sites of luminescence.

The foregoing discussion may suggest an illustrative picture of the fungus tissue. In luminous fungi, the cells producing SOD and those containing the light-emitting substance are located remotely or arranged in a manner that impedes the diffusion of SOD. On the other hand, in the non-luminous forms, the two types of cells are located more closely or arranged in a manner that promotes the diffusion of SOD. In such a situation, SOD will be able to diffuse rapidly into the light-emitting cells to destroy O_2^- to block the luminescence reaction in the non-luminous forms. In the luminous form, young fungus is brightly luminescent because the light-emitting cells are scarcely affected by SOD, although the luminescence will gradually diminish with time due to the slow diffusion of SOD into the light-emitting cells.

The finding by several investigators more than 60 years ago that crushing a fungus always destroys the light (Harvey, 1952) can be easily understood in the picture described above. The significant increase in the luminescence intensity of the mycelium of *Armillaria mellea* after soaking it in water (Table 9.2) can be explained by the leaching out of SOD.

9.4   Studies on *Panellus stipticus*

The luminous fungus *P. stipticus* (Fig. 9.1) is very common in North America, and is also found in Europe and Japan. The fruiting body, with a cap size of 1–2 cm, is almost white with a tint of yellowish tan when young, and becomes brown by aging in 2–3 weeks. The fungus continuously produces light (λ_{max} 530 nm) from its young growing fruiting body and mycelium. The luminescence of the fruiting bodies continues for many days until they get darkly pigmented because of aging or they are dried by dry weather. A young, small specimen is always the brightest. Dried specimens are not luminous, but they regain the ability to luminesce when they are made wet by rain. Frozen specimens, when thawed, produce very little luminescence. When fresh specimens (both fruiting body and mycelium) are ground in a mortar, the light is instantly extinguished. Airth's luciferin-luciferase reaction is negative with both fruiting body and mycelium.

Our study on *Panellus* luminescence was started when the fruiting bodies of this species were found in the backyard of my residence in the early 1980s. This fungus commonly grows on dead oak trees, and we observed a rich growth of this fungus on the fallen trees in the Beebe Wood, Falmouth, MA, over a period of 3–4 years after Hurricane Bob struck the area in 1991.

Cultivation of *P. stipticus*. Cultured fruiting bodies are better in quality than those collected from nature, because they can be harvested at their best. In producing the cultured fruiting bodies of *P. stipticus*, we utilized the cultivation technique of the Shiitake mushroom. The initial mycelial culture is grown on a YM agar plate from the spores collected from a brightly luminescent fruiting body. Small cylindrical pieces of wood (8 mm diameter, 12 mm long) are autoclaved in YM agar medium, and the excess medium is drained off prior to the hardening of the agar. After cooling to room temperature, the wood pieces are inoculated with the fungus mycelium. In two or three weeks, the growth of mycelium spreads and completely covers the surface of the wood pieces. The logs of oak wood (cut down two months in advance; 50–70 cm long) are drilled with a 10 mm bit at intervals of about 7–8 cm, making holes of 2 cm depth. The holes are plugged with the mycelium-covered wood pieces, and the logs are left in the shade of trees with occasional watering. The logs start to produce the fruiting bodies in about 3–4 months under a moderate temperature of 15–25°C.

Chemiluminescent compounds and their precursors in *P. stipticus*. Although *P. stipticus* is negative in the luciferin-luciferase reaction, crude extracts of this fungus are chemiluminescent, like the luciferin obtained from *Omphalia flavida* by Kuwabara and Wassink (1966). The chemiluminescence is elicited by the addition of H_2O_2 and Fe^{2+} under a mild condition of pH 5–8, and the luminescence is strongly

Table 9.3 Designation of the compounds related to the *Panellus* luminescence

Designation	Description
Panal	Obtained from PS-A and PS-B by hydrolysis
PS	Fatty acid ester of panal
PS-A	Decanoic acid ester of panal
PS-B	Dodecanoic acid ester of panal
PS-A/MA	Methylamine adduct of PS-A
PS-B/MA	Methylamine adduct of PS-B
PM	Methylamine-activation product of PS
PM-1	Methylamine-activation product derived from PS-A
PM-2 and -3	Methylamine-activation product derived from PS-A plus PS-B
PM-4	Methylamine-activation product derived from PS-B
K-1	A synthetic model compound of PS
KM-1 and KM-2	Methylamine activation product of K-1

enhanced by a cationic surfactant, such as cetyltrimethylammonium bromide (CTAB). Moreover, it was accidentally discovered that both the light intensity and the total amount of light were greatly increased when the extract had been pretreated with ammonium sulfate or a primary amine, suggesting the presence of luciferin precursor(s) in the extract. Actually, we were able to isolate three kinds of precursors from the fruiting bodies of *P. stipticus*, namely, panal, PS-A and PS-B (Table 9.3), which yield strongly chemiluminescent substances upon activation with the salts of ammonia or amines (Nakamura *et al.*, 1988a; Shimomura, 1989, 1991b, 1993).

9.4.1 *Panal*

Extraction and purification (Shimomura, 1989). Fruiting bodies are extracted with 30% methanol, and the extract is concentrated under reduced pressure. The residual solution (pH 6.0) is first washed with ethyl acetate, then acidified to pH 2.0 with 1 M HCl and extracted with ethyl acetate. The extract is evaporated nearly to dryness. The residue is dissolved in 30% methanol and purified by the following 3-step procedure.

(1) Anion-exchange chromatography on a column of TSK DEAE-650 M (EM Science) in 30% methanol. Elution with a linear gradient of NaCl concentration from 0 to 1 M.

(2) Heat-treatment: the active fractions from step 1 are combined and evaporated to a small volume ($\sim$4 ml), and mixed with a 1/4 volume of 1 M H_3PO_4. The mixture is heated in a bath of boiling water for 2 minutes, then quickly cooled. The precipitate formed is removed by centrifugation. The panal in the clear solution is extracted with ethyl acetate, and the extract is evaporated to dryness. The residue is redissolved in 30% methanol. During this process, the hydrated form of panal is converted into its basic structure.

(3) HPLC on a C18-silica column, with 27% (v/v) acetonitrile (pH 2.6 with H_3PO_4). The active fraction eluted is concentrated, and the panal in the residue is extracted with ethyl acetate. After evaporating ethyl acetate, panal is redissolved in 30% methanol, and stored at $-30°$C.

The yield of panal is about 0.75% of fresh fruiting bodies or 2–3% of the dried material, based on weight. In the fruiting bodies, most of the panal probably exists in the forms of PS-A and PS-B.

Properties of panal (Nakamura *et al.*, 1988a). Purified panal is a colorless, amorphous solid, soluble in alcohols, water, ethyl acetate, and chloroform. The absorption spectrum (Fig. 9.3) shows a single peak (λ_{max} 217 nm, ε 15,300). Optical rotation $[\alpha]_D$ $-17°$ (c 0.9, methanol). Mass spectrometry and NMR analysis showed that panal is a sesquiterpene aldehyde, $C_{15}H_{18}O_5$ (M_r 278.30), with the structure shown below.

Panal   R = H
PS-A   R = $CO(CH_2)_8CH_3$
PS-B   R = $CO(CH_2)_{10}CH_3$

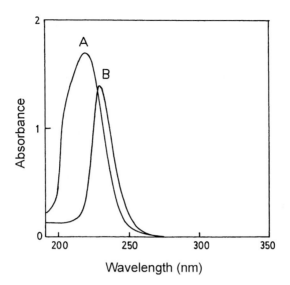

Fig. 9.3   Absorption spectra of panal in methanol (**A**), and in 18% acetonitrile/water (v/v) containing 20 mM morpholine and acetic acid, pH 4.3 (**B**). The absorption peak in A is broad due to the partial enolization of the keto group forming a conjugated structure C=C–C=C–OH. In B, the C=O group is completely converted into an enamine (C=C–C=C–NR$_2$) by morpholine, giving a sharp absorption peak. From Shimomura, 1989, with permission from the American Society for Photobiology.

Panal is stable in aqueous media (pH 1–5) at room temperature, except for a gradual hydration that can be reversed with an aqueous acid. Panal is activated into chemiluminescent compounds upon treatment with the salts of ammonia or primary amines.

9.4.2   Activation Products of Panal

Activation procedures (Shimomura, 1989). To activate trace amounts of panal, several granules of the crystals of ammonium sulfate are added to 5–50 μl of a sample made in 30% methanol, making sure that $(NH_4)_2SO_4$ is in excess to saturate the solution. The mixture is left standing at room temperature for 1 h (85% activation) or one night (complete activation). After the activation, 1 μg of panal emits about 1.5×10^{10} photons using the assay method given below.

To activate 1–2 mg of panal with ammonium sulfate, 1 ml of 3 M $(NH_4)_2SO_4$ (pH 3.7 with acetic acid) and 2–3 mg of CTAB are added, and the mixture is incubated at 25°C for 12–16 hours. The product is deep yellow or orange.

To activate 1–2 mg of panal with a primary amine, 1 ml of 2–3 M primary amine hydrochloride (pH 1.2, with HCl), and 2–3 mg of CTAB are added, and the mixture is incubated at 45°C for 2–5 days. The product is dark orange and fluoresces yellow.

The yield of luminescence activity is increased by including CTAB, Tergitol 7 (a sulfate detergent), or *p*-toluenesulfonyl chloride to various extents.

Assay methods. Activity can be measured at pH 5.5 or at pH 8.0. With the same sample, the pH 5.5 method gives a much higher luminescence intensity than the pH 8.0 method (and, shorter reaction time), although the total amounts of light emitted by the two methods are practically equal (Fig. 9.4). The pH 5.5 method is susceptible to inhibition by various salts, whereas the pH 8.0 method is not affected.

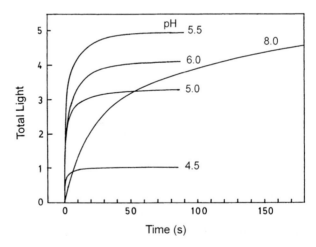

Fig. 9.4 Time course of the chemiluminescence reaction of $(NH_4)_2SO_4$-activated panal at pH values 4.5, 5.0, 5.5, and 6.0, in 3 ml of 10 mM acetate buffer in the presence of 10 mg of CTAB, 20 µl of 0.1 M $FeSO_4$, and 20 µl of 10% H_2O_2; and at pH 8.0, in 3 ml of 50 mM Tris-HCl buffer containing 0.18 mM EDTA, 10 mg of CTAB, 10 mg of $NaHCO_3$, 20 µl of 0.1 M $FeSO_4$, and 20 µl of 10% H_2O_2. All at 25°C. From Shimomura, 1989, with permission from the American Society for Photobiology.

The pH 5.5 method: To an activated sample ($<10\,\mu l$), 3 ml of 10 mM sodium acetate buffer (pH 5.5), 5–10 mg of CTAB, 20 μl of 0.1 M $FeSO_4$, and 20 μl of 10% H_2O_2 are added in that order. The light emission is triggered when H_2O_2 is injected with a constant rate syringe (Hamilton CR-700-20).

The pH 8.0 method: To an activated sample ($<100\,\mu l$), 3 ml of 50 mM Tris-HCl buffer containing 0.18 mM EDTA (pH 8.0), 5–10 mg of CTAB, about 10 mg of $NaHCO_3$, 20 μl of 0.1 M $FeSO_4$, and 30 μl of 10% H_2O_2 are added in that order. The last component, H_2O_2, triggers the light emission.

Properties of activated products. The luminescence of the activation products of panal is elicited by Fe^{2+} and H_2O_2, and the light emission is significantly enhanced by the presence of various detergents, particularly cationic detergents, such as CTAB. The influence

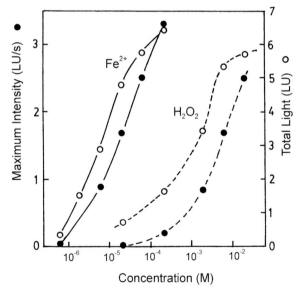

Fig. 9.5 Influence of the concentrations of Fe^{2+} (solid lines) and H_2O_2 (broken lines) on the light intensity (●) and total light (○) of the luminescence of $(NH_4)_2SO_4$-activated panal in 3 ml of 10 mM acetate buffer (pH 5.5) containing 10 mg of CTAB, with various concentrations of $FeSO_4$ plus 30 μl of 10% H_2O_2, or with 20 μl of 0.01 M $FeSO_4$ plus various concentrations of H_2O_2. From Shimomura, 1989, with permission from the American Society for Photobiology.

of the concentrations of Fe^{2+} and H_2O_2 on the light intensity and total light in the presence of CTAB is shown in Fig. 9.5. The emission spectrum may differ with the detergent used, as well as with the amine used. With CTAB, the emission peak of $(NH_4)_2SO_4$-activation product was 570 nm, and the emission peaks of the primary amine-activation products were 485–495 nm. When DOAB (dioctadecyldimethylammonium bromide; contains two hydrophobic chains) was used as the detergent, a weak emission peak at 530–540 nm was observed from the primary amine-activation products.

9.4.3 *PS-A and PS-B*

Extraction and purification (Shimomura, 1991b). The luciferin precursors PS-A and PS-B are extracted from the dried fruiting bodies of *Panellus stipticus* (5 g) with methanol, and the extract is evaporated under reduced pressure to remove most of the methanol. The residue (pH 6.3) is diluted with a small amount of water and extracted with ethyl acetate, and the extract is evaporated to dryness. The precursors in the residue are purified by three steps of silica gel chromatography and one step of HPLC, while monitoring the absorbance of eluate at 220–230 nm. The procedure is summarized below:

(1) The dried residue is dissolved in a chloroform/ethyl acetate mixture (95:5, v/v) and applied to a silica gel column (1.5 × 7 cm) that has been packed with chloroform. The column is washed with chloroform. Then the precursors are eluted with chloroform/ethanol (98:2, v/v).

(2) A repetition of (1), except that the elution is carried out with chloroform/ethyl acetate mixture (first 90:10, then 80:20).

(3) The precursors dissolved in a hexane/ethyl acetate mixture (90:10, v/v) are adsorbed on a silica gel column that has been packed with hexane. The precursors were eluted with hexane/ethyl acetate mixtures, 80:20, 70:30, and 60:40 (v/v), in that order.

(4) HPLC on a PRP-1 column (0.7 × 30.5 cm, Hamilton), with 65% acetonitrile containing 0.05% acetic acid, at a flow rate of 4 ml/min. To reverse the possible hydration of the molecules, the material obtained in step 3 is first evaporated to dryness, and redissolved in chloroform. Immediately before injection onto the HPLC column, an

aliquot of the chloroform solution is evaporated to dryness, and the residue is redissolved in 0.2 ml of the HPLC solvent and injected at once. Two precursors, PS-A and PS-B, are cleanly separated (eluting at 28 ml and 50 ml, respectively). In a typical experiment, 5 g of dried *P. stipticus* yielded 27 mg of PS-A and 23 mg of PS-B.

Properties of PS-A and PS-B (Shimomura, 1991b; Shimomura et al., 1993b). Both PS-A and PS-B are colorless viscous liquid, and their absorption spectra resemble that of panal (Fig. 9.6). By NMR analysis and mass spectrometry, PS-A and PS-B are found to be 1-O-decanoylpanal and 1-O-dodecanoylpanal, respectively. As a minor component, 1-O-tetradecanoylpanal has also been isolated. PS-A and PS-B gain chemiluminescence activity when treated with the salt of primary amines (see below for the conditions). Taking the activity obtained with methylamine as 100%, the activities obtained with other amines were: ethylamine, 38%; ethanolamine, 10%; propylamine, 20%; hexylamine, 3%; and decylamine, 1%.

Both PS-A and PS-B have a tendency to hydrate like panal, and they also form adducts with methylamine. The adducts, PS-A/MA and PS-B/MA, are prepared by incubating PS-A or PS-B in 75% methanol containing an excess amount of methylamine hydrochloride plus some sodium acetate to neutralize the HCl, at 45°C for 30 min. The adducts can be purified by HPLC on a PRP-1 column (80% acetonitrile containing 0.05% acetic acid). Their chemical structures have been determined by NMR and mass spectrometry as shown in Fig. 9.8 (p. 288). Both adducts are colorless and show an absorption maximum at 218 nm.

9.4.4 Activation of PS-A and PS-B

The activation by methylamine is carried out by one of the following two methods (Shimomura et al., 1993b).

(a) A sample of precursor (0.1–0.3 mg in 0.1 ml of methanol) is mixed with 0.4 ml of 2 M methylamine (pH 1.2 with H_2SO_4). The cloudy mixture is incubated overnight at 45°C to complete the activation.

(b) A sample of precursor (0.1–0.3 mg in 0.05 ml of methanol) is mixed with 0.45 ml of 0.2 M acetic acid/NaOH buffer (pH 3.5)

containing 50 mM methylamine hydrochloride and 0.1% (v/v) Tergitol 4 (a sulfate-type anionic surfactant), and the mixture is incubated at 25–35°C. The activation is complete usually in less than 12 h
at 35°C.

The activation product is extracted with ethyl acetate, and the
extract (orange colored with a strong yellow fluorescence) is evaporated to dryness.

Purification of the activation products (PMs). The methylamine
activation product dissolved in methanol is purified by chromatography, first on a column of silica gel using a mixed solvent of
chloroform/ethanol, followed by reversed-phase HPLC on a column of divinylbenzene resin (such as Jordi Reversed-Phase and
Hamilton PRP-1) using various solvent systems suitable for the target
substance (for example, acetonitrile/water containing 0.15% acetic
acid).

The activation product of an equal-amount mixture of PS-A
and PS-B gave four major chemiluminescent compounds, designated
PM-1, PM-2, PM-3 and PM-4 in an approximate ratio of 1:2:2:1. The
activation product of PS-A gave only PM-1, and that of PS-B gave only
PM-4.

Assay of luminescence activity. A methanolic solution of the activation product (5–50 μl) is mixed with 3 ml of 50 mM Tris-HCl buffer,
pH 7.8, containing 0.18 mM EDTA and about 5 mg of CTAB. After
leaving the solution for a few minutes, 15 μl of 50 mM $FeSO_4$ and
30 μl of 10% H_2O_2 are added in this order. The light emission begins
when H_2O_2 is added.

Properties of the activation products (Shimomura, 1991b). All
PMs are orange-colored, with an absorption maximum at 488 nm
(Fig. 9.6). The absorption characteristics and chemiluminescence
activities of those compounds are shown in Table 9.4. All PMs are
brightly fluorescent in yellow in organic solvents and also in aqueous solutions containing a surfactant (emission maxima 520–530 nm).
The chemiluminescence spectra of PMs are significantly affected by the

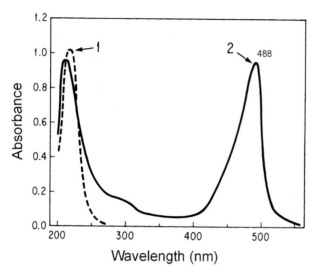

Fig. 9.6   Absorption spectra of PS-A and PS-B (superimposable; Curve 1), and PM-1, PM-2, PM-3, and PM-4 (superimposable; Curve 2), all in methanol. From Shimomura, 1991b, with permission from Oxford University Press.

Table 9.4   Absorption Characteristics and Chemiluminescence Activities of Various Compounds derived from the Fruiting Bodies of *Panellus stipticus* (Shimomura, 1991b)

Compound	Absorption Maximum (nm)	$A_{1\%, 1cm}$ (in methanol)	Chemiluminescence Activity (10^{12} quanta/mg) after Activation with:		
			None	Methylamine	$(NH_4)_2SO_4$
Panal	217	550		4	10
PS-A	215	340		130	1.8
PS-B	215	305		130	1.2
PM-1*	210	261	3,000		
	488	256			
PM-2*	210	255	3,000		
	488	250			
PM-3*	210	245	3,000		
	488	240			
PM-4*	210	245	3,000		
	488	240			

*The compound has been prepared by activation with methylamine, and requires no further activation.

Table 9.5 Compositions of the Chemiluminescent Compounds, PM-1, -2, -3 and -4

Compound*	m/z	Molecular Formula	Made from:
PM-1	1286	$C_{77}H_{110}O_{14}N_2$	3PS-A + 2MeNH$_2$ − 4H$_2$O
PM-2	1314	$C_{79}H_{114}O_{14}N_2$	2PS-A + PS-B + 2MeNH$_2$ − 4H$_2$O
PM-3	1342	$C_{81}H_{118}O_{14}N_2$	PS-A + 2PS-B + 2MeNH$_2$ − 4H$_2$O
PM-4	1370	$C_{83}H_{122}O_{14}N_2$	3PS-B + 2MeNH$_2$ − 4H$_2$O

*The samples of PM-2 and PM-3 probably consist of at least 3 kinds of isomers with the same molecular formula according to high-resolution chromatography.

type of surfactant used, as in the case of the activation products of panal.

The molecular formulas of PMs (Table 9.5) obtained by high-resolution mass spectrometry in collaboration with Prof. Y. Kishi, Harvard University, indicate that PMs are formed by the condensation of three molecules of PS and two molecules of methylamine, with the removal of four water molecules. No study has been made on their conformational isomers.

The bioluminescence spectrum of *P. stipticus* and the fluorescence and chemiluminescence spectra of PM are shown in Fig. 9.7. The fluorescence emission maximum of PM-2 (525 nm) is very close to the bioluminescence emission maximum (530 nm), but the chemiluminescence emission maximum in the presence of a cationic surfactant CTAB (480 nm) differs significantly. However, upon replacing the CTAB with the zwitter-ionic surfactant SB3-12 (3-dodecyldimethylammonio-propanesulfonate), the chemiluminescence spectrum splits into two peaks, 493 nm and 530 nm, of which the latter peak coincides with the emission maximum of the bioluminescence. When PM-1 is heated at 90°C for 3 hr in water containing 10% methanol, about 50% of PM-1 is converted to a new compound that can be isolated by HPLC; the chemiluminescence spectrum of this compound in the presence of SB3-12 (curve 5, Fig. 9.7) is practically identical with the bioluminescence spectrum.

The quantum yields of PMs are calculated at approximately 0.0065 from the data of Tables 9.4 and 9.5. The chemical yields of PMs from PS-A and PS-B are calculated at about 4.3%, based on the

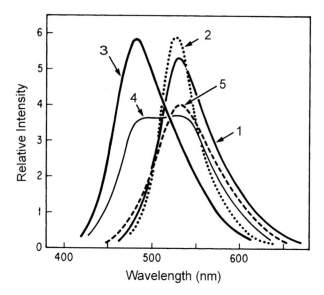

Fig. 9.7 Bioluminescence of *Panellus stipticus* fruiting body (**1**); fluorescence of PM-2 in the presence of CTAB upon excitation at 440 nm (**2**); chemiluminescence of PM-2 in the presence of CTAB (**3**); chemiluminescence of PM-2 in the presence of 3-(dodecyldimethylammonio)propanesulfonate (SB3-12) (**4**); and chemiluminescence of the hot-water treatment product of PM-1 in the presence of SB3-12 (**5**). Curves 2–5 were measured in 2 ml of 50 mM Tris-HCl buffer (pH 8.0) containing 0.18 mM EDTA. Chemiluminescence was elicited by the addition of 5 μl of 50 mM $FeSO_4$ and 10 μl of 10% H_2O_2. From Shimomura, 1991b, with permission from Oxford University Press.

comparison between the luminescence activities of PMs and those of PS-A and PS-B after the methylamine activation.

Requirement for superoxide anion in the luminescence of PMs. It is difficult to investigate the role of superoxide anion in the presence of a large amount of added H_2O_2, which readily produces a trace amount of superoxide anion. Fortunately, PMs do chemiluminesce by the xanthine-xanthine oxidase system that produces O_2^- without adding H_2O_2, although the luminescence intensity is only one-tenth of that elicitable with the Fe^{2+}-H_2O_2 system.

Table 9.6 shows the results of the tests carried out to examine the possible involvement of O_2^- in the luminescence reaction of PMs

Table 9.6   Examination of Superoxide Involvement in the Chemiluminescence of PM-2 Elicited by the Fe^{2+}-H_2O_2 System and the Xanthine-Xanthine Oxidase System[a] (Shimomura, 1991b)

(A) Fe^{2+}-H_2O_2 system

Fe^{2+}	H_2O_2	Substance Added	Initial Intensity	Total Light
0.1 mM	10 mM	None (control)	100% (4×10^9 quanta/s)	100% (8×10^{11} quanta)
None	None	None	0%	0%
0.1 mM	None	None	0%	0%
None	10 mM	None	10%[b]	2.2%[b]
None	None	$FMNH_2$ (10 μM)[c]	12%	0.3%
None	None	$Na_2S_2O_4$ (0.1 mM)[c]	14%	0.4%
None	10 mM	Peroxidase (10 μg)[d]	80%	80%
0.1 mM	10 mM	Argon gas[e]	45%	

(B) Xanthine-xanthine oxidase system (50 μM xanthine, 5 μg xanthine oxidase)

Substance added	Initial maximum intensity
None (control)	100% (3×10^8 quanta/s)
SOD (5 μg; from bovine erythrocytes, 3,600 units/mg)	4%
Catalase (2 μg)	75%
Bilirubin (1 μM)	70%
Cytochrome c (1 μM)	Strong inhibition for initial 30 s.
Argon gas[e]	0%

[a]Tested in 3 ml of 50 mM Tris-HCl buffer, pH 8.0, containing 0.18 mM EDTA, 0.27 μg PM-2, 2 mg of CTAB.
[b]Superoxide anions are spontaneously formed.
[c]Superoxide anions are known to be formed in the air-oxidation of this compound.
[d]Peroxidase is nearly as effective as Fe^{2+} in this chemiluminescence system.
[e]Solution was bubbled with argon gas for 10 min from a capillary tubing.

(Shimomura, 1991b). In the experiments of part A, $FMNH_2$ and $Na_2S_2O_4$ produce O_2^- when they are oxidized by air. The effect of peroxidase resembles that of Fe^{2+}, apparently due to the presence of an iron atom in this enzyme. In the experiments of part B, H_2O_2 is absent in the test solutions, at least at the beginning of the reaction. The data show that SOD strongly inhibits the luminescence reaction. Catalase that decomposes H_2O_2, and bilirubin that quenches singlet oxygen show only small effects, whereas cytochrome c that oxidizes O_2^- resulted in a strong inhibition of luminescence until the reagent is exhausted. The bubbling argon removes oxygen O_2, thus O_2^- cannot be formed. These results strongly support that the luminescence reaction requires O_2^- as well as O_2.

9.4.5   Mechanism of the in vivo Bioluminescence of P. stipticus

According to the information available, it would be reasonable to consider that *P. stipticus* emits light when its natural luciferin is oxidized with molecular oxygen in the presence of O_2^- and a suitable surfactant (Shimomura *et al.*, 1993b). Also, it seems almost certain that the natural luciferin is formed from PS-A, PS-B and a simple primary amine by the addition and condensation reactions.

The chemiluminescent compounds PMs are produced from PS-A, PS-B and methylamine under surprisingly mild conditions that are apparently compatible with the intracellular environment of living fungus. For example, the formation of PMs from PS-A, PS-B and methylamine takes place in a pH 3.6 buffer solution containing a trace amount of Tergitol 4 (an anionic surfactant) even at room temperature (Fig. 9.8). PMs are also produced by reacting PS-A and PS-B with the adducts PS-A/MA and PS-B/MA under the same condition, but not produced from the adducts alone. In addition to Tergitol 4, various types of fatty alcohol sulfate are also effective in catalyzing the addition and condensation. It is intriguing that the light emission from the growing mycelium of *P. stipticus* has been reported to be optimum at pH 3.5 (Bermudes *et al.*, 1990).

The chemiluminescence reaction of PMs also takes place under mild conditions. The luminescence occurs in aqueous media at pH 5.5–8.5, and is triggered by O_2^- in the presence of O_2 and a cationic surfactant. The O_2^- required is probably abundantly generated in growing fungus. Regarding the cationic surfactant, the highly effective surfactant CTAB and its homologues are man-made and not expected in nature. However, it has been recently discovered that some choline esters, such as hexadecanoylcholine and tetradecanoylcholine, are as effective as CTAB. Moreover, these surfactants shift the chemiluminescence spectrum of PM very close to the luminescence spectrum of a fresh fruiting body of *P. stipticus* (Fig. 9.9). The findings suggest

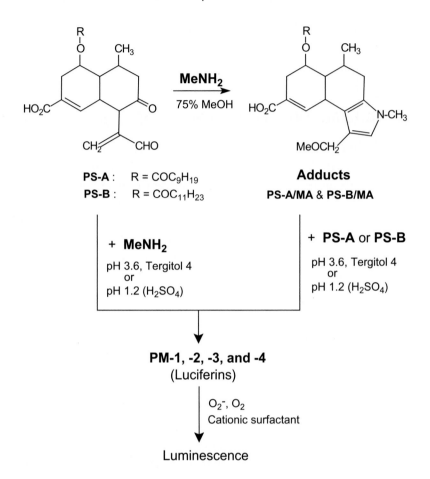

PS-A :   R = COC$_9$H$_{19}$
PS-B :   R = COC$_{11}$H$_{23}$

Adducts

PS-A/MA & PS-B/MA

+ MeNH$_2$

pH 3.6, Tergitol 4
or
pH 1.2 (H$_2$SO$_4$)

+ PS-A or PS-B

pH 3.6, Tergitol 4
or
pH 1.2 (H$_2$SO$_4$)

PM-1, -2, -3, and -4
(Luciferins)

O$_2^-$, O$_2$
Cationic surfactant

Luminescence

Fig. 9.8   A scheme showing the formation of PMs from PS-A and PS-B.

that fatty acid choline esters might actually be the cationic surfactants functioning in the bioluminescence of fungi.

The role of a cationic surfactant must be to provide a necessary hydrophobic and polarized environment for the molecule of luciferin for its luminescence reaction. In the case of a common luciferin-luciferase reaction, such an environment is provided by the enzyme luciferase. The chemical structures of PMs, as well as that of the natural luciferin, have not been determined yet (see the next section).

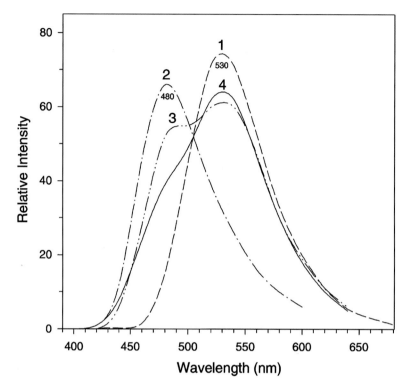

Fig. 9.9 Luminescence spectrum of a young fruiting body of *Panellus stipticus* (**1**); the chemiluminescence spectra of PM-1 in the presence of: CTAB (**2**); hexadecanoyl-choline iodide (**3**); and tetradecanoylcholine chloride (**4**). Chemiluminescence was elicited with Fe^{2+} and H_2O_2 in 50 mM Tris buffer, pH 8.0, containing 0.18 mM EDTA.

9.4.6 *Synthetic Studies of Panellus Luciferin*

The structures of PMs are trimeric, consisting of three molecules of PS and two molecules of methylamine that are condensed together (Table 9.5). In the structures of PMs, however, the bonds created by the condensation lack adjacent hydrogen atoms, making the connectivity assignment in ^{1}H-NMR studies virtually impossible (see the 9-membered rings in Fig. 9.11). To circumvent this problem, a model compound of panal, K-1, having ^{13}C-labeles at the C12 and C13 positions, has been synthesized at Kishi's laboratory (Harvard University) to make a model compound of PM (Stojanovic, 1995).

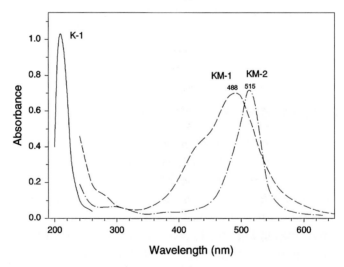

In the activation of PS-A and PS-B, treatment with methylamine resulted in a considerably higher chemiluminescence activity than with other amines. In the case of K-1, however, a significantly higher chemiluminescence activity was obtained with $(NH_4)_2SO_4$ or with hexylamine than with methylamine. In spite of this finding, methylamine was used in the activation of K-1 to prepare a model compound of PM.

Thus, ^{13}C-labeled K-1 (2 mg) was treated with 4 mM methylamine in a mixture of methanol (6 ml) and 0.2 M sodium acetate buffer pH 3.6 (1.5 ml) at 50°C for 2 hr, and the product was purified by HPLC on a column of PRP-1 (50–85% acetonitrile gradient, in the presence of 0.1 M acetic acid and 10 mM NaCl). Two weakly chemiluminescent compounds, KM-1 (5% yield) and KM-2 (1% yield), were obtained.

Fig. 9.10 Absorption spectra of K-1 (a model compound) and its chemiluminescent methylamine-activation products KM-1 and KM-2. All in methanol.

Both compounds were non-fluorescent. Compound KM-1 was similar to PMs in the absorption maximum (λ_{max} 488 nm) but not in the spectral shape, whereas KM-2 was similar to PMs in the spectral shape but not in the absorption maximum at 515 nm (Fig. 9.10). The chemical structures of KM-1 and KM-2 have been determined by high-resolution mass spectrometry and NMR spectrometry (Fig. 9.11; Stojanovic, 1995).

The compound KM-1 is composed of three molecules each of K-1 and methylamine in its structural formula, differing from PM that is made from three molecules of PS and two molecules of methylamine. However, KM-2 is composed of three molecules of K-1 and two molecules of methylamine, corresponding to the formula of PM.

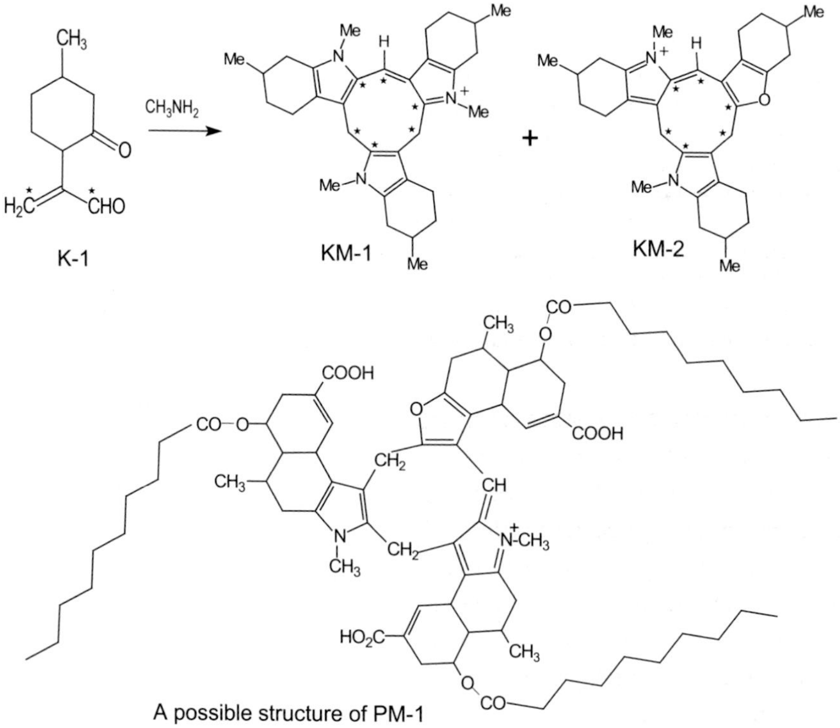

Fig. 9.11   Structures of the model compounds KM-1 and KM-2, and a possible structure of PM-1 drawn on the assumption that PMs contain the same chromophore structure as in KM-2.

Thus, on the assumption that KM-2 and PM contain an identical chromophore, PM-1 may have a structure as shown in Fig. 9.11. Although this is only a possible structure at present, it is intriguing to see a luciferin structure quite different from other known luciferins.

9.5 Studies on *Mycena citricolor*

9.5.1 Luciferin Obtained by Kuwabara and Wassink

Kuwabara and Wassink (1966) extracted, purified and crystallized the luciferin of *Omphalia flavida* (a synonym of *Mycena citricolor*) from 15 kg of the mycelium. The luciferin was extracted from the mycelium with boiling water at pH 5.5. The aqueous extract was acidified to pH 3, and extracted with ethyl acetate. The luciferin in the ethyl acetate extract was purified by partition chromatography on a column of Celite, using water as the stationary phase and chloroform-butanol as the moving phase. The luciferin obtained (approx. 12 mg) was a microcrystalline solid of brownish-orange color. It was stated rather ambiguously: "This active substance may produce light with either its fungal enzymes or with the particulate luciferase itself which has been characterized by Airth and his colleagues (1959, 1960, 1962). Moreover, like luminescence produced by luminol, this substance may also produce light non-enzymatically merely upon the addition of sodium hydroxide and hydrogen peroxide, with or without a suitable catalyst such as ferrous sulfate."

According to the Kuwabara-Wassink paper, the purified luciferin in aqueous neutral buffer solution showed an absorption maximum at 320 nm, and a fluorescence emission peak at 490 nm. The luminescence emission maximum measured with Airth's fungal luciferase system was 524 nm at pH 6.5, whereas the chemiluminescence emission maximum of the luciferin with H_2O_2 plus a droplet of strong NaOH plus ferrous sulfate was 542 nm. No information was reported on the chemical nature of the luciferin.

9.5.2 Studies on the Mycena citricolor Luminescence by the Author

Extraction and purification of luciferin and its precursors. The luciferin and its precursors are extracted from the mycelium of

Mycena citricolor with methanol or aqueous methanol at room temperature. The advantage of using methanol instead of the boiling water used by Kuwabara and Wassink is not clear, because methanol extracts luciferin more efficiently but it also extracts more impurities. The extracted material was dissolved in 16% methanol and added on a column of a reversed phase adsorbent (Jordi gel, a polydivinylbenzene polymer). Then, the column was eluted stepwise with aqueous solvents containing increasing concentrations of methanol or acetonitrile. Natural luciferin with an orange-yellow color was eluted first, followed by the fractions of colorless precursors. The luminescence activity of the luciferin obtained was $2 - 6 \times 10^{12}$ quanta per one gram of mycelium scraped from culture plates by the assay method described below; however, this luciferin was not investigated in my study. The luminescence activity of precursors after decylamine-activation (described below) was more than twice that of the natural luciferin.

The precursor fractions were purified on an anion-exchange column (Q HyperD, from Biosepra); the material in 16% methanol was adsorbed on the column, and then eluted by a linear increase of methanol, from 20% to 25%, in 30 mM Tris-HCl buffer, pH 7.5, containing 0.35 M NaCl. A further purification was carried out by HPLC on a column of PRP-1 (Hamilton), by elution with a linear increase of acetonitrile from 20% to 40% in 0.1% acetic acid or in 5 mM tributylamine-H_2SO_4 aqueous solution (pH 7.5). This resulted in the separation of the precursor into more than a dozen peaks (Fig. 9.12). When HPLC was repeated with a fraction of each peak, the resulting chromatogram showed one, two or three extra peaks in addition to the original peak, each coinciding exactly with a peak in the original chromatogram. Apparently, the precursors have a strong tendency to isomerize. Therefore, the number of the parent species of the precursors should be considerably less than that of the peaks actually observed. Thus, it is extremely important to prevent isomerization in the study of the *M. citricolor* precursors.

Assay methods. Luciferin is measured by mixing 5–20 µl of a sample, 1–2 mg of CTAB, 1–2 mg of $NaHCO_3$, 3 ml of 50 mM Tris-HCl (pH 8), 10 µl of 10% H_2O_2, and 15 µl of 50 mM $FeSO_4$, in this order. The addition of $FeSO_4$ triggers the luminescence reaction.

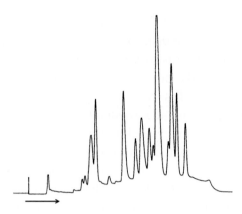

Fig. 9.12 An example of HPLC separation of the isoforms of *Mycena citricolor* luciferin precursors. The sample adsorbed on a PRP-1column was eluted by a linear increase of acetonitrile from 20% to 40% in a pH 7.5 buffer containing 5 mM tributylamine sulfate. Practically all the peaks have the precursor activity.

To measure the luciferin precursors, 0.1 ml of 0.4 M decylamine hydrochloride (made in 75% methanol) is added to the above mixture before the addition of $FeSO_4$. This mixture is left at room temperature for 20–30 min, then 15 µl of 50 mM $FeSO_4$ is added to trigger luminescence.

Properties of luciferin precursors. About one dozen of the luciferin precursors of *M. citricolor* isolated by HPLC had a strong tendency of isomerization, as mentioned above. Their molecular weights could not be established by mass spectrometry, which is probably due to isomerization, although they appear to be in a range of 300–600. The precursors showed an absorption peak at about 369 nm in methanol and aqueous acetonitrile (Fig. 9.13). According to an NMR study, all precursors probably contain the following common partial structure (personal communication from Dr. H. Nakamura, 1998).

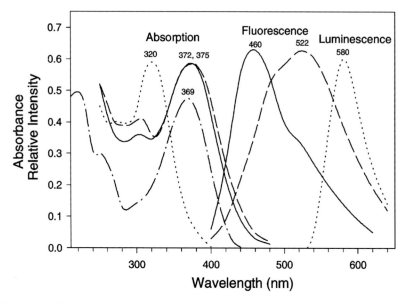

Fig. 9.13   Absorption spectrum of one of the luciferin precursors of *Mycena citricolor* in methanol (dash-dot line, λ_{max} 369 nm). The absorption and fluorescence emission spectra of the decylamine-activation product of the same precursor in neutral aqueous solution (solid lines; abs. λ_{max} 372 nm and fl. λ_{max} 460 nm), and in ethanol (broken lines; abs. λ_{max} 375 nm and fl. λ_{max} 522 nm). The chemiluminescence spectrum of the same activation product (dotted line, λ_{max} 580 nm). The dotted line (λ_{max} 320 nm) is the absorption spectrum of *M. citricolor* natural luciferin reported by Kuwabara and Wassink (1966).

Treatment of the precursors with decylamine resulted in a high level of chemiluminescence activity. Taking the activity obtained with decylamine as 100%, the activities obtained with other amines were: methylamine, 5%; hexylamine, 23%; octylamine, 39%; and dodecylamine, 55%. For comparison, the *Panellus* precursors, PS-A and PS-B, are best activated with methylamine, and a synthetic model compound, K-1, is best activated with hexylamine.

Activation of the *M. citricolor* luciferin precursor with decylamine. The following mixture was incubated overnight at room temperature (23~24°C): luciferin precursors before the separation of isomers by HPLC (50 μl, activity 2×10^{12} photons by the assay method described above), 30 mM Bis-tris buffer (pH 6.4, 4.5 ml); acetonitrile

(5.5 ml), 0.4 M decylamine hydrochloride in 75% methanol (0.4 ml), and 10% H_2O_2 (0.2 ml).

Purification of the decylamine-activation product. The orange-colored reaction mixture was evaporated under reduced pressure to remove most of the acetonitrile, and the residual solution was added on a column of anion-exchanger Q HyperD (Biosepra). The column was washed with a small amount of 20 mM HEPES, pH 7.5, containing 25% methanol, and then eluted with the same buffer containing 33% methanol. The yellowish fractions eluted earlier had higher specific activities than the orange-colored fractions eluted later. The active fractions were further purified by HPLC on an Asahi ODP-50 column (vinylalcohol copolymer with C18 functional group) using 14% acetonitrile containing 50 mM NaCl and 10 mM HEPES, pH 7.5. Two yellow-colored substances (luciferins) were obtained.

Properties of the activation product. The two decylamine-activation products (luciferins) showed similar absorption characteristics (λ_{max} 372 nm in water, and 375 nm in ethanol), which clearly differ from the absorption peak of the natural luciferin (320 nm) reported by Kuwabara and Wassink (1966). The fluorescence emission of the activation products varied significantly by solvents, showing a peak at 460 nm in neutral aqueous solution and a broad peak at 485–522 nm in ethanol. They emitted chemiluminescence (λ_{max} 580 nm) in the presence of CTAB, H_2O_2 and Fe^{2+} (Fig. 9.13), in resemblance to the $(NH_4)_2SO_4$-activation product of panal (λ_{max} 570 nm).

9.6 Summary on the Chemistry of Fungal Luminescence

Components required for fungal bioluminescence. According to our studies, fungal luciferins emit light in the presence of superoxide anion (O_2^-), molecular oxygen (O_2), and a cationic surfactant. Superoxide anion should be abundant in growing fungus. The cationic surfactant CTAB is highly effective and convenient for laboratory use but does not occur in nature. However, the fatty acid esters of choline, such as hexadecanoylcholine and tetradecanoylcholine, are as effective as CTAB and may occur in fungus. The characteristics of the luminescence are modulated by cationic surfactants. It appears that a cationic surfactant functions in the role of luciferase in the luminescence of fungi.

The precursors of luciferins. We have isolated the precursors of luciferins from two species of luminous fungus, *Panellus stipticus* and *Mycena citricolor*. The precursors possess highly reactive carbonyl functions, and they can be readily converted (activated) into chemiluminescent luciferins by treating them with a salt of primary amine. The amine that gives the highest chemiluminescence activity is methylamine in the case of the *P. stipticus* precursors, and decylamine in the case of the *M. citricolor* precursors. Both the activation of precursors and the chemiluminescence of luciferins take place under very mild conditions compatible with a cellular environment, and both reactions are strongly influenced by the environment created by surfactants.

The light emitter. The luminescence spectra of most species of luminous fungi are practically identical, with a peak at about 530 nm (Van der Burg, 1943), suggesting that the chromophores of the light emitters involved are the same or nearly identical. The light-emitter might be the excited state of an oxyluciferin, or a fluorescent compound that functions as the light-emitter after being excited by energy transfer from the excited state oxyluciferin. Luminous fungi contain a variety of fluorescent substances, and some of the substances show a fluorescence emission spectrum that matches the luminescence of live luminous fungi. Such a coincidence may support but does not prove that the fluorescent substance is the light-emitter in a luminous fungus. The author believes that a flavin chromophore has not been completely ruled out as a candidate for the light emitter, despite the rejection of free riboflavin as the light emitter of luminous fungi (O'Kane *et al.*, 1994a), because riboflavin *in vivo* might not be in its free form in the presence of various proteins, lipids, and surfactants.

Future research. For the elucidation of the mechanism of fungal luminescence, it would be necessary to determine the structures of the natural luciferins occurring in luminous fungi. Considering recent advances in technology, it appears feasible to achieve this goal with the luciferins existing in the mycelium of *Mycena citricolor* and the young fruiting bodies of *Panellus stipticus*; both contain a relatively large amount of natural luciferins (see Table 9.2).

10

OTHER LUMINOUS ORGANISMS

10.1   Ophiuroidea: Brittle Stars

The phylum Echinodermata contains bioluminescent species in four classes: Crinoidea (sea lilies and feather stars), Holothuroidea (sea cucumbers), Asteroidea (starfish), and Ophiuroidea (brittle stars); no luminous organism is found in the class Echinoidea (the sea urchins) (Herring, 1978a). Bioluminescent species are particularly abundant in Ophiuroidea. Harvey (1952) wrote that the luminescence of the ophiuroid *Amphiura squamata* requires molecular oxygen and is negative in the luciferin-luciferase reaction. He also noted that both *Amphiura squamata* and *Ophiopsila aranea* are fluorescent in yellowish green under ultraviolet light, which is enhanced after they have been stimulated to luminesce, suggesting that an oxidation product formed in the luminescence reaction is fluorescent. Herring (1974) studied more than a dozen kinds of deep-sea echinoderms and reported that the extracts of four species, the crinoid *Thaumatocrinus jungerseni* and the holothurians *Kolga kyalina*, *Peniagone théeli*, and *Laetmogone violacea*, gave flashes of light when Fe^{2+} and H_2O_2 were added.

Despite the relatively common occurrence of luminous echino-derms, only two species of ophiuroid, *Ophiopsila californica* and *Amphiura filiformis*, have been biochemically investigated. It is sur-prising that the former species luminesces with a photoprotein system, whereas the latter emits light with a luciferin-luciferase system.

10.1.1   *The Brittle Star Ophiopsila californica*

The brittle star *Ophiopsila californica* (Fig. 10.1.1) is abundant around Catalina Island, off the coast of Los Angeles. An average-sized specimen weighs about 3–4 g, and has five arms of about 10 cm long, which emit greenish light upon stimulation (λ_{max} 510 nm).

Extraction and purification of photoprotein (Shimomura, 1986b). The live animals are chilled to 5°C, and the five arms of each specimen are cut off and dropped into cold 3.5% magnesium acetate solution. After 10 min, the arms are drained from the solution, and stored at

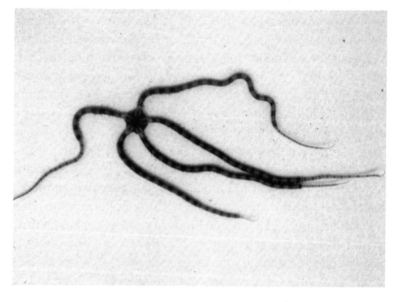

Fig. 10.1.1   The brittle star *Ophiopsila californica*.

−75°C until use. The photoprotein is extracted and purified by a five-step procedure as summarized below, at 2–4°C except as noted:

Step 1. Frozen arms are ground with cold water, and centrifuged. The pellets are further ground with water and centrifuged. The supernatants are combined, saturated with ammonium sulfate, and the precipitate formed is collected by centrifugation.

Step 2. The pellets are extracted with 10 mM sodium phosphate buffer, pH 6.7, containing 2 mM EDTA and 0.2 M NaCl, and chromatographed on a gel-filtration column (Ultragel AcA 54; LKB) using the same pH 6.7 buffer. The photoprotein is eluted slightly before a brownish substance.

Step 3. The photoprotein fractions are combined and the NaCl concentration of the solution is reduced to 50 mM by gel filtration through a column of Sephadex G-25. Then, the solution is poured onto a column of DEAE cellulose. The photoprotein adsorbed on the column is eluted by a linear increase of NaCl concentration from 50 mM to 0.4 M in the same pH 6.7 buffer. Active fractions are combined and concentrated by ammonium sulfate precipitation.

Step 4. A repetition of gel filtration on a smaller column of Ultragel AcA 54, followed by the concentration of the eluted photoprotein by ammonium sulfate precipitation.

Step 5. HPLC on a gel filtration column (Zorbax GF250, Du Pont Instruments) at 20°C, using 10 mM sodium phosphate, pH 6.5, containing 2 mM EDTA and 0.1 M Na_2SO_4. A major impurity is eluted immediately before the photoprotein peak. The purified photoprotein is stored at −75°C.

The purification method given above may be improved by utilizing various new chromatographic media, and the yield of the active photoprotein may be increased by utilizing catalase to remove traces of H_2O_2 in solutions.

Assay of *Ophiopsila* photoprotein. The luminescence reaction is initiated by the injection of 0.1 ml of 1% H_2O_2 into 3 ml of Tris-HCl buffer, pH 7.5, containing 0.5 M NaCl and 10–100 μl of a sample solution, and the total light emission is measured. When assaying the photoprotein in crude extracts, 10 µl of 0.1% horseradish peroxidase is included in the assay mixture to prevent erratic results caused by various inhibitory impurities, although the mechanism involved is unclear.

Properties of *Ophiopsila* photoprotein (Shimomura, 1986b). The molecular weight of the *Ophiopsila* photoprotein was estimated at about 45,000 by gel-filtration. The photoprotein is unstable even after purification (half-life: 12–18 hr at 0°C). The specific luminescence activity was 3×10^{12} quanta/mg, and the quantum yield is calculated at 0.02%. The unusually low values of specific activity and quantum yield must be due to the inactivation of the photoprotein during the extraction and purification.

The absorption spectrum of the photoprotein showed a small peak (λ_{max} 423 nm, with a shoulder at about 450 nm) in addition to the protein peak at 280 nm (Fig. 10.1.2). The peak at 423 nm decreased slightly upon the H_2O_2-triggered luminescence reaction. The photoprotein is fluorescent in greenish-blue (emission λ_{max} 482 nm), which coincides exactly with the luminescence spectrum of the photoprotein

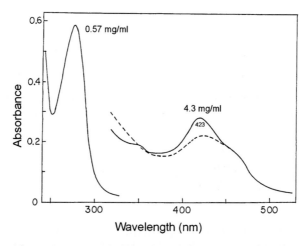

Fig. 10.1.2 Absorption spectra of the photoprotein of the brittle star *Ophiopsila californica* (solid line) and the spent product of the photoprotein after H_2O_2-triggered luminescence (dashed line). From Shimomura, 1986b, with permission from the American Society for Photobiology.

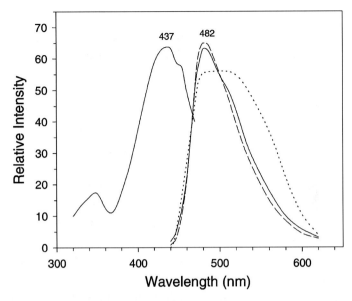

Fig. 10.1.3 Fluorescence excitation and emission spectra (solid lines) and H_2O_2-triggered luminescence spectrum (dashed line) of *Ophiopsila* photoprotein (Shimomura, 1986b, revised). The dotted line indicates the *in vivo* bioluminescence spectrum of *Ophiopsila californica* plotted from the data reported by Brehm and Morin (1977).

in the presence of H_2O_2 (Fig. 10.1.3), suggesting that the fluorescent chromophore is the light-emitter. However, the fluorescence of the photoprotein did not change noticeably after the H_2O_2-triggered luminescence reaction. Thus, it is not clear whether the chromophore of greenish-blue fluorescence has been formed by the light-emitting reaction or whether it originally existed.

Live specimens of *O. californica* are green fluorescent and their bioluminescence spectrum shows a broad peak at about 510 nm (Fig. 10.1.3). The green fluorescence is visually identical with the luminescence (Brehm and Morin, 1977), indicating the involvement of a green fluorescent substance in the *in vivo* luminescence. During the purification of the photoprotein, most of the green fluorescent substance is removed, although a trace amount is recognized in the fluorescence emission spectrum of the purified photoprotein (Fig. 10.1.3).

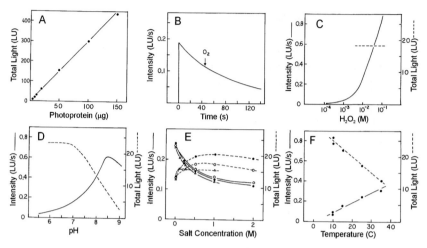

Fig. 10.1.4 Characteristics of the H_2O_2-triggered luminescence of *Ophiopsila* photoprotein. (A) Relationship between the amount of photoprotein and total light emission. (B) An experiment showing the lack of oxygen requirement for luminescence; luminescence reaction was started in the complete absence of O_2 (in argon gas) and air was introduced at 40 s. (C) Effect of the concentration of H_2O_2 on luminescence. (D) Effect of pH on luminescence. (E) Effect of the concentration of salts on luminescence: NaCl (o), KCl (•), $(NH_4)_2SO_4$ (Δ). (F) Effect of temperature on luminescence. The luminescence reactions were started generally by injecting 0.1 ml of 1% H_2O_2 into 3 ml of 10 mM Tris-HCl buffer, pH 7.5, containing 0.5 M KCl and 6 μg of photoprotein, with necessary modifications depending on the purpose of experiment. From Shimomura, 1986b, with permission from the American Society for Photobiology.

Some characteristics of the H_2O_2-triggered luminescence of the photoprotein are illustrated in Fig. 10.1.4. Panel A shows the proportionality between the amount of photoprotein and the total light emission. The data in Panel B show that the luminescence reaction does not require the presence of air (O_2), although the data do not necessarily indicate the non-involvement of O_2 molecule because of the presence of a relatively high concentration of H_2O_2 that may spontaneously generate a trace amount of O_2. Panels C–F show the influence of the H_2O_2 concentration, pH, salt concentration and temperature on the light intensity and total light. It is intriguing, and somewhat puzzling, to find that the luminescence reaction requires relatively high concentrations of H_2O_2.

10.1.2   The Brittle Star Amphiura filiformis

This luminous brittle star has been briefly studied recently (Mallefet and Shimomura, 2004, unpublished). The animal contained a high level of coelenterazine luciferase activity (4×10^{12} photons$\cdot$s^{-1}g^{-1}), which is comparable to those in the luminous anthozoans such as the sea pansy *Renilla* and sea pen *Ptilosarcus* (Shimomura and Johnson, 1979b). There is no evidence for the presence of a photoprotein in this brittle star. Thus, the luminescence system of *Amphiura filiformis* is considered to be a coelenterazine-luciferase system, differing from that of *Ophiopsila californica*. The luciferase has a molecular weight of 23,000 on the basis of gel filtration on Superdex 200 Prep, and catalyzes the luminescence reaction of coelenterazine in the presence of oxygen; the light emission (λ_{max} 475 nm) is optimum at pH 7.2.

10.2   Millipede Luminodesmus sequoiae (Diplopoda)

The phylum Arthropoda includes the classes Diplopoda (millipedes), Chilopoda (centipedes), Crustacea (see Chapter 3), and Insecta (see Chapter 1). All luminous arthropods other than crustaceans are terrestrial, and not very many luminous millipedes and centipedes are known. The luminescence of millipedes is usually intracellular, whereas luminous centipedes discharge luminous secretion. Substantial chemical studies have been made only with the millipede *Luminodesmus sequoiae* and the centipede *Orphaneus brevilabiatus*, of which the latter is discussed in the Section 10.3.

The luminous millipede *Luminodesmus sequoiae* (Fig. 10.2.1) was first described by Loomis and Davenport (1951). The animal measures about 25 mm long and 6 mm wide, and emits greenish-blue light continuously from the whole body, including the antenna and legs. The fluorescence of *L. sequoiae* appears visually identical with the bioluminescence. This species is found in the Sequoia National Forest, California, at an altitude of approximately 1,500 m, usually under redwood trees. They crawl out from their hiding places at night (9–10 pm) during the period of mid-April to early May after the ground snow has melted away. The animals emit strong fluorescence when illuminated

Fig. 10.2.1 The Sequoia millipede *Luminodesmus sequoiae* illuminated by a flashlight (*left*) and by ultraviolet light (*right*).

with a portable UV light; thus they can be readily located from a distance in the dark. Because of the ease of collecting them, there is a high risk of annihilating their population. Therefore, careful attention must be paid when collecting this millipede.

Davenport *et al.* (1952) were unsuccessful in their attempts to restore the luminescence of the filtered aqueous extract of *Luminodesmus*. Hastings and Davenport (1957) saw a weak luminescence in their filtered aqueous extracts made from the acetone powder of the millipedes. They found that the luminescence is dependent on pH, with an optimum at about pH 8.9, and that the light intensity could be increased by 10–30% by adding ATP. Hastings and Davenport also measured the luminescence spectrum of live animals, finding an emission peak at 495 nm.

Later, an active photoprotein was extracted and purified from *L. sequoiae* and its properties were investigated (Shimomura, 1981, 1984). The results are summarized below.

Extraction and purification of the *Luminodesmus* photoprotein. Specimens of *L. sequoiae* were collected in the vicinity of Camp Nelson, California, in May 1980 and April 1982 and 1983. The specimens were anesthetized with chloroform vapor, and their guts were pulled out from the head or tail and discarded. The body shells (terga) were frozen with dry ice, and kept at $-75°C$ until use.

The frozen shells were ground in a cold mortar with 50 mM sodium acetate buffer, pH 5.8, containing 10 mM EGTA and 0.2 M NaCl, then the mixture was centrifuged. The pellets were re-extracted with the same buffer, and centrifuged. All supernatants were combined, and the photoprotein was precipitated with ammonium sulfate. The photoprotein in the precipitate was purified by four steps of column chromatography at near $0°C$. Due to the instability of the photoprotein, efforts were made to reduce the time required for purification.

Step 1. Gel filtration on Sephadex G-100, in 10 mM sodium phosphate buffer, pH 6.5, containing 5 mM EGTA and 0.2 M NaCl, which was the basic buffer used throughout the purification process.

Step 2. Hydrophobic interaction chromatography on a column of Phenyl-Sepharose CL-4B. The sample was adsorbed on the column in the basic buffer containing 0.5 M $(NH_4)_2SO_4$. The photoprotein adsorbed was first washed with the same buffer, then eluted with the basic buffer.

Step 3. Gel filtration on Ultrogel AcA 44, using the basic buffer.

Step 4. Anion-exchange chromatography on a column of DEAE-Sephadex A-50. The column had been equilibrated with the basic buffer containing 0.12 M NaCl. The NaCl in the photoprotein solution was removed by gel filtration, and then the solution was added onto the column. The photoprotein adsorbed on the column was eluted with the equilibration buffer.

Assay of *Luminodesmus* photoprotein. A sample solution (10~100 µl) is mixed with 2 ml of 10 mM Tris buffer, pH 8.5, containing 0.1 mM ATP and 1 mM $MgCl_2$, and the peak intensity of the luminescence is measured. After mixing, the light intensity reaches its maximum in a few seconds, then gradually diminishes in a period of about 15 min. By this method, 1 mg of a purified photo-protein showed a peak intensity of approximately 7×10^{11} quanta/s, at 23°C.

Properties of *Luminodesmus* photoprotein. The molecular weight of the photoprotein was estimated at about 60,000 by gel filtration. The protein emits light (λ_{max} 496 nm) in the presence of ATP, Mg^{2+} or Ca^{2+}, and molecular oxygen. The effects of the concentrations of ATP, Mg^{2+} and Ca^{2+} are illustrated in Fig. 10.2.2, which indicates the extremely high sensitivities of this photoprotein to ATP and Mg^{2+}; 5 nm ATP or 1 μM Mg^{2+} can be detected with this luminescence sys-tem. The luminescence intensity is affected by the pH and temperature of the medium, and is optimum at pH 8.3 and at 30°C (Figs. 10.2.3 and 10.2.4).

The photoprotein is unstable; the half-life of the activity measured in 10 mM phosphate buffer, pH 6.5, containing 5 mM EGTA and

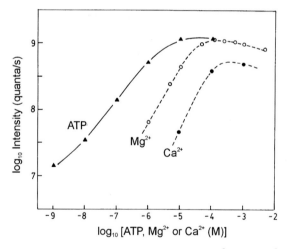

Fig. 10.2.2 Influence of the concentrations of ATP, Mg^{2+} and Ca^{2+} on the maxi-mum luminescence intensity of the photoprotein of the millipede *Luminodesmus*. The luminescence reaction was started by mixing a solution of the photoprotein (A_{280} 0.3, 10 μl) with 2 ml of 10 mM Tris-HCl buffer, pH 8.3, containing either 1 mM $MgCl_2$ plus various concentrations of ATP or 0.05 mM ATP plus various concentrations Mg^{2+} or Ca^{2+}. From Shimomura, 1981, with permission from the Federation of the European Biochemical Societies.

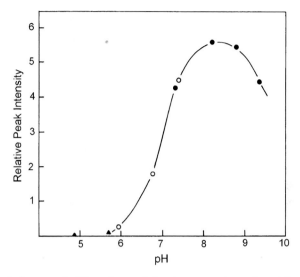

Fig. 10.2.3 Influence of pH on the peak luminescence intensity of *Luminodesmus* photoprotein, when 0.5 ml of a solution of the photoprotein was added to 2 ml of 10 mM sodium acetate buffers (▲), 10 mM sodium phosphate buffers (o) or 10 mM Tris-HCl buffers (•), each containing 1 mM MgCl₂ and 0.05 mM ATP. From Shimomura, 1981, with permission from the Federation of the European Biochemical Societies.

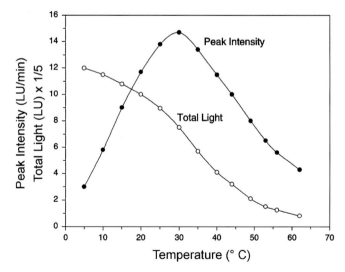

Fig. 10.2.4 Influence of temperature on the peak light intensity and total light of *Luminodesmus* photoprotein, when 25 μl of a solution of the photoprotein was injected into 2 ml of 10 mM Tris-HCl buffer, pH 8.3, containing 0.1 mM ATP and 1 mM MgCl₂, that had been equilibrated at various temperatures.

0.2 M NaCl was 15 hr at 0°C and 45 min at 23°C. Based on this inactivation rate, it is estimated that the purified protein contained only 5–10% of the active form. No method has been found to stabilize the photoprotein.

The photoprotein is non-fluorescent. The absorption spectrum of purified photoprotein shows a very small peak at 410 nm, in addition to the protein peak at 280 nm (Fig. 10.2.5). The peak height at 410 nm appears to be proportional to the luminescence activity of the protein. The protein also shows extremely weak absorption peaks at about 497, 550 and 587 nm (not shown). These absorption peaks, except the 280 nm peak, might be due to the presence of a chromophore that is functional in the light emission.

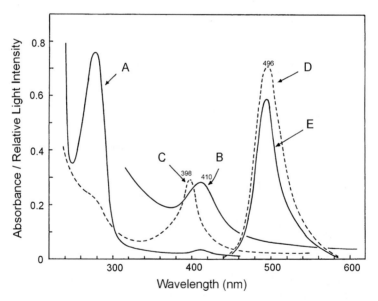

Fig. 10.2.5 (A) Absorption spectrum of a purified *Luminodesmus* photoprotein having a luminescence activity of 2.5×10^{11} quanta $\cdot$ $s^{-1}ml^{-1}$. (B) Curve A plotted with 8 times expanded absorbance. (C) Absorption spectrum of the chromophore isolated from the photoprotein, in methanol in molar concentration roughly equivalent to A, plotted with 8 times expanded absorbance. (D) Luminescence spectrum measured with a live specimen anesthetized with chloroform. (E) Luminescence spectrum of the photoprotein in 30 mM Tris-HCl buffer, pH 7.5, containing 0.3 mM $MgCl_2$ and 50 µM ATP. From Shimomura, 1984, with permission from Elsevier.

When a photoprotein solution (1.3 ml) was shaken with ethanol (0.7 ml) containing one drop of concentrated HCl and then the mixture was extracted twice with 2 ml each of ethyl acetate, about 75% of the chromophore was extracted into the ethyl acetate extract. The chromophore isolated showed an absorption peak at 398 nm in neutral methanol (Fig. 10.2.5). The isolated chromophore was practically non-fluorescent, like the native photoprotein. However, the acidification of a methanolic solution with HCl resulted in a sharpening and two-fold increase of the 398 nm absorption peak, accompanied by the appearance of fluorescence. In aqueous 0.1 M HCl, two fluorescence emission peaks (595 nm and 650 nm) were found, together with a corresponding excitation peak (400 nm). Treatment of the 398 nm absorbing chromophore with 0.1 M NaOH resulted in a rapid loss of the 398 nm absorption peak. Dithionite did not affect the 398 peak, suggesting that the chromophore does not contain Fe^{3+}.

Porphyrin as a possible chromophore of *Luminodesmus* photoprotein. The oxidation of the isolated chromophore with CrO_3 in 2 M H_2SO_4 yielded 3-methyl-4-vinylmaleimide and hematinic acid, but not a fused ring diketoaldehyde derived from euphausiid luciferin (the compound F) and dinoflagellate luciferin (Shimomura, 1980; Dunlap *et al.*, 1981; Nakamura *et al.*, 1988, 1989). These results, upon consideration of the spectroscopic properties noted above, seem to suggest that the chromophore of *Luminodesmus* photoprotein might be a porphyrin, bile pigment or chlorophyll. Such a possibility is consistent with the narrow bioluminescence spectrum of this photoprotein (Fig. 10.2.5, curve E), with a full-width at half maximum (FWHM) of only 38 nm, which is close to the FWHM values of the bioluminescence spectra of the compound F of euphausiid (FWHM 40 nm) and the dinoflagellate luciferin (FWHM 38 nm). The FWHM values of the luminescence spectra of other bioluminescence reactions are generally considerably wider (65 nm to 100 nm), unless a GFP is involved as the light emitter.

Kuse *et al.* (2001) isolated a fluorescent compound from *L. sequoiae* (fluorescence emission λ_{max} 505 nm upon excitation at

390 nm; HWHM 75 nm) and determined its chemical structure to be 7,8-dehydropterin-6-carboxylic acid. This compound may be involved in the bioluminescence of *Luminodesmus* as an *in vivo* light emitter to some extent, although its direct involvement in a light-emitting chemical reaction is unlikely.

10.3   Centipede *Orphaneus brevilabiatus* (Chilopoda)

According to Haneda (1955), the centipede *Orphaneus brevilabiatus* is found widely in Micronesia, the East Indies, the Malay Peninsula, Indochina, Formosa and Okinawa. The centipedes (about 6 cm long) are often found in native houses, on the walls and inside furniture, and they secrete a greenish luminous slime when irritated. When exposed to a chloroform vapor, they discharge strikingly luminous slime from every segment of the body.

Anderson (1980) reported that the luminous slime of *O. brevilabiatus* shows a positive luciferin-luciferase reaction and that the luminescence reaction requires oxygen. The same author measured the luminescence spectrum of the slime, and found two emission peaks, at 480 nm and 510 nm (Fig. 10.3.1), which might indicate the involvement of energy transfer from the 480 nm light-emitter to the 510 nm light-emitter. Anderson noted that the optimum pH for light emission

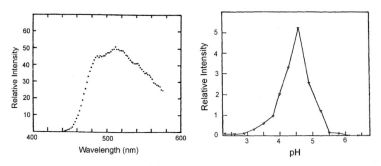

Fig. 10.3.1 Bioluminescence spectrum of the centipede *Orphaneus brevilabiatus* (*left* panel), and the influence of pH on the luminescence of the exudate of the same centipede in 0.1 M potassium citrate/phosphate buffers (*right* panel). From Anderson, 1980, with permission from the American Society for Photobiology.

is at 4.6 (Fig. 10.3.1), an unusually acidic condition in the bioluminescence systems, although the luminescence of certain luminous fungi
also takes place under acidic conditions.

10.4   Hemicordata

The phylum Hemicordata contains bioluminescent enteropneusts
(acorn worms) *Ptychodera*, *Glossobalanus* and *Balanoglossus*. They
are relatively large worm-like organisms (length 10–100 cm) and
inhabit shallow water, burrowing in mud and sand. They produce
a bluish luminous slime from every portion of the body upon stimulation (Herring, 1978a). According to Harvey (1952), a *Ptychodera*
species from Bermuda and a *Balanoglossus* species from Naples were
both negative in the classic luciferin-luciferase reaction. He also noted
that, when the whole animal of *Balanoglossus minutus* is shaken in
seawater through which pure hydrogen is passing, no luminescence
can be observed, but upon admission of air to the container, both the
animal and seawater luminesce.

Dure and Cormier (1961) demonstrated a luciferin-luciferase reaction for the first time in the extracts of the acorn worm *Balanoglossus
biminiensis*, and also discovered that the luminescence reaction is stimulated by H_2O_2, of which the details are described below. Recently,
Kanakubo and Isobe (2005) reported the chemical structure of a probable luciferin of another acorn worm *Ptychodera flava*.

10.4.1   *The Acorn Worm Balanoglossus biminiensis*

According to Dure and Cormier (1961, 1963) and Cormier and
Dure (1963), they made the preparations of luciferase and luciferin
from *Balanoglossus biminiensis*, collected on Sapelo Island, Georgia,
and investigated the luciferin-luciferase reaction, as summarized
below.

Preparation of luciferase. Organisms were freeze-dried, powdered, and washed with ethyl acetate to destroy the majority of catalase activity. The washed residue was extracted with 50 mM potassium
phosphate buffer, pH 6. The extract was fractionated by ammonium

sulfate precipitation, taking a fraction of 35–56% saturation. The precipitate containing both luciferase and luciferin was dissolved in the same pH 6 buffer, and luciferin was removed by gel filtration on a column of Sephadex G-25 using 50 mM potassium phosphate, pH 8. The dark brownish-green colored protein fraction obtained was used as the luciferase preparation.

Preparation of *Balanoglossus* luciferin. The residue of the first pH 6 extraction above was re-extracted with 50 mM potassium phosphate buffer, pH 8. After centrifugation, the supernatant was used as the standard luciferin preparation. Luciferin was highly labile and easily inactivated at an extreme pH, by heat, and also by freezing and thawing. The instability resembled that of certain proteins.

Cormier and Dure (1963) found another type of luciferin and called it "protein-free luciferin." Protein-free luciferin was found in the vapor condensate of freeze-drying whole animals, and also in the 35–56% ammonium sulfate fraction of the crude extract noted above. The protein-free luciferin behaved like an aromatic or heterocyclic compound and it was strongly adsorbed onto Sephadex and other chromatography media, requiring a considerable amount of solvent to elute it. The luminescence reaction of protein-free luciferin in the presence of luciferase required a 500-times higher concentration of H_2O_2 compared with the standard luciferin preparation. Both types of the luciferin preparation had a strong odor of iodoform.

Assay of luminescence activity. Luciferin solution (1 ml) is mixed with 1.2 ml of 0.5 M Tris buffer (pH 8.2) and 0.3 ml of luciferase solution. The luminescence reaction is initiated by the injection of 0.5 ml of 0.176 mM H_2O_2 to the luciferin-luciferase mixture. The light emission is characterized by a flash of light, followed by a rapid decay to a much lower steady-state level (Fig. 10.4.1). The maximum light intensity of the flash is taken as the measure of the luminescence activity.

Characteristics of the *Balanoglossus* bioluminescence. The luciferin of *Balanoglossus* emits light in the presence of *Balanoglossus* luciferase and H_2O_2. In the luminescence reaction, the apparent K_m for

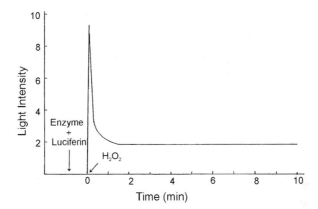

Fig. 10.4.1 Light emission profile of the luminescence reaction of the acorn worm *Balanoglossus biminiensi*, when H_2O_2 is injected into a mixture of the luciferin and luciferase. From Dure and Cormier, 1961, with permission from the American Society for Biochemistry and Molecular Biology.

H_2O_2 was 2.9×10^{-6} M. A requirement for O_2 could not be detected, suggesting that O_2 is not involved in the luminescence reaction. The luciferase was found to be a peroxidase that catalyzes peroxidation of the luciferin, and it can be substituted with horseradish peroxidase.

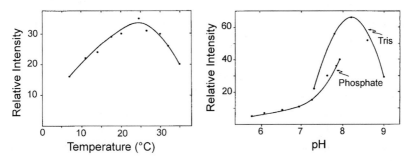

Fig. 10.4.2 The effects of temperature (*left* panel) and pH (*right* panel) on the peak intensities of the *Balanoglossus* luminescence reaction. In the measurements of the temperature effect, 0.5 ml of 0.176 mM H_2O_2 was injected into a mixture of 1.2 ml of 0.5 M Tris buffer (pH 8.2), 0.3 ml of luciferase, and 1 ml of luciferin, at various temperatures. For the pH effect, the Tris buffer (pH 8.2) was replaced with the Tris buffers and phosphate buffers that have various pH values, and the measurements were made at room temperature. From Dure and Cormier, 1963, with permission from the American Society for Biochemistry and Molecular Biology.

The luminescence reaction is optimum pH 8.2 and at a temperature of 25°C (Fig. 10.4.2).

10.4.2   The Luciferin of Ptychodera flava

According to Kanakubo and Isobe (2005) and Kanakubo *et al.* (2005), the acorn worm *Ptychodera flava* from Okinawa emits green light upon stimulation with H_2O_2, but does not show a luciferin-luciferase reaction. The bioluminescence spectrum of a live specimen shows a peak at 528 nm (Fig. 10.4.3). Putative luminescence substance was extracted from the freeze-dried organisms with ethyl acetate, and purified by repeated chromatography on C18 silica columns, yielding 2,3,5,6-tetrabromohydroquinone (TBHQ, the structure shown at the end of this section). The yield was surprisingly high: 647 mg from 1.4 g of the dried material of the initial ethyl acetate extract. The residue of the ethyl acetate extraction contained green fluorescent riboflavin that could be extracted with methanol. A mixture of TBHQ and riboflavin in an alkaline 70% dioxane solution (pH 12) emitted a chemiluminescence when H_2O_2 was added (Fig. 10.4.3). Based on the

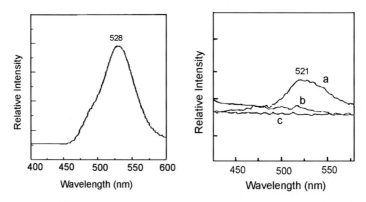

Fig. 10.4.3   *Left* panel: Bioluminescence spectrum of the acorn worm *Ptychodera flava* stimulated with H_2O_2. *Right* panel: (a) The spectrum of the chemiluminescence emitted when 70% dioxane containing 1.7% H_2O_2 (5 ml) was added to a mixture of a solution of 2,3,5,6-tetrabromohydroquinone (TBHQ) in ethyl acetate (2.5 ml), 50 mM glycine buffer (pH 12.0; 2.5 ml), and riboflavin; (b) when riboflavin was omitted; and (c) when TBHQ was omitted. Dioxane was included to solubilize the ethyl acetate solution containing TBHQ. From Kanakubo *et al.*, 2005, with permission from Elsevier.

closeness between the observed emission maximum (521 nm) and the bioluminescence maximum (528 nm), TBHQ was postulated to be the light-producing substance of *P. flava* in the presence of riboflavin. The mechanism of the *in vivo* luminescence involving TBHQ remains unknown.

$$\begin{array}{c} \text{Br} \quad \text{Br} \\ \text{HO} - \bigcirc - \text{OH} \\ \text{Br} \quad \text{Br} \end{array}$$

10.5   Tunicates (Phylum Chordata)

The luminous tunicate *Pyrosoma* has been well known for its strikingly beautiful, sustained bright luminescence, ever since it was first reported by Péron (1804). Harvey (1952) remarked, "The single genus, *Pyrosoma*, are undoubtedly the most extraordinary colonies of luminous animals in the living world." He noted that the animals are abundant as the plankton in warm seas, ranging from the surface to a depth of 200 m. A colony, usually cylinder-shaped (Fig. 10.5.1), may attain a length of 4 m, although it is usually 3 to 10 cm. Single animals are usually 4 to 5 mm, sometimes 20 mm, in length. If a *Pyrosoma*

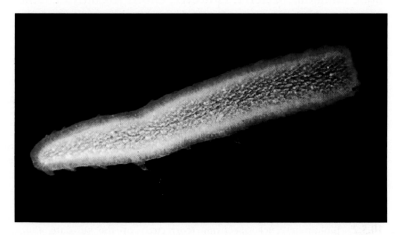

Fig. 10.5.1   The tunicate *Pyrosoma atlanticum*. Photo by Herb Gruenhagen.

colony is pressed through a cloth, the resultant filtrate remains luminous for a short while, but soon becomes dark. However, on mixing with fresh water, the dark extract again gives a brilliant light. The luciferin-luciferase reaction with *Pyrosoma* was negative.

Luminescence of *Pyrosoma*. All species of the genus *Pyrosoma* (about 10 species) are bioluminescent. *Pyrosoma* is one of the few organisms reported to luminesce in response to light (Bowlby *et al.*, 1990). The luminescence emission spectrum of *Pyrosoma atlantica* is bimodal according to Kampa and Boden (1957), with the primary peak near 482 nm, and the secondary near 525 nm. Swift *et al.* (1977) reported the emission maxima of two *Pyrosoma* species at 485 and 493 nm, respectively, and Bowlby *et al.* (1990) found an emission peak at 475 nm with *P. atlantica*. A corrected bioluminescence spectrum of *P. atlantica* (λ_{max} 485 nm) reported by Herring (1983) is shown in Fig. 10.5.2.

The source of light emission. Each individual animal in a colony of *Pyrosoma* has two groups of luminous cells at the entrance to the branchial sac (Herring, 1978a). The luminous cells contain tubular inclusions whose identity has been a matter of dispute for nearly a century. Buchner (1914) originally suggested that the inclusions were symbiotic luminous bacteria. However, efforts to cultivate luminous bacteria from the luminous cells of *Pyrosoma* consistently failed. In

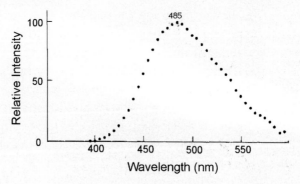

Fig. 10.5.2 Bioluminescence spectrum (corrected) of the tunicate *Pyrosoma atlanticum*. From Herring, 1983, with permission from the Royal Society.

addition, on the basis of symbiosis it is difficult to explain the phenomenon in which the luminescence elicited locally by mechanical or other stimulations spreads a wave of light all over the colony; such a characteristic of the *Pyrosoma* luminescence distinctly differs from that of luminous bacteria and bacterial symbionts that are continuously luminescent. Pierantoni (1921) suggested that the inclusions in the luminous cells are modified luminous bacteria. Later, Mackie and Bone (1978) proposed the presence of intracellular symbiotic bacteria based on observations by electron microscopy. Furthermore, Leisman *et al.* (1980) have detected bacterial luciferase activity in the extract of the light organs of *Pyrosoma*, as the evidence supporting the presence of symbiotic luminous bacteria. However, the evidence presented seems insufficient to prove the bacterial origin of the *Pyrosoma* luminescence. To prove the involvement of luminous bacteria, it would be necessary to show that the luciferase occurs in a sufficient quantity to account for the bright luminescence of the specimen used; the mere detection of the luciferase activity would be insufficient to conclude the bacterial origin considering the ubiquitous occurrence of luminous bacteria. No additional information is available on the biochemical side of the *Pyrosoma* luminescence, except that a luciferin-luciferase reaction has never been positive.

Luminescence of *Clavelina*. The luminescence of the colonial ascidian *Clavelina miniata* was first reported by Aoki *et al.* (1989). This organism is a sessile tunicate, and emits a strong green luminescence upon stimulation by a mechanical means, hypotonic treatment, and increasing the concentration of K^+. No evidence was found to support the bacterial origin of its bioluminescence (Aoki *et al.*, 1989). Hirose *et al.* (1996) suggested that the tunic phagocytes in the light organs might be the source of luminescence on the basis of structural evidence and Chiba *et al.* (1998) isolated tunic phagocytes capable of emitting light, thus confirming that they are the origin of luminescence; no other types of cells in the light organ have a capability to emit light. The luminescence spectrum of *C. miniata* showed an emission maximum at 535 nm (Fig. 10.5.3).

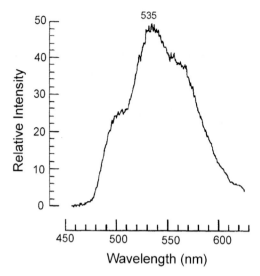

Fig. 10.5.3 Luminescence spectrum of the light emitted from freeze-dried, powdered specimens of the ascidian *Clavelina miniata* upon addition of water. The same luminescence spectrum was obtained when living specimens were stimulated. From Chiba *et al.*, 1998, with permission from John Wiley & Sons Ltd.

10.6   The Luminous Fishes

The fishes and tunicates belong to the same phylum Chordata although their appearances are quite different each other. There are many kinds of luminous fishes, possessing diverse types of light organs (reviews: Haneda, 1970, 1985; Herring and Morin, 1978; Herring, 1982). Luminous fishes are all marine inhabitants, and can be loosely categorized into two groups, the fishes of coastal shallow-water and the fishes of the oceanic deep-sea. Most luminous fishes of the former group utilize symbiotic luminous bacteria as their light sources; the flashlight fishes *Anomalops* and *Photoblepharon* are the most well known of them (Haneda and Tsuji, 1971; Morin *et al.*, 1975). A few members of the same group are self-luminous and utilize *Cypridina* luciferin as their source of light emission (for example, *Parapriacanthus* and *Apogon*). On the other hand, the majority of the luminous oceanic and deep-sea fishes are self-luminous and appear to utilize coelenterazine in their luminescence reaction, although some members of this group harbor luminous bacteria for their luminescence (for example anglerfishes: only the females have light organs). Several luminous fishes are illustrated in Fig. 10.6.1. In this chapter, the luminous fishes that utilize symbiotic luminous bacteria are excluded from further discussion.

10.6.1   Coastal and Shallow-water Fishes that Utilize Cypridina Luciferin

In 1958, it was discovered that the pempherid *Parapria-canthus beryciformes* and the apogonid *Apogon ellioti* show a

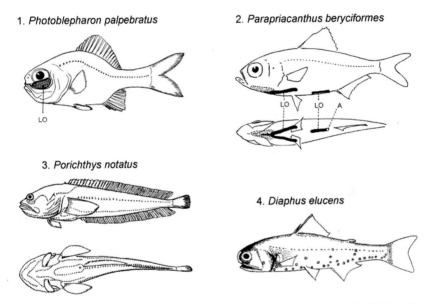

1. *Photoblepharon palpebratus*

2. *Parapriacanthus beryciformes*

3. *Porichthys notatus*

4. *Diaphus elucens*

Fig. 10.6.1 Some examples of luminous fishes. (**1**) The flashlight fish *Photoblepharon* emits light from symbiotic luminous bacteria in the light organs (LO) located on the eyelids. The light can be blocked by raising a black curtain. (**2**) The pempherid fish *Parapriacanthus* has two light organs (LO), a Y-shaped anterior organ and a small rod-like posterior organ at the front of anus. The light is produced by *Cypridina* luciferin, stored in the pyloric caeca and portions of the intestine (see Fig. 10.6.2), and a luciferase. (**3**) The midshipman fish *Porichthys* emits luminescence also with *Cypridina* luciferin and a luciferase. The light is emitted from several hundreds of tiny photophores distributed along the four pairs of lateral lines. (**4**) The deep-sea lantern fishes *Diaphus* emit blue light from many photophores distributed on the ventral and lateral surfaces of the body. Some species, such as *Diaphus elucens* shown here, possess an additional large nasal light organ (a white patch at the nose tip). Light is emitted with coelenterazine and a luciferase.

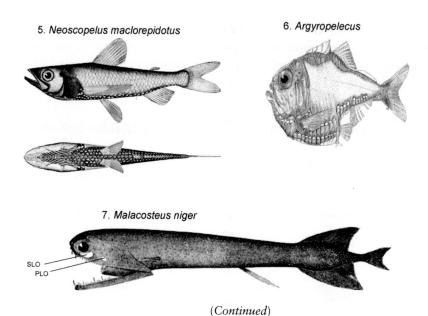

5. *Neoscopelus maclorepidotus*

6. *Argyropelecus*

7. *Malacosteus niger*

SLO
PLO

(*Continued*)

(5) The deep-sea neoscopelid *Neoscopelus* has photophores on the ventral surface and also on the tongue, an unusual area, which may lure small animals directly into the mouth. (6) The hatchet fish *Argyropelecus aculeatus* has a complex system of conspicuous light-producing photophores distributed predominantly on the ventral surface. (7) The loosejaw *Malacosteus nigar* has two pairs of light organs, suborbital light organs (SLO) that emit red light and postorbital light organs (PLO) that emit blue light. Drawings 1 and 3 from Herring and Morin, 1978 (with permission from Elsevier); 2 from Haneda and Johnson, 1962; 4 from Tsuji and Haneda, 1971a; 5 and 7 from NOAA Photo Library; 6 from Weitzman, 1974.

luciferin-luciferase reaction and the luciferins and luciferases of these fishes cross-react with those of the crustacean ostracod *Cypridina hilgendorfii* (Haneda and Johnson, 1958; Haneda *et al.*, 1958; Iwai and Asano, 1958; Johnson and Haneda, 1958). The discovery constitutes the first examples of the luciferin-luciferase cross-reaction between different types of luminous organisms. The absorption spectra of the luciferins obtained from *Cypridina* and *Apogon* were identical (Sie *et al.*, 1961), and various properties of the crystalline luciferin preparations obtained from *Cypridina* and *Parapriacanthus* were indistinguishable (Johnson *et al.*, 1961a), indicating that the luciferins from the two fish species are chemically identical with *Cypridina* luciferin. Following this discovery, a large amount of luciferin was found in the pyloric caeca of *Parapriacanthus* and some dead *Cypridina* were discovered in the stomach of the fish, implying that the fish had obtained the luciferin from the *Cypridina* in food (Haneda and Johnson, 1962; Johnson *et al.*, 1961a). In the photograph of a dissected specimen of *Parapriacanthus* (Fig. 10.6.2), the pyloric caeca and the gut leading to the anus are easily distinguishable by the yellow fluorescence of *Cypridina* luciferin, in support of the food origin hypothesis.

The batrachoidid *Porichthys porosissimus* (the midshipman fish) from the Gulf of Mexico shows a luciferin-luciferase reaction and cross-reacts with the luciferin and luciferase of *Cypridina*, although the amount of the luciferase found in the photophores of this fish was extremely small (Cormier *et al.*, 1967). The fish *Porichthys* has several hundred tiny photophores on the skin, differing from *Apogon* and *Parapriacanthus* that have a small number of larger light organs. Strum (1969) found that a Pacific species *Porichthys notatus* from Pacific Grove, California, has photophores and luminesces upon the injection of adrenalin, whereas the same species from Puget Sound, Washington, with the same photophores does not luminesce upon adrenalin injection.

The reason for the difference in the luminescence capability between the specimens of *Porichthys* from the two areas, Pacific Grove and Puget Sound, was clarified by Tsuji *et al.* (1972). The specimens of *Porichthys notatus* from Puget Sound did not contain a detectable amount of luciferin in the photophores, although they contained luciferase. On the other hand, specimens of the same species from the Santa Barbara area, further south of Pacific Grove, were found to contain both luciferin and luciferase, and luminesced upon the injection of noradrenalin, even after 150 days of starvation. When the Puget Sound *Porichthys* had been injected intraperitoneally with a solution of *Cypridina* luciferin, the specimen gained the ability to luminesce by adrenalin injection within a week, thus confirming a deficiency of luciferin in the Puget Sound *Porichthys*. An oral administration, instead of the intraperitoneal injection, of dried *Cypridina* or

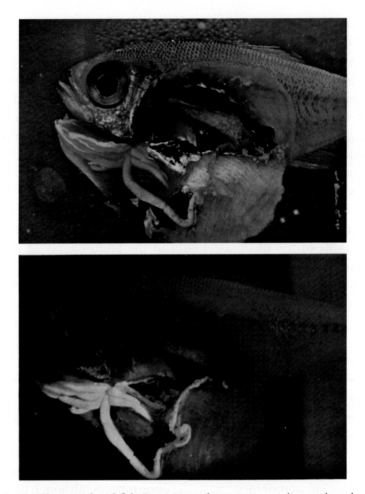

Fig. 10.6.2 The pempherid fish *Parapriacanthus ransonneti* dissected to show the pyloric caeca and portions of the visceral organs, photographed by daylight (*top*) and by ultraviolet light (*bottom*). The strong yellowish fluorescence in ultraviolet light is due to the presence of a large amount of *Cypridina* luciferin in the pyloric caeca and portions of the intestine. From Haneda and Johnson, 1962.

purified *Cypridina* luciferin showed the same effect. Thus, it is clear that the luminescence is induced by exogenous *Cypridina* luciferin, although it is not yet clear whether the specimens of normally luminous *Porichthys* have any capability to synthesize luciferin in their bodies or whether they obtain all of their luciferin from dietary sources.

The chemical mechanism of the light-emitting reaction involved in *Parapriacanthus*, *Apogon* and *Porichthys* is considered the same as in *Cypridina* (see Chapter 3).

10.6.2 Oceanic Deep-sea Luminous Fishes

There are numerous kinds of luminous deep-sea fishes, of which myctophidae (lantern fishes) is probably the largest family. Herring (1978b) lists bioluminescent fishes of 30 genera of myctophidae, 19 genera of gonostomatidae, and 16 genera of melanostomiatidae, together with many others. The spectral characteristics of luminescence for more than 35 species of luminous fishes have been reported (Herring, 1983; Widder *et al.*, 1983), of which a majority are deep-sea fishes. Despite the large number of deep-sea species known, chemical knowledge on their luminescence is rather meager at present. Only a small number of species have been tested for the presence of luciferin and luciferase. Remarkably, however, almost all of the luminescent species tested contained coelenterazine when tested with a coelenterazine-specific enzyme, such as *Oplophorus* luciferase and apoobelin (see Table 5.1). The chemical identity of the coelenterazine has been confirmed by TLC or HPLC when the species contained a relatively large amount of this compound, i.e. in the cases of the neoscopelid *Neoscopelus microchir* (Inoue *et al.*, 1977c) and the myctophid lantern fishes *Diaphus elucens* and *Diaphus coeruleus* (Shimomura *et al.*, 1980). It should be noted that the liver of *Neoscopelus microchir* contained an unusually large amount of coelenterazine; the livers from 25 specimens (5 g dry weight in total) yielded over 2 mg of purified coelenterazine (Inouye *et al.*, 1977c). Rees *et al.* (1992) identified coelenterazine in the hatchet fish *Argyropelecus hemigymnus* by HPLC, and reported that one specimen contained 0.044 ng of coelenterazine, in contrast to the data of 20 ng per specimen reported by Shimomura *et al.* (1980); the cause for the large difference is unclear.

The two species of *Diaphus* mentioned above contained relatively large amounts of coelenterazine and also exhibited weak but unmistakable luciferase activities (Shimomura *et al.*, 1980). Thus, it is most likely that these species have luminescence systems of luciferin-luciferase type, although a further confirmation would be desirable. On the other hand, the neoscopelid *N. microchir* contained a large amount of coelenterazine but the amount of luciferase found was too small (Shimomura *et al.*, 1980) to postulate a luciferin-luciferase system in this fish. Clearly a further study is needed to identify the luminescence system of this species. The amounts of coelenterazine and the activities of coelenterazine luciferase found in *Argyropelecus* and the gonostomatid *Yarrella* (Table 5.1) are both too small for discussing the type of luminescence systems involved; they might utilize a luciferin-luciferase system, a photoprotein system, or other type of luminescence system that is not yet known. Nevertheless, these data still strongly suggest the involvement of coelenterazine or its derivative in their luminescence.

The stimulation of luminescence by H_2O_2 has been observed with the photophores of various luminescent fishes: the batfish *Dibranchus* (Crane, 1968), *Diaphus* (Anctil and Gruchy, 1970), the sternopty-chids *Argyropelecus* and *Sternoptyx*, the scopelarchid *Benthalbella*, the shark *Isistius* (Herring and Morin, 1978), and the searsiidae fishes (Herring, 1983). However, the stimulation of luminescence by H_2O_2 is not necessarily an indication of the involvement of H_2O_2 or an active oxygen species in the bioluminescence reaction of these fishes. Further investigation is clearly needed.

Red-emitting luminous fishes. Most deep-sea luminous fishes emit bluish light. However, the stomiatoid fishes *Aristostomias scintillans*, *Malacosteus niger*, and *Pachystomias* have two types of light organs, and they emit red light from the suborbital light organs and blue light from the postorbital light organs (see Fig. 10.6.1). According to Widder *et al.* (1984), the luminescence emission maxima of the pos-torbital light organs from *A. scintillans* and *M. niger* are 479 nm and 469 nm, respectively, and those of the suborbital light organs are 703 nm and 702 nm, respectively (Fig. 10.6.3).

In the case of *M. niger*, the removal of the brown surface layer (apparently a filter) of the suborbital light organ and cutting open the organ resulted in a shift of the emission maximum from 702 nm to

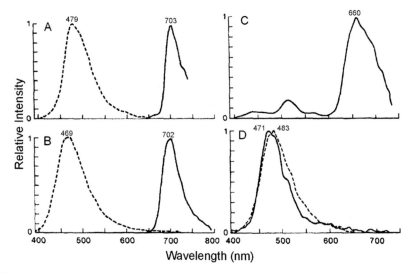

Fig. 10.6.3   Luminescence emission spectra of: (**A**) *Aristostomias scintillans* postorbital organ (λ_{max} 479 nm) and suborbital organ (λ_{max} 703 nm); (**B**) *Malacosteus niger* postorbital organ (λ_{max} 469 nm) and suborbital organ (λ_{max} 702 nm); (**C**) *M. niger* suborbital organ after the superficial tissue was removed and the organ was cut open (λ_{max} 660 nm); and (**D**) *M. niger* postorbital organ before isolation (λ_{max} 471 nm) and after isolation (λ_{max} 483 nm). From Widder *et al.*, 1984, Copyright 1984 AAAS.

660 nm. It would be unlikely that two different light-emitting reactions occur in one organism. Thus, it would be reasonable to consider the occurrence of an energy transfer from the 469 nm light-emitter to the 660 nm light-emitter.

Considering the occurrence of an energy transfer from the 469 nm light-emitter to the 660 nm light-emitter, there is a need of a further red shift of the emission peak from 660 nm to 702 nm to explain the *in vivo* luminescence. This red shift is apparently done by the brown surface layer of the suborbital light organs. According to Denton *et al.* (1985), the brown filter heavily absorbs light for all wavelengths of the visual spectrum below about 650 nm, resulting in the red shift.

Campbell and Herring (1987) isolated and partially purified a red fluorescent protein from the suborbital light organs of *M. niger*. The absorption spectrum of this red fluorescent protein had a peak at 612 nm, a shoulder at 555 nm, and a secondary peak at 490 nm.

Its fluorescence showed two emission peaks, 564 nm and 626 nm; the excitation peaks for the former emission peak were at 305 nm and 374 nm, and those for the latter were at 332 nm and 392 nm, indicating the presence of two different components in the preparation. It appears that the characteristics of the fluorescent proteins were altered by isolation and purification.

Currently no information is available on the chemical nature of the luminescence reaction that causes the 479 and 469 nm emission peaks from *A. scintillans* and *M. niger*.

10.6.3   *Future Research on Fish Bioluminescence*

Some luminous fishes harbor symbiotic luminous bacteria as their source of light emission, and a few kinds of shallow-water luminous fishes emit light utilizing *Cypridina* luciferin. Concerning the numerous kinds of self-luminous deep-sea fishes, however, we know very little about their light-emitting reactions, although it is commonly believed that most of them utilize coelenterazine. In fact, we have no knowledge except that some of the fishes use coelenterazine as a luciferin in the traditional luciferin-luciferase reaction. In some deep-sea luminous fishes, coelenterazine might be utilized in a photoprotein system or in a bioluminescence system that is not yet known.

Chemical studies on fish luminescence have been hampered by difficulties in obtaining specimens and the minute sizes of the luminous organs and photophores. Despite the setbacks, it might be possible to find out the basic nature of a luminescence reaction when coelenterazine (or *Cypridina* luciferin) is utilized in the luminescence. Once the basic nature of the luminescence reaction is found, then further details might become available by comparison with an organism having the same type of luminescence system.

The first step of a chemical study should be the quantitative measurements of coelenterazine, dehydrocoelenterazine, and a coelenterazine-specific luciferase, in the light organs, liver, digestive tract (with empty stomach), and eggs if available (see Section C5 of Appendix for the method). A clear, unequivocal presence of a coelenterazine luciferase indicates the involvement of a luciferin-luciferase system, even if only a trace amount of coelenterazine is found. If a significant amount of coelenterazine is found but the coelenterazine luciferase activity found is very low, or very weak, an involvement of a coelenterazine luminescence system other than the luciferin-luciferase type should be investigated. If both coelenterazine and luciferase are found in very small amounts, the possibility of a bioluminescence system unrelated to coelenterazine must be considered.

APPENDIX

A. Taxonomic Classification of Selected Luminous Organisms[a]

Phylum	Class	Sub-class	Order	Family	Luminous genus
BACTERIA					
				Bacteriaceae	*Photobacterium, Beneckea, Vibrio,* etc.
FUNGI					
	Basidiomycetes				
		Agaricales			
				Agaricaceae	*Panellus, Mycena, Pleurotus, Armillaria, Omphalia,* etc.
PROTOZOA					
	Radiolaria				*Collozoum, Thalassicolla,* etc.
	Dinoflagellata				*Noctiluca, Dissodinium, Gonyaulax (Lingulodinium), Peridinium, Pyrodinium, Pyrocystis,* etc.
CNIDARIA					
	Hydrozoa				
		Hydroida			
				Campanulidae	*Obelia, Phialidium, Campanularia,* etc.

(*Continued*)

(*Continued*)

Phylum	Class	Sub-class	Order	Family	Luminous genus
			Siphonophora	Aequoridae	*Aequorea*
				Mitrocomidae	*Mitrocoma* (*Halistaura*)
					Vogtia, Hippopodius,
					Abyla, Apolemia,
					Diphyes, Praya, etc.
	Scyphozoa				
				Periphyllidae	*Periphylla*
				Atollidae	*Atolla*
				Pelagiidae	*Pelagia*
	Anthozoa				
			Pennatulacea		*Cavernularia, Renilla,*
					Pennatula, Ptilosarcus,
					Virgularia, Stylatula,
					etc.
			Zoanthiniaria		*Parazoanthus,*
					Epizoanthus
CTENOPHORA					
				Mnemiopsidae	*Mnemiopsis*
				Bolinopsidae	*Bolinopsis*
				Pleurobrachiidae	*Pleurobrachia,*
					Callianira, etc.
				Cestidae	*Cestum*
				Beroidae	*Beroe*
MOLLUSCA					
	Gastropoda				*Latia, Planaxis,*
					Quantula (*Dyakia*),
					Plocamopherus,
					Kalinga, etc.
	Bivalvia				*Pholas, Rocellaria*
	Cephalopoda				
		Sepioidea			
				Spirulidae	*Spirula*
				Sepiolidae	*Semirossia,*[b]
					Heteroteuthis,
					Sepiola,[b] *Euprymna,*
					Sepiolina, etc.
		Teuthoidea			
			Myopsida		
				Loliginidae[b]	*Loligo, Uroteuthis,*
					etc.

(*Continued*)

(Continued)

Phylum	Class	Sub-class	Order	Family	Luminous genus

Oegopsida
Ommastrephidae *Symplectoteuthis, Eucleoteuthis, Ommastrephes, Ornithoteuthis,* etc.
Enoploteuthidae *Enoploteuthis, Watasenia, Pyroteuthis, Abralia, Pterygioteuthis,* etc.
Onychoteuthidae *Onychoteuthis, Chaunoteuthis,* etc.
Chiroteuthidae *Chiroteuthis, Chiropsis*
Cranchiidae *Cranchia, Leachia, Pyrgopsis, Liocranchia,* etc.
Vampyromorpha *Vampyroteuthis*
Octopoda *Japetella, Eledonella, Stauroteuthis*

ANNELIDA
Polychaeta
Chaetopteridae *Chaetopterus, Mesochaetopterus*
Polynoidae *Harmothoë, Polynoë, Acholoë,* etc.
Syllidae *Odontosyllis, Syllis, Pionosyllis, Eusyllis*
Tomopteridae *Tomopteris*
Terebellidae *Polycirrus, Thelepus*
Oligochaeta
Enchytraeidae *Henlea*
Megascolecidae *Microscolex, Pontodrilus, Diplocardia, Lampito, Octochaetus,Ramiella,* etc.
Lumbricidae *Eisenia*

ARTHROPODA
Crustacea
Ostracoda

(*Continued*)

Phylum	Class	Sub-class	Order	Family	Luminous genus
				Cypridinidae	*Cypridina, Vargula, Pyrocypris*
				Halocypridae	*Conchoecia*
		Copepoda			
				Augaptilidae	*Euaugaptilus, Centraugaptilus*
				Metridiidae	*Metridia, Gaussia, Pleuromamma*
				Heterorhabdidae	*Heterorhabdus, Hemirhabdus, Disseta*, etc.
				other families	*Oncaea, Megacalanus Lucicutia*, etc.
		Malacostraca			
			Mysidacea		*Gnathophausia*
			Amphipoda		*Scina, Acanthoscina, Parapronoë, Cyphocaris, Thoriella, Chevreuxiella*, etc.
			Euphausiacea		*Euphausia, Thysanoessa, Meganyctiphanes*, etc.
			Decapoda		
				Oplophoridae	*Oplophorus, Systellaspis, Acanthephyra, Ephyrina*, etc.
				Pandalidae	*Heterocarpus, Parapandalus*
				Penaeidae	*Hymenopenaeus, Mesopenaeus, Hadropenaeus*
				Sergestidae	*Sergestes, Sergia*
	Diplopoda (millipede)				*Luminodesmus (Motyxia), Spirobolellus*
	Chilopoda (centipede)				*Orphaneus, Orya, Stigmatogaster*, etc.
	Insecta				
			Coleoptera		

(*Continued*)

(*Continued*)

Phylum	Class	Sub-class	Order	Family	Luminous genus
				Lampyridae	*Lampyris, Photinus, Photuris, Luciola, Aspisoma*, etc.
				Elateridae	*Pyrophorus, Nyxophyxis, Noxlumenes, Cryptolampros, Sooporanga*, etc.
				Phengodidae	*Phengodes, Zarhipis, Phrixothrix, Diplocladon*, etc.
			Diptera		
				Mycetophilidae	*Orfelia (Platyura), Arachnocampa, Keroplatus*
ECHINODERMATA					
	Ophiuroidea				*Amphiura, Ophiopsila, Ophioscolex, Ophiacantha*, etc.
	Crinoidea				*Thaumatocrinus, Annacrinus*, etc.
	Holothuroidea				*Kolga, Peniagone, Paroriza, Laetmogone, Paelopatides*, etc.
	Asteroidea				*Plutonaster, Dytaster, Freyella, Brisinga*, etc.
HEMICHORDATA					
	Enteropneusta				
				Ptychoderidae	*Balanoglossus, Ptychodera, Glossobalamus*
CHORDATA					
TUNICATA (UROCHORDATA)					
	Larvacea				
		Copelata			*Oikopleura*
	Thaliacea				
		Pyrosomatida			*Pyrosoma*
		Salpida			*Cyclosalpa*

(*Continued*)

(*Continued*)

Phylum	Class	Sub-class	Order	Family	Luminous genus
		Ascidiacea			*Clavelina*
VERTEBRATA — PISCES (Fishes)					
	Chondrichthyes (Sharks)				
				Squalidae	*Isistius, Etmopterus, Squaliolus, Centroscyllium*
	Osteichthyes				
		Salmoniformes			
				Bathylagidae[b]	*Opisthoproctus, Winteria, Rhynchohyalus*
				Gonostomatidae	*Yarrella, Gonostoma, Cyclothone, Danaphos, Maurolicus, Sonoda*, etc.
				Sternoptychidae	*Argyropelecus, Polyipnus, Sternoptyx*
				Astronesthidae	*Astronesthes, Heterophotus, Cryptostomias*
				Melanostomiatidae	*Melanostomias, Eustomias, Bathophilus, Photonectes, Echistoma*, etc.
				Malacosteidae	*Malacosteus, Aristostomias, Photostomias*
				Chauliodontidae	*Chauliodus*
				Stomiatidae	*Stomias, Macrostomias*
				Idiacanthidae	*Idiacanthus*
				Alepocephalidae	*Photostylus, Rouleina, Xenodermichthys*, etc.
				Searsiidae	*Searsia, Holtbyrnia, Persparsia, Maulisia*, etc.

(*Continued*)

(*Continued*)

Phylum	Class	Sub-class	Order	Family	Luminous genus
				Scopelarchidae	*Scopelarchoides, Benthalbella*
				Neoscopelidae	*Neoscopelus*
				Myctophidae (Lantern fishes)	*Lampadena, Myctophum Protomyctophum, Diaphus, Benthosema,* etc.
			Batrachoidiformes		
				Batrachoididae	*Porichthys*
			Lophiiformes		
				Ogcocephalidae	*Dibranchus*
				Oneirodidae[b]	*Oneirodes, Chaenophryne, Danaphryne, Puck,* etc.
				Himantolophidae[b]	*Himantolophus*
				Linophrynidae[b]	*Linophryne, Photocorynus, Borophryne,* etc.
			Gadiformes		
				Moridae[b]	*Physiculus, Brosmiculus, Gadella, Tripterophycis*
				Macrouridae[b]	*Coelorhynchus, Nezumia, Malacocephalus,* etc.
			Beryciformes		
				Monocentridae[b]	*Monocentris, Cleidopus*
				Anomalopidae[b]	*Anomalops, Photoblepharon, Kryptophanaron*
			Perciformes		
				Apogonidae	*Apogon, Archamia, Epigonus, Howella, Siphamia,*[b] etc.
				Pempheridae	*Parapriacanthus, Pempheris*

[a] Compiled and then condensed from the tables given by Harvey (1952), Herring (1978b) and Haneda (1985).
[b] Symbiotic luminous bacteria found.

B.   Lists of Luciferins, Luciferases and Photoproteins Isolated

The structures of known luciferins

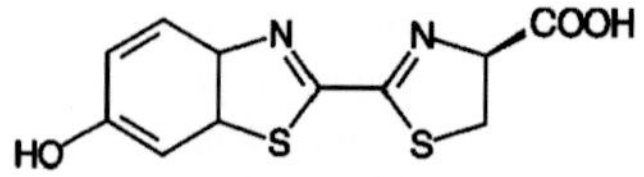

Firefly

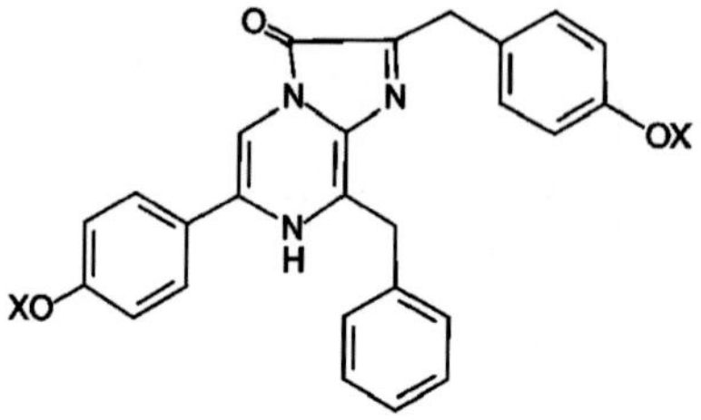

Coelenterazine (X = H)
Squid *Watasenia* (X = SO$_3$H)

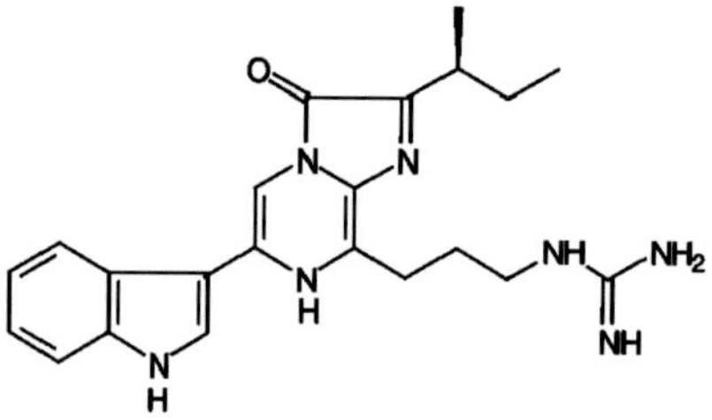

Cypridina

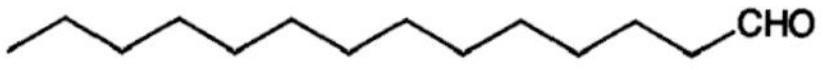

Luminous bacteria

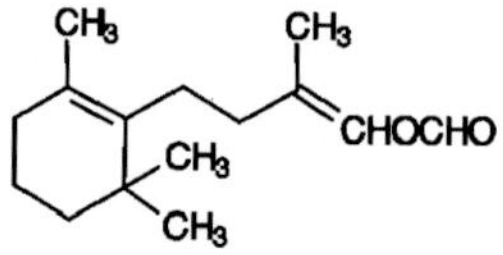

Limpet *Latia*

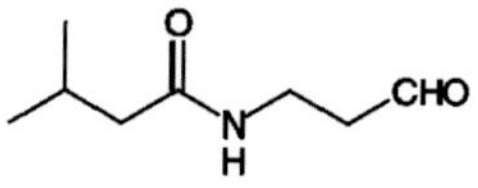

Earthworm *Diplocardia*

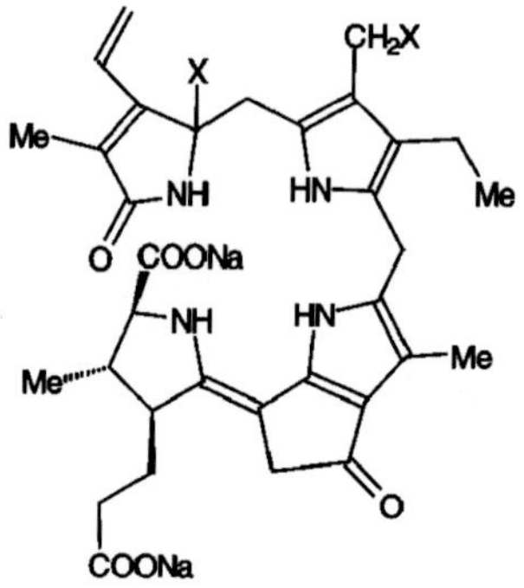

Krill (X = OH)
Dinoflagellate (X = H)

LUCIFERINS

Name of Luciferin	Molecular Formula (M_r)	Absorption Max. (nm) (ε value)	Fluorescence Emission Max. (nm)	Luminescence Max. (nm)	Quantum Yield
Firefly luciferin[a] (D-form)	$C_{11}H_8N_2O_3S_2$ (280.33)	328 (pH 5), 384 (pH 11) (both 18,200)	537	560 (pH 7.1) 615 (pH 5.4)	0.88 ± 0.25 (pH 7.6)
Bacterial luciferin (tetradecanal)[b]	$C_{14}H_{28}O$ (212.36)	290 (16)	None	490	0.10–0.16
Cypridina luciferin[c]	$C_{22}H_{29}O_2N_7 \cdot$ 2HCl (496.44) $C_{22}H_{27}ON_7 \cdot$ 2HBr (567.33)	432 (9,000)	540 (weak)	450–460	0.30 ± 0.04
Latia luciferin[d]	$C_{15}H_{24}O_2$ (236.35)	207 (13,700)	None	536	0.009
Odontosyllis luciferin[e]	Unknown	330	None	507	

(Continued)

(Continued)

Name of Luciferin	Molecular Formula (M_r)	Absorption Max. (nm) (ε value)	Fluorescence Emission Max. (nm)	Luminescence Max. (nm)	Quantum Yield
Diplocardia luciferin[f]	$C_8H_{15}O_2N$ (157.21)		None	490	0.03 (H_2O_2 adduct)
Krill luciferin[g] (Compound F)	$C_{33}H_{38}O_8N_4Na_2$ (664.67)	388 (15,400)	476	476	0.6 (0°C)
Dinoflagellate luciferin[h]	$C_{33}H_{38}O_6N_4Na_2$ (632.67)	390 (14,000)	474	474	
Coelenterazine	$C_{26}H_{21}O_3N_3$ (423.47)	435 (9,800)	530 (weak)	450–480 (anion); 400 (neutral)	0.1–0.3

[a] Bitler and McElroy, 1957; Seliger and McElroy, 1960; Morton et al., 1969.
[b] Uritzur and Hastings, 1978, 1979a; Lee, 1972; McCapra and Hysert, 1973; Shimomura et al., 1972, 1974.
[c] Kishi et al., 1966a,b; Johnson et al., 1962; Shimomura and Johnson, 1970a.
[d] Shimomura and Johnson, 1968b,c, 1972a.
[e] Shimomura et al., 1963d, 1964.
[f] Bellisario et al., 1972; Ohtsuka et al., 1976; Rudie et al., 1981.
[g] Shimomura, 1995a; Nakamura et al., 1988.
[h] Nakamura et al., 1989.

LUCIFERASES

Name of Luciferase (Lase)	M_r	Required Components[*]	Specific Activity, Quanta $\cdot$ s^{-1}mg^{-1} (Turnover No) K_m	Remarks[†]
Firefly luciferase[a] (*Photinus*)	62,000	ATP, Mg^{2+} and O$_2$	7.8×10^{14} (10/min)	Em. max. 560 nm
Phrixothrix Lase[b]		ATP, Mg^{2+} and O$_2$		Em. max. 542 nm and 638 nm
Bacterial Lase[c]	76,000	FMNH$_2$ and O$_2$		Em. max. 490 nm
Cypridina Lase[d]	62,200	O$_2$	7.7×10^{16} (1,600/min) K_m 0.52 μM	Inhibitor: EDTA
Latia luciferase[e]	173,000, 180,000	O$_2$	(8.4/min)	Activator: purple proteine, ascorbic acid, NADH
Odontosyllis luciferase[f]		Mg^{2+} and O$_2$		Activator: HCN
Diplocardia luciferase[g]	300,000	H$_2$O$_2$	1.9×10^{14}	Contains 4 Cu atoms
Krill luciferase[h] (*Euphausia* and *Meganyctiphanes*)	600,000	O$_2$	5×10^{12} (0.5/min)	Partially inactivated sample

(*Continued*)

(Continued)

Name of Luciferase (Lase)	M_r	Required Components[*]	Specific Activity, $Quanta \cdot s^{-1} mg^{-1}$ (Turnover No) K_m	Remarks[†]
Dinoflagellate luciferase[i]	35,000 (pH 6) 130,000 (pH 8)	O_2	8.5×10^{14} at pH 8	From *Gonyaulax*
Coelenterazine luciferases				
Periphylla Lase[j]	20,000, 32,000, 40,000, 80,000	O_2	$1–4 \times 10^{16}$ K_m 0.2, 1.2 μM	Em. max. 465–470 nm
Renilla reniformis luciferase[k]	34,000	O_2	1.8×10^{15} (111/min) K_m 0.44 μM	Em. max. 480 nm, Q_{clz} 0.10–0.16
Ptilosarcus gruneyi Lase	35,000	O_2	K_m 0.81 μM	Em. max. 475 nm, Q_{clz} 0.12
Oplophorus Lase (*gracilirostris*)[l]	106,000 (active subunit 19,000)	O_2	1.7×10^{15} (55/min) K_m 0.18 μM	Em. max. 452 nm, Q_{clz} 0.30
Heterocarpus sibogae Lase	110,000	O_2	K_m 0.12 μM	Em. max. 453 nm, Q_{clz} 0.32
Heterocarpus ensifer Lase	140,000	O_2	K_m 0.17 μM	Em. max. 452 nm, Q_{clz} 0.28

(Continued)

(Continued)

#Gaussia princeps luciferase[m]	19,900	O_2	1.5×10^{15} K_m 1.4 µM	Em. max. 470 nm, Q_{clz} 0.16
#Metridia longa luciferase[n]	23,900	O_2		Em. max. 480 nm
Amphiura filiformis Lase[o]	23,000	O_2		Em. max. 475 nm

*In addition to luciferin and luciferase.

†Em., luminescence emission; Q_{clz}, quantum yield of coelenterazine (unpublished data included).

#Recombinant protein.

[a]Bitler and McElroy, 1957; Woods et al., 1984; Shimomura et al., 1977; Branchini et al., 2005.

[b]Viviani et al., 1999.

[c]Friedland and Hastings, 1967; Hastings et al., 1969a.

[d]Shimomura et al., 1961, 1969; Thompson et al., 1989.

[e]Shimomura and Johnson, 1968c; Kojima et al., 2000a.

[f]Shimomura et al., 1963d, 1964.

[g]Bellisario et al., 1972; Rudie et al., 1976.

[h]Shimomura, 1995a.

[i]Hastings, 1978.

[j]Shimomura et al., 2001.

[k]Matthews et al., 1977a.

[l]Shimomura et al., 1978; Inouye et al., 2000.

[m]Ballou et al., 2000; Verhaegen and Christpoulos, 2002.

[n]Markova et al., 2004.

[o]Mallefet and Shimomura, 2004 (unpublished).

PHOTOPROTEINS

Source Organism	Name	M_r	Requirements for Luminescence	Luminescence Maximum, nm (Quantum Yield)
PROTOZOA				
Thalassicolla sp.[a]	Thalassicollin		Ca^{2+}	440
COELENTERATA				
Aequorea aequorea[b]	Aequorin	21,500	Ca^{2+}	465 (0.16)
Halistaura cellularia	Halistaurin		Ca^{2+}	470
(Mitracoma cellularia)[c]	Mitrocomin	21,800	Ca^{2+}	470
Phialidium gregarium	Phialidin[d]	23,000	Ca^{2+}	474
	Clytin[e]	21,600	Ca^{2+}	
Obelia geniculata	Obelin	21,000[f]	Ca^{2+}	475[h] 485[i]
Obelia longissima	Obelin	22,200[g]	Ca^{2+}	495[i]
CTENOPHORA				
Mnemiopsis sp.[j]	Mnemiopsin-1	24,000	Ca^{2+}	485
	Mnemiopsin-2	27,500	Ca^{2+}	485
Beroe ovata[i]	Berovin	25,000	Ca^{2+}	485

(Continued)

(Continued)

Source Organism	Name	M_r	Requirements for Luminescence	Luminescence Maximum, nm (Quantum Yield)
ANNELIDA				
Chaetopterus variopedatus[k]		120,000 and 184,000	Fe^{2+}, hydroperoxide and O_2	455
Harmothoë lunulata[l]	Polynoidin	500,000	Fe^{2+}, H_2O_2 and O_2	510
MOLLUSCA				
Pholas dactylus[m]	Pholasin	34,600	Peroxidase or Fe^{2+}, plus O_2	490 (0.09)
Symplectoteuthis oualaniensis[n]	Symplectin	60,000	Alkaline pH? and O_2	470–480[o]
Symplectoteuthis luminosa[p]		50,000	Catalase, H_2O_2 and O_2	
DIPLOPODA				
Luminodesmus sequoiae[q]		60,000	ATP, Mg^{2+} and O_2	496

(Continued)

(Continued)

Source Organism	Name	M_r	Requirements for Luminescence	Luminescence Maximum, nm (Quantum Yield)
ECHINODERMATA *Ophiopsila californica*[r]		45,000	H_2O_2	482

[a] Campbell and Herring, 1990.
[b] Shimomura, 1986a.
[c] Shimomura et al., 1963a; Fagan et al., 1993.
[d] Levine and Ward, 1982.
[e] Inouye and Tsuji, 1993.
[f] Stephanson and Sutherland, 1981.
[g] Illarionov et al., 1995.
[h] Morin and Hastings, 1971a.
[i] Markova et al., 2002.
[j] Ward and Seliger, 1974a,b.
[k] Shimomura and Johnson, 1968d.
[l] Nicolas et al., 1982.
[m] Michelson, 1978.
[n] Fujii et al., 2002.
[o] Takahashi and Isobe, 1993, 1994.
[p] Shimomura unpublished.
[q] Shimomura, 1981, 1984.
[r] Shimomura, 1986b.

C.   Miscellaneous Technical Information

C 1.   *Basic Principle of the Isolation of Bioluminescent Substances*

The physiology of living organisms involves various metabolic reactions that are interrelated in a highly complex manner. In bioluminescent organisms, light emission is caused by one such reaction liberating a part or most of the generated energy in the form of light. In studying bioluminescence reactions, it is crucially important to use the purified forms of the components necessary for light emission. The results obtained with crude or impure materials are difficult to interpret and often lead to erroneous interpretations. Therefore, the isolation and purification of the essential components of light emission, such as luciferin, luciferase and photoprotein, are extremely important. Since bioluminescence systems are highly diverse in their mechanisms and required components, individual methods to isolate and purify luminescent substances need to be designed specifically for each luminescence system. Thus, only general strategies are discussed in the following sections.

The substances involved in bioluminescence reactions are usually unstable. Thus, the extraction and purification of bioluminescent substances should be carried out in the shortest possible period of time, usually at a low temperature. It is known through experience that luminescent substances are almost always more stable in the original animal tissues than in extracts when preserved at a low temperature. Therefore, before starting extraction and purification, the stability of the extracts and purified substances should be investigated by carrying out a small-scale pilot experiment. A pilot experiment is also essential in the course of purification to avoid an unexpected loss of the target substance. If a component of the luminescence system is insoluble in common buffer solutions, it must be solubilized to purify it (see C1.3).

Before extracting a luminescent substance, it is desirable to find a condition under which the luminescence is reversibly inhibited. This step may not always be simple, but it is extremely important and useful; the condition found is often directly related to the basic nature of the luminescence system. For example, in the case of extracting the photoprotein aequorin, it must be first found out that Ca^{2+} causes the luminescence of the photoprotein aequorin; this information leads to the use of EDTA to inhibit the luminescence, a step which is easily reversible by removing EDTA. A method for reversible inhibition is crucial in the study of bioluminescence; without this information, the extraction and purification of luciferins and photoproteins are often difficult. It should be noted that the extraction and purification of the photoproteins of the millipede *Luminodesmus* and the brittle star *Ophiopsila* were carried out without efficient inhibitors, resulting in purified photoproteins of very low specific activities.

Sometimes an aqueous extract of a luminous organism is nonluminous although it contains all the components necessary for light emission. If that is due to the presence of various inhibitors that are extracted together with the luminescent substances, the extract will become luminous by merely diluting the extract with water or a buffer solution. A means of reversible inhibition will be needed to purify such a luminous extract.

C 1.1. *Reversible Inhibition of Bioluminescence*

The reversible inhibition of luminescence can be achieved by a variety of methods that block luminescence or remove one of the components essential for luminescence, as explained below.

Effect of pH. The light emission from most bioluminescence systems is affected by the pH of the medium, and some luciferases and photoproteins can be made inactive at certain pH ranges without resulting in permanent inactivation. For example, the luminescence of euphausiids can be quenched at pH 6, the luminescence of aequorin can be suppressed at pH 4.2–4.4, and the luciferase of the decapod shrimp *Oplophorus* becomes inactive at about pH 4. In the case of *Cypridina* luminescence, however, the acidification of an extract to below pH 5 results in an irreversible inactivation of the luciferase.

Elimination of a cofactor needed for luminescence. The chelators EDTA and EGTA efficiently quench the luminescence of Ca^{2+}-activated systems such as aequorin, obelin and mnemiopsin. The luminescence systems that require ferrous ions, such as extracts of the polychaete *Chaetopterus*, can be inhibited by 8-hydroxyquinoline and certain other chelators (such as 2,2′-dipyridyl and o-phenanthroline) but not significantly by EDTA or EGTA. The effect of magnesium ions is suppressed by EDTA but not by EGTA (Log stability constants, K_1: Mg-EDTA 8.69, and Mg-EGTA 5.21; Fabiato and Fabiato, 1979). In the isolation of pholasin, ascorbate was used to inhibit *Pholas* luciferase, a copper-containing enzyme; and a copper chelator diethyldithiocarbamate was used to inactivate the luciferase.

When H_2O_2 is a necessary component of a luminescence system, it can be removed by catalase. If a luminescence system involves superoxide anion, the light emission can be quenched by destroying O_2^- with superoxide dismutase (SOD). The ATP cofactor usually present in the fresh extracts of the fireflies and the millipede *Luminodesmus* can be used up by their spontaneous luminescence reactions, eventually resulting in dark (nonluminous) extracts containing a luciferase or photoprotein. The process is, however, accompanied by a corresponding loss in the amount of luciferin or photoprotein. The use of ATPase and the elimination of Mg^{2+} in the extract may prevent such a loss.

Anaerobic condition. Molecular oxygen is an essential requirement in a majority of bioluminescence systems. Thus, the light emission from those systems stops if molecular oxygen is completely eliminated. The methods for removing oxygen from biological fluids were discussed in detail by Harvey (1926b), which contains useful information on various procedures that are valuable even today.

There are several ways to achieve an anaerobic condition.

(1) *Argon gas.* This inert gas is heavier than air and most frequently used. In the case of a solution, argon gas can be bubbled to remove oxygen. It is probably the most convenient way to achieve an anaerobic condition. However, argon gas contains a trace of oxygen. Thus, the light emission of the *Cypridina* luminescence system cannot be stopped completely with high purity argon (99.999%). Luminous bacteria begin to luminesce at an oxygen pressure of 0.0053 mm Hg (Harvey, 1926b), thus bubbling a bacterial suspension with 99.999% argon will stop the luminescence according to calculation.

(2) CO_2 *gas.* The use of dry ice (i.e. adding small pieces of dry ice during extraction) has the merit of cooling the sample in addition to displacing oxygen with CO_2 gas. The method cannot be used if the

luminescent substance is sensitive to the acidity of carbon dioxide as in the case of the luciferins of euphausiids and dinoflagellates (note that these luciferins in the organisms before extraction are not affected by dry ice). However, commercial dry ice contains a small amount of oil, which may contaminate purified substances; thus, dry ice should not be used in the final stages of purification.

(3) *Other anaerobic methods.* Carrying out experiments in an atmosphere of pure hydrogen is a highly reliable way to eliminate a loss due to oxidation, but it has some shortcomings. In addition to a risk of accidental explosion, hydrogen gas must be freed from any trace of contaminating oxygen before use, by passing the gas through a heated quartz tube containing a catalyst, such as platinized asbestos and copper fragments. Experiments in a high vacuum (<10 μm Hg) are possible but technically very difficult due to a greater possibility of air leakage.

A small amount of sodium dithionite added to an aqueous medium depletes oxygen almost instantly. Due to the acidity of the reagent, a sufficiently strong buffer must be included in the medium. However, the use of dithionite is not recommended in the stages of extraction and purification because the reagent and its decomposition products are highly reactive and can cause undesirable reactions.

If a bioluminescence system is not inhibited by anaerobic conditions, the luminescence system probably does not require oxygen. In such a case, the luminescence system might require hydrogen peroxide, or it might involve an aequorin-like photoprotein that does not require oxygen for luminescence.

Other inhibition techniques. The ionic strength of solution (0–1 M NaCl, KCl or $(NH_4)_2SO_4$) influences the light emission of many bioluminescence systems. Adjusting ionic strength alone is usually insufficient to stop or strongly inhibit luminescence, but it might be sufficiently effective if used in conjunction with other techniques, such as the adjustment of pH. There are many substances that nonspecifically inhibit bioluminescence systems by a variety of mechanisms, but the inhibition is often incomplete and irreversible (for example, thiol-modification reagents, ascorbic acid, sulfite, cyanide, hydroquinone, and various detergents).

C 1.2.   *Extraction of Luciferin-luciferase Systems*

When the luminescence of an organism is a simple luciferin-luciferase type reaction involving a luciferin of relatively small size (M_r < about 1,000), the luciferin can usually be extracted with methanol. Methanol inactivates the luciferase and instantly quenches luminescence. It is often a preferred shortcut method to extract luciferin, despite the unavoidable loss of the luciferase. The advantage of using methanol, instead of boiling water, which is used in the classical method, is that (a) methanol does not dissolve luciferase and most proteins, (b) methanol dissolves a wide range of organic, low molecular weight compounds more efficiently than water, and (c) the extraction can be carried out at a temperature lower than $0°C$, thus decreasing the loss of luciferin due to oxidation.

The luciferase can be extracted with cold water or an aqueous buffer solution, as in the classical method of preparing luciferase solution. In some instances, the extraction can be carried out without causing luminescence by adjustments of the pH and ionic strength of the extraction buffer. If the extract is luminous, however, the luminescence will gradually diminish due to the exhaustion of a luciferin or a cofactor (if involved), eventually yielding a non-luminous solution containing luciferase. To confirm the luciferin-luciferase reaction, the crude luciferase thus prepared is tested for its capability of catalyzing light emission by adding a small amount of the methanolic extract of luciferin in various buffer solutions having different pH values (6–9) and salt compositions (0.05–0.5 M NaCl or KCl). In the test, the final concentration of methanol should not exceed 5% to avoid inhibition by methanol. If light emission cannot be observed, then the possible involvement of a cofactor should be investigated. The cofactors so far found in bioluminescence systems include $FMNH_2$, NADH, ATP, Ca^{2+}, Mg^{2+}, cyanide ions, H_2O_2, and ferrous ions.

C 1.3.   *Solubilization of Proteins*

In some luminous organisms, luciferases and photoproteins exist in particulate forms that are insoluble in water or common buffer solutions, resembling membrane proteins. The protein is probably highly aggregated or bound to an insoluble material; thus, the protein must be solubilized in order to extract it into a solution. Generally the solubilization is accomplished by the action of a suitable detergent (surfactant) in an appropriate buffer solution.

Detergents. The detergent chosen should not interfere with the assay of luminescence activity; namely, the detergent used should not inhibit the luminescence reactions when the extract is sufficiently diluted with an assay buffer. In addition, it should be taken into account that the detergent might need to be removed at the end of purification. The solubilization procedures of proteins and the properties of various detergents are discussed in detail in *Methods in Enzymology,* Volume 182, Section V (1990). Although there are many kinds of detergents to choose from, the author would first test only a few of them, for example, Triton X-100 (nonionic, at 0.1%), SB3-12 (zwitterionic, at 0.05%), lauroylcholine chloride (cationic, at 0.02%), and CHAPS (zwitterionic, at 0.1%). If the test indicates possible usefulness with any of them, the detergent will be further tested at various different concentrations, and also the usability of homologous detergents will be examined. Although Triton X-100 is an excellent detergent and widely used, it shows some absorption at 280 nm due to the presence of aromatic rings, which is a disadvantage for monitoring at this wavelength. All the other detergents mentioned above do not absorb at 280 nm.

It must be mentioned here that detergents have a tendency to destabilize luciferins and photoproteins and sometimes cause spontaneous luminescence from them, resulting in a partial loss of active luciferins and photoproteins. Such an effect is strongest with CTAB, a cationic detergent.

Buffer composition. To solubilize luciferases and photoproteins, a high concentration of salt (0.5–1.0 M NaCl or KCl) is frequently useful. In the case of membrane proteins, 0.1–0.5 M phosphate buffer was reported to be highly effective for solubilization (Dey *et al.,* 1981). Guanidine hydrochloride (1–2 M) is occasionally useful, but the author has not obtained good results using urea. Glycerol (10–30%) is worth trying; it affects not only the solubility but also markedly increases the stability of proteins. The use of a high concentration of glycerol (>20%) causes a significant increase in viscosity; if the viscosity becomes a problem, it is recommended that sucrose be used as it increases viscosity much less than glycerol does.

The result of solubilization can be evaluated after centrifugation of the solubilization mixture at 20.000 g for 10 minutes, by measuring the luminescence activity of the supernatant.

C 1.4.   Purification

With or without reversible inhibition of luminescence, a bioluminescent substance is extracted from a luminous organism or its tissue and then purified. Purification is usually done through various forms of chromatography, supplemented by other techniques such as fractional precipitation, electrophoresis and ultrafiltration. The techniques of chromatography commonly used are column chromatography, TLC, HPLC and FPLC, utilizing various principles of separation, such as adsorption, ion-exchange, gel filtration, phase partition, reversed phase, affinity, and hydrophobic interaction. The use of low concentrations of EDTA and/or 2-mercaptoethanol (about 1 mM each) in solvent is often, but not always, beneficial to stabilize unstable substances.

In the purification of a luciferin, a combination of adsorption chromatography (on silica gel or alumina), reversed phase chromatography, and ion-exchange chromatography (if luciferin has an ionic group) is usually sufficient to obtain pure substance if the luciferin is not very unstable. The purification of a luciferase and photoprotein is more complex and often requires considerable effort to obtain a pure protein. For purification efficiency, the purification sequence should include at least two, preferably three, different separation principles, such as gel filtration, ion-exchange and hydrophobic interaction. The available chromatographic media are being improved continuously; therefore, the best medium available should be chosen at the time. The author's preferences at present are: Superdex Prep grade for gel filtration, Q Sepharose Fast Flow for anion exchange, SP Sepharose Fast Flow for cation exchange, and Butyl Sepharose 4 Fast Flow for hydrophobic interaction, all from Pharmacia.

Purification is not routine work in the study of bioluminescence, and the purification method for a new substance must be devised specifically for each situation. General information on the available techniques is described in various publications (e.g., for natural products: Cannell, 1998; for proteins: Deutscher, 1990, and Scopes, 1993).

C 2. *Storage of Samples*

Raw materials. Most luminous organisms can be stored at $-70°C$ or below under aerobic conditions, or with dry ice, without a significant loss of luminescence activity for a period of several months or more, although a trial is always recommended. Even if a substance already extracted is unstable when stored with dry ice (like the luciferase of *Cypridina* and the luciferins of euphausiids and dinoflagellates), the same substance in the organisms before extraction can be safely stored at $-70°C$ or with dry ice. The material can also be stored with liquid nitrogen for added safety, but the quantity storable in a laboratory setup (e.g., Dewar flask) is limited.

Extracts and purified materials. Luminescent substances extracted or purified are usually less stable than the substances in the original organisms, and some proteins extracted might denature by freezing. Although many purified luminescent substances can be safely stored in an ultracold freezer at below $-70°C$, some substances, such as the purified luciferins of euphausiids and dinoflagellates, are rapidly oxidized even at such a low temperature. These highly oxygen-sensitive substances can be stored only in a completely evacuated sealed glass container (see Method II below) at below $-20°C$, or with liquid nitrogen.

Oxygen-sensitive substances. Substances that are moderately sensitive to oxygen (such as crystalline coelenterazine and *Cypridina* luciferin) can be stored aerobically at below $-70°C$ in a desiccated container for many years. They can be permanently stored in an evacuated fuse-sealed glass container even at room temperature, in darkness (for the technique, see Method II below).

Alternatively, moderately oxygen-sensitive substances can be stored under vacuum or in high purity argon gas, at a temperature

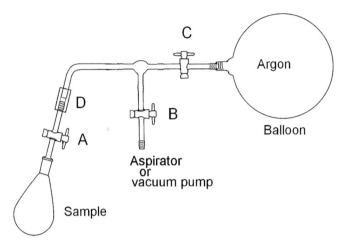

Fig. A.1.   A setup for replacing the air in the sample container with argon, using a balloon. At the connection D, a short piece of thick-walled silicon-rubber tubing is used.

of 0 to −20°C. In the case of storage under vacuum, the material or a vial containing the material is placed in a flask equipped with a ground-glass joint, and a stopcock is attached to the ground-joint (see the sample flask in Fig. A.1). The joint and stopcock should be lubricated with high-vacuum grease. The flask is evacuated with a vacuum pump (<10 μm Hg), and then stored at a temperature not lower than −20°C. At a temperature lower than −20°C, the risk of air leakage increases. The safe period of storage varies depending on the quality of the ground-glass joints, stopcocks and grease used, as well as the stability of the material itself; it might last more than one year or it might be only a few days before air leaks in. At −70°C, the seal of the ground-glass joint will fail probably within one hour. The safe storage period can be significantly extended by filling the flask with argon gas (see Method I below).

For the storage of a substance dissolved in a volatile solvent, the solution must be placed directly in the flask, not within a vial. During the evacuation, gently sway the flask and adjust the evacuation speed to avoid sample loss due to a sudden boiling of the solution. Replace the inside space of the flask completely with the solvent vapor (typically methanol) by boiling the sample solution for a period of 1–2 minutes before closing the stopcock. In using this method, however, the vacuum pump should be protected with a vapor trap dipped in a dry ice/acetone bath; if the substance is unharmed at a temperature of 45–50°C, a water aspirator can be used instead of a vacuum pump (in the case of methanol solvent).

Storage under argon gas (Method I). A simple setup for replacing the inside space of a flask with argon gas is shown in Fig. A.1. The sample flask contains a solid or liquid sample. The balloon used is an ordinary one available at toy stores. The balloon is filled with argon gas at least 10 times the volume of the sample flask. The balloon is quickly attached to stopcock C that is in open position. Immediately, stopcock C is closed when the air inside the stopcock is washed out with argon gas by the balloon's own pressure. Now stopcocks A and B are opened, and the system is evacuated, using a water aspirator or a vacuum pump. After the evacuation, stopcock B is closed, and then stopcock C is opened to fill the flask with argon gas. The steps of evacuation and filling with argon are repeated four or five times, by the manipulation of stopcocks B and C. After the last filling, the aspirator or vacuum pump is disconnected from stopcock B. Then, stopcock B is opened to equalize the atmospheric pressure by releasing some argon gas, and then stopcock A is closed immediately. Finally, the sample container is disconnected from the system at connection D and stored in a freezer or refrigerator. The strains at the seals are very little due to the small difference of the atmospheric pressures inside and outside the flask, thus significantly reducing the possibility of a seal failure and extending the safe storage period.

Storage under vacuum in a sealed tube (Method II). Substances that are extremely oxygen-sensitive, such as the fluorescent compound F of euphausiids and dinoflagellate luciferin, have to be stored in an evacuated sealed container at a low temperature. For long-term storage, they must be fuse-sealed in an evacuated glass vial using the method outlined below.

For a small amount of material (<0.1 g), the method is simple and easy, using the basic technique of glass blowing. A piece of borosilicate glass tube, 5–6 mm inside diameter and about 30 cm long, is washed and dried in advance (Fig. A.2, (A)). First, one end of the tube is sealed with a narrow flame of a gas-oxygen torch. This is done by heating

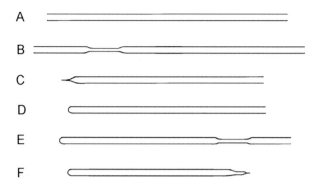

Fig. A.2.   The sequence of making a sealed glass tube, from top to bottom.

and melting a portion of the tube about 7 cm from the end, while rotating the tube at a constant speed, followed by a rapid pull-out motion to make a short length of capillary (B). In the flame, the hole of the capillary fuses and the tube is separated into two pieces. The sealed end of the longer tube, which has a sharp projection (C), is further heated and softened in the flame. Then, a slight air pressure is applied from the mouth (through a rubber tubing connected to the open end of the tube) to transform the sealed end into a round shape of even thickness (D). The product is allowed to cool. A sample (solid or liquid) is placed into the rounded bottom of the tube, and the tube is again heated at a position at least 10 cm above the sample to avoid heating the sample, and the softened glass is pulled slowly to make a constriction of about 3 mm outside diameter and 2–3 cm long (E). After cooling, the tube is connected to a vacuum line and evacuated (<10 μm Hg), and then the tube is sealed by heating the constricted part with a small flame (F). In the case of a liquid sample, the portion of the tube containing the sample should be cooled with dry ice (or other means) during evacuation to prevent boiling. The constricted part should not be cooled, and cooling is usually not necessary during sealing, which takes only 2–3 seconds.

For storage of a very small amount of highly oxygen-sensitive substance, it is recommended that an empty tube (D) is continuously evacuated at 100–120°C for several hours before use, to remove the oxygen adsorbed in the glass.

C 3.   *Measurement of Luminescence*

In the chemical study of bioluminescence, an instrument for measuring light intensity and total (integrated) light emission is an absolute necessity. The apparatus is commonly called a "light meter" or "luminometer." There are many commercially available luminometers (Stanley, 1997, 1999). Some of them are specialized for certain purposes, such as ATP assay, but most instruments are capable of measuring both light intensity and total light. The intensity mode is convenient for kinetic studies, and the total light mode is essential in the studies of luciferins and photoproteins to quantitate the amount of these substances; the total light emitted is proportional to the amount of a luciferin or photoprotein luminesced. When purchasing a luminometer, it is advisable to choose one that is equipped with a photomultiplier tube with a spectral response curve as flat as possible in the wavelength range to be measured. This is important to avoid cumbersome corrections of the data. It is also recommended to calibrate the light meter in units of photons (see Section C4).

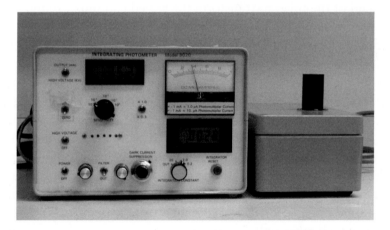

Fig. A.3.   Light meter used by the author (Model 8020, Pelagic Electronics). For total light measurements, the signals are integrated with capacitors. Milliammeter reading is automatically reset at full-scale position, and the number of resets is digitally indicated below the meter. The box at the right contains a photomultiplier and sample compartment.

Anderson *et al.* (1978) described a light meter they designed. The author uses Model 8020 Integrating Photometer (Pelagic Electronics, East Falmouth, MA 02536) shown in Fig. A.3, which was custom-made in 1991 based on the design made by Dr. Edward F. McNichol (Chase, 1960). The light meter uses capacitors to integrate the photomultiplier signals in order to quantify the total light, which is simple and highly suitable for continuous recording of integrated total light.

C4. *Calibration of Luminometer and the Measurement of Quantum Yield*

In bioluminescence and chemiluminescence reactions, light is emitted when the energy level of light emitter molecules falls from the excited state to the ground state. The quantum yield Q of the substance A is given by:

$$Q = E_c Q_f$$

where E_c is the yield of the excited state molecules of the substance A in its light-emitting chemical reaction, namely, the number of singlet excited state light-emitter molecules formed divided by the total number of the A molecules reacted (inclusive of any side reactions), and Q_f is the fluorescence quantum yield of the light emitter molecules.

In the measurements of quantum yield, a calibration of the light measuring apparatus in an absolute unit is necessary. The quantum yield of firefly luciferin catalyzed by *Photinus* luciferase was reported at 0.88 ± 0.25 (Seliger and McElroy, 1960), whereas that of *Cypridina* luciferin catalyzed by *Cypridina* luciferase was reported to be 0.30 ± 0.04 at 4°C (Johnson *et al.*, 1962; Shimomura and Johnson, 1970a). The quantum yield of luminol luminesced with H_2O_2 and hemin at pH 11.6 was found to be approximately 0.012 at various temperatures between 3°C and 54°C (Lee *et al.*, 1966; Lee and Seliger, 1972). These data were obtained with photomultiplier systems that had been calibrated with Standard Lamps, involving highly complicated methods. For practical purposes, however, a light meter can be calibrated and the quantum yield easily measured by utilizing the chemiluminescence of luminol as the secondary standard (Lee *et al.*, 1966; O'Kane and Lee, 2000). The adequacy of using a luminol standard to calibrate light meters is supported by the fact that the calibration results with the luminol luminescence and the *Cypridina* luminescence showed a close agreement (Shimomura and Johnson, 1967).

Calibration with luminol luminescence (Lee *et al.*, 1966). Calibration with luminol can be performed in aqueous solution in the presence of H_2O_2 and a suitable catalyst (luminescence range 380–550 nm, λ_{max} 430 nm), or in DMSO in the presence of potassium *t*-butoxide (λ_{max} 486 nm). The calibration method in aqueous medium requires the following three reagents:

(1) Luminol stock solution (approx. 20 mg luminol in 1 liter of 10 mM NaOH, to give an A value of 0.80 at 347 nm). Working solution: dilute the stock solution 10 times with pH 11.6 buffer (3 g of $Na_2HPO_4 \cdot 7H_2O$ and 15 g of $Na_3PO_4 \cdot 12H_2O$ in 1 liter of water).
(2) Hydrogen peroxide, 0.1% in water.
(3) Hemoglobin aqueous solution, A 0.2 at 414 nm, and a 1:3 dilution.

The luminescence reaction is started by injecting 0.1 ml of hydrogen peroxide solution into 1 ml of luminol solution, to begin a low intensity luminescence. Injection of hemoglobin 1:3 dilution (0.1 ml) to the mixture raises the intensity by a factor of 100. After the reaction is essentially over, 0.1 ml of A_{414} 0.2 hemoglobin is injected to ensure the complete utilization of luminol. If the spectral response curve of the photomultiplier used is not flat enough for the wavelength range involved, a correction of the data may be needed (Lee and Seliger, 1965; Wampler, 1978). Luminol emits light in a wavelength range of 380–550 nm with the maximum at 430 nm.

Under the conditions described above, 1 ml of luminol solution of A_{347} 1.0 will emit a total of $9.75 \pm 0.7 \times 10^{14}$ photons.

For monitoring the day-by-day fluctuation of the sensitivity of luminometer, the radioactive luminescence standard described by Hastings and Weber (1963), and Hastings and Reynolds (1966) is most useful and convenient (see also Section 2.6).

C 5.   Measurements of Coelenterazine, its Derivatives, and other Important Substances in Bioluminescence

In investigating a new bioluminescent organism, one of the initial experiments is to examine the presence of coelenterazine and some other substances, in addition to testing the luciferin-luciferase reaction. The information obtained might give important firsthand knowledge on the luminescence system involved. Concerning the luciferin-luciferase reaction, a brief discussion of the method is given in the Section C1.2. The information given below is to assist the detection and measurement of coelenterazine and several other substances that might play an important role in the luminous organism being studied.

C 5.1.   Assay of Coelenterazine

Coelenterazine can be detected and measured with a coelenterazine luciferase, i.e. a luciferase specific to coelenterazine. As the coelenterazine luciferase, the luciferases from the sea pansy *Renilla* and the copepods *Gaussia* and *Pleuromamma* are commercially available. Certain kinds of decapod shrimps, such as *Oplophorus* and *Heterocarpus*, contain a large amount of luciferase, and the luciferases purified from them are most satisfactory for the assay of coelenterazine considering their high activities and high quantum yields. Even partially purified preparations of these luciferases are satisfactory for most measurements. The author routinely uses purified *Oplophorus* luciferase.

To measure the amount of coelenterazine in a tissue, the sample is ground in a small agate mortar or homogenized mechanically with about 10 volumes of methanol at -5–$0°C$ (preferably under argon gas), and then the homogenate is centrifuged ($12,000\,g$, 5 min, $0°C$). A small amount ($\sim 10\,\mu l$) of the supernatant is placed in a test tube. Then 1 ml of 10 mM Tris-HCl buffer, pH 7.8, containing 0.5 M NaCl, 0.05% BSA and a certain amount of luciferase is injected into the test tube at room temperature, measuring the resulting light emission in terms of total light (quanta or light units). The amount of luciferase should be sufficient to complete the luminescence reaction within 2–3 minutes. The amount of coelenterazine in the sample is calculated from the total number of photons emitted (N), the quantum yield of coelenterazine (Q) that is specific to the luciferase used, and the Avogadro constant (6.02×10^{23} mole^{-1}); the amount of coelenterazine will be $N/(Q \times 6.02 \times 10^{23})$ mole. When *Renilla* luciferase is employed, the use of a pH 7.4 buffer instead of the pH 7.8 buffer will optimize the reaction rate (see Section 4.6). The BSA included (crystalline, Calbiochem, Cat. No. 12657) prevents the inactivation of luciferase, and the amount can be increased up to 0.2% depending on the luciferase used.

In tissues, most coelenterazine exists in a protein-bound stabilized form, which liberates free coelenterazine when extracted with methanol. Thus, the amount of coelenterazine measured by this method is the sum of free coelenterazine and its protein-bound form.

Minute amounts of coelenterazine can also be measured utilizing apoaequorin or apoobelin (Campbell and Herring, 1990; Thompson *et al.*, 1995). In this method, a sample containing coelenterazine is treated with an excess amount of apophotoprotein (apoaequorin or apoobelin) to convert it to a Ca^{2+}-sensitive photoprotein (aequorin or obelin). The photoprotein formed is assayed by luminescing it with Ca^{2+} to determine the amount of coelenterazine originally existed. With this method, the luminescence reaction is fast and usually complete in a few seconds, in contrast to the slower luminescence reactions with luciferases that sometimes require a few minutes to complete. However, the formation of photoprotein from apoaequorin is slow and not necessarily quantitative, and the overall accuracy of the photoprotein method does not compare favorably with that of the luciferase method that directly measures coelenterazine. The author recommends using a luciferase if the enzyme is available.

C 5.2. Assay of the Coelenterazine Luciferase Activity

The activity of a coelenterazine luciferase is assayed by measuring the intensity of light emitted upon the addition of coelenterazine. A sample tissue is ground or homogenized with 10–20 volumes of pure water at 0–3°C. To measure the luciferase activity, 10 µl of the homogenate is placed in a measuring tube. Then 1 ml of 10 mM Tris-HCl buffer, pH 7.8, containing 0.5 M NaCl, 0.05% BSA and 5 µl of methanolic coelenterazine solution ($A_{430, 1\,cm}$ 1.0) is injected into the tube at room temperature; the methanolic coelenterazine solution must be mixed with the buffer solution immediately before the injection. The resulting light emission is measured in terms of light intensity (photons/s or light units/s). Then, the rest of the homogenate is centrifuged ($15,000\,g$, 10 min, $0°C$), and the luciferase activity of 10 µl of the supernatant is measured by the same method as above. Normally the results of the two measurements are nearly identical, reflecting the amount of luciferase in the sample. If the result of the first measurement is significantly larger than that of the second measurement, it indicates the presence of bound luciferase in the pellet.

In order to solubilize the bound luciferase, the pellet is homogenized with 20–30 volumes of cold 10 mM Tris-HCl buffer, pH 7.5, containing 1 M NaCl, and the activity of the homogenate is measured. The homogenate is centrifuged and the activity of supernatant is also measured. A close agreement between the two measurements indicates that 1 M NaCl has solubilized the bound luciferase. If the two values differ significantly, a further effort of solubilization is needed (see Section C1.3).

C 5.3. Assay of the Stabilized Forms of Coelenterazine

Two types of stabilized coelenterazine are known to exist in bioluminescent organisms, i.e. the protein-bound form (usually bound to a calcium-binding protein) and the enol ester form (usually enol-sulfate; see Structure I, Fig. 5.5).

The assay result of coelenterazine described in Section C5.1 includes the amount of the protein-bound form of coelenterazine in addition to free coelenterazine. The assay of only the protein-bound form, or only the free coelenterazine, is complicated because the protein-bound form tends to liberate free coelenterazine by various stimuli, not only by Ca^{2+}. An example of the measurement of protein-bound coelenterazine is given by Shimomura and Johnson (1979b).

To measure the amount of the enol-sulfate of coelenterazine in a tissue, the methanol homogenate prepared in Section C5.1 is centrifuged, and 10 µl of the supernatant is heated with 0.1 ml of 0.5 M HCl at $95°C$ for one minute under argon gas, in order to hydrolyze the enol-sulfate group. The material is quickly cooled in ice water, neutralized with a small amount of solid $NaHCO_3$, and then the amount of coelenterazine in the sample is measured in the same pH 7.8 buffer as used in Section C5.1. The amount of coelenterazine obtained in this manner represents the sum of the enol-sulfate form, free coelenterazine and the protein-bound form. The amount of the enol-sulfate form is obtained by subtracting the amount of coelenterazine measured in Section C5.1 from the amount of coelenterazine obtained after the acid hydrolysis.

C 5.4.   *Assay of Dehydrocoelenterazine*

Dehydrocoelenterazine was first found in the liver of the luminous squid *Watasenia* (20 μg per specimen; Inouye *et al.*, 1977b). Later it was found that dehydrocoelenterazine is involved in the light emission of the luminous squid *Symplectoteuthis oualaniensis* (Isobe *et al.*, 1994; Takahashi and Isobe, 1994; see Section 6.3). More recently, large amounts of this compound have been found in the livers of the luminous squid *Symplectoteuthis luminosa* and certain luminous deep-sea fishes at the author's laboratory. These findings strongly suggest the importance of dehydrocoelenterazine in bioluminescence, as the precursor or storage form of coelenterazine. Dehydrocoelenterazine can be easily reduced into coelenterazine with a trace amount of sodium borohydride ($NaBH_4$) in both aqueous and methanolic solutions.

To measure dehydrocoelenterazine, a sample tissue is homogenized in methanol and then centrifuged, as in Section C5.1. A small amount of the supernatant (20–50 μl) is placed in a test tube, to which about 1 mg of $NaBH_4$ is added and stirred. After leaving it for five minutes, 1 ml of 20 mM Tris-HCl buffer, pH 7.8, containing luciferase, 0.5 M NaCl and 0.05% BSA is injected into the test tube, and the resulting light emission is measured in terms of total light.

C 5.5.   *Assay of Cypridina Luciferin*

The amount of *Cypridina* luciferin is measured with *Cypridina* luciferase. To measure the content of *Cypridina* luciferin in a tissue, the sample is extracted with methanol in the same manner as in the case of coelenterazine (Section C5.1). However, the extracted *Cypridina* luciferin is extremely unstable in air. Therefore, it is necessary to use methanol that has been bubbled with argon gas, and the extraction must be carried out as rapidly as possible under argon gas to prevent the oxidation of the luciferin. To assay *Cypridina* luciferin, 1 ml of 10 mM phosphate buffer, pH 7.0, containing 0.1 M NaCl and *Cypridina* luciferase is injected to a test tube containing 10 µl of the methanol extract, and the total light emission is measured.

Cypridina luciferase is not available commercially at present. However, *Cypridina* luciferase can be readily extracted from both live and dried *Cypridina*, and the crude extract, after dialysis, can be used in the measurement of *Cypridina* luciferin. Live *Cypridina* can be collected in Japan, and the ostracod can be cultivated in laboratory (see the last part of Section 3.1.2). Dried *Cypridina* is available from certain sources, including the author's laboratory.

C 5.6. *Assay of Cypridina Luciferase*

The amount of *Cypridina* luciferase is measured with *Cypridina* luciferin. However, the detection and measurement of a trace amount of the luciferase is extremely difficult, for the reason explained below.

Both *Cypridina* luciferin and coelenterazine spontaneously chemiluminesce in aqueous solutions in the absence of luciferase, and the chemiluminescence of *Cypridina* luciferin is much stronger than that of coelenterazine. The chemiluminescence of these substances is strongly enhanced when the molecules are in a hydrophobic environment created by association with various proteins, lipids and surfactants. For example, *Cypridina* luciferin in a pH 8 buffer (0.05 µg/3 ml) emits a low level of luminescence (1×10^7 quanta/s) at 25°C, which is increased to a higher level when 0.1 mg of BSA is added (3×10^7 quanta/s), or to a much higher level when a trace of CTAB is added ($3-8 \times 10^8$ quanta/s). The luminescence of *Cypridina* luciferin is also enhanced to very significant levels when 2.5% (w/w) of chicken egg yolk or mayonnaise is added (3×10^8 and 1.3×10^8 quanta/s, respectively). Therefore, it is necessary to take great care when detecting and measuring the activity of *Cypridina* luciferase with *Cypridina* luciferin in a crude tissue extract that contains proteins and lipids.

To assay *Cypridina* luciferase, 1 ml of 10 mM pH 7.0 phosphate buffer containing 0.1 M NaCl and 5–10 µl of an aqueous extract of sample tissue is injected into a test tube containing $5\,\mu l$ of methanolic solution of *Cypridina* luciferin ($A_{430,\,1\,cm}$ 1.0), and the resulting luminescence is measured in terms of light intensity. *Cypridina* (*Vargula*) luciferin is commercially available.

If a trace activity is indicated by the luminescence intensity measurement, the following two methods can be used to determine whether the light emission is due to the luciferase or it is an artifact: (1) Measure the luminescence intensity with a buffer that contains 1 mM EDTA (add luciferase to this buffer and wait 1 min before mixing with luciferin). If the luminescence was caused totally by luciferase, the light intensity will be decreased to about 20% by EDTA (see Section 3.1.7). (2) Inactivate luciferase by acidifying the sample to pH about 2.0, followed by neutralization with $NaHCO_3$. Inactivated luciferase should not show any luciferase activity.

C 5.7.   Assay of Ca^{2+}-sensitive Photoproteins

A sample tissue is ground or homogenized with 10–20 volumes of a cold neutral 10–20 mM buffer solution containing 10–20 mM EDTA (do not use phosphate buffer that might precipitate Ca^{2+}). The photoprotein activity of the homogenate is assayed by injecting 1 ml of 10 mM calcium acetate into 0.05–0.1 ml of the homogenate, measuring the emitted light in terms of total light. Then, the rest of the homogenate is centrifuged ($15,000\,g$, 10 min), and the activity of the supernatant is assayed similarly. Nearly equal activities obtained from the two assays indicate that the photoprotein is extracted into the supernatant. A significantly smaller activity in the supernatant than in the homogenate indicates that part or most of the photoprotein is in an insoluble particulate matter. In such a case, a sample tissue is first homogenized with 10–20 volumes of cold neutral solution of saturated ammonium sulfate containing 10–20 mM EDTA. After centrifugation, the precipitate obtained is mixed with the cold neutral buffer solution containing EDTA, as described at the beginning of this section, and the activity of the mixture is measured. Then, centrifuge the mixture, and measure the activity of the supernatant obtained. A good agreement between the last two measurements indicates that the photoprotein is solubilized by the pre-treatment with saturated ammonium sulfate solution. If the photoprotein still resists solubilization, see Section C1.3.

C 5.8.   *Measuring Bioluminescence in the Field*

The availability of an instrument to detect and measure the luminescence of organisms in the field is a great advantage in the study of bioluminescence. However, a portable luminometer of a small size, preferably weighing 1 lb or less, is difficult to find in the commercial market. It will also be quite expensive if it is available due to the low demand for this type of instrument. Anderson *et al.* (1978) described a portable luminometer. We have been using hand-made light meters since about 1970 (Johnson and Shimomura, 1972, 1978). These light meters are equipped with a phototube and are considerably less sensitive than a luminometer equipped with a photomultiplier tube, but we found them very useful. One of them, in use since 1980, is illustrated in Fig. A.4. This light meter was assembled from a phototube of low operating voltage, one Op-amp, and a milliammeter. The signal from the phototube is amplified by the Op-amp, and read on the

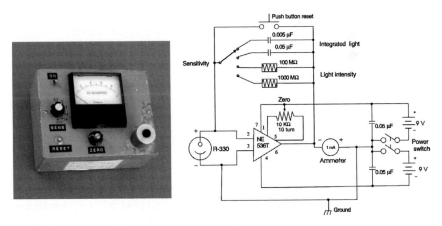

Fig. A.4.   A portable light meter hand-made in 1980, and its circuit diagram. This inexpensively made instrument contains a phototube (Hamamatsu R-330) and one Op-amp, and is powered with two 9-volt batteries. Signal integration for total light measurement is made by polystyrene capacitors of low-leakage type. Reagents can be injected into the test tube in the sample compartment (at lower right in the photo) through a syringe needle that has a small bend to prevent leakage of light. The R-330 phototube is now discontinued, but probably replaceable with R645 or R414.

milliammeter in the units of light intensity or integrated total light. In this light meter, integration is performed with capacitors.

When using a light meter in the field, measurements should not be done under sunlight or at brightly lit places to avoid the effects of the hysteresis of phototube or photomultiplier and the phosphorescence-like luminescence of various substances. Materials such as green plants and *Cypridina* luciferin emit a strong after-glow when illuminated by sunlight. A bathroom can be an ideal place for measurements during fieldwork.

Coelenterazine and the corresponding luciferase can be easily tested in the field. A small piece of tissue sample is put in a test tube with methanol (for coelenterazine) or water (for luciferase), and crushed with a spatula. To measure coelenterazine, a buffer solution containing a coelenterazine luciferase is injected into a small amount of the fluid part of the crushed sample mixture. Similarly, luciferase can be measured with a buffer solution containing coelenterazine. The presence of *Cypridina* luciferin can be tested in the same fashion, with the methanol extract of samples and crude *Cypridina* luciferase. However, the detection of a very weak *Cypridina* luciferase activity in the field is not recommended (see Section C5.6). To test the presence of a Ca^{2+}-sensitive photoprotein, crush a sample in a neutral buffer solution containing 20–50 mM EDTA, and then add 10 mM calcium acetate to a small portion of the fluid part of the crushed sample to detect any light emission.

C 6. ^{18}O-Labeling of the Reaction Product CO_2

When the bioluminescence reaction of a luciferin produces a CO_2 molecule from the carbonyl group of the luciferin, the identification of the origin of an O atom in the product CO_2 — whether it originates in molecular oxygen or in solvent water — often provides an important clue in deducing the mechanism of luminescence reaction. DeLuca and Dempsey (1970) were the first to use ^{18}O to investigate the origin of the O atom in the CO_2 produced by the firefly bioluminescence reaction, though their experiments unfortunately resulted in an incorrect conclusion (see Section 1.8). The source of the error was identified and the method was improved (Shimomura and Johnson, 1973a, 1975a; Shimomura et al., 1977). There are four factors that affect the result in the labeling of the O atom in CO_2 (Shimomura, 1982):

(1) Exchange of oxygen atom between the C=O group of luciferin and the solvent water. The exchange is slow with carboxylic acids but fairly rapid with aldehydes and ketones (Samuel, 1962).

(2) Exchange of oxygen atom between the product CO_2 and solvent water. This exchange is quite rapid (Mills and Urey, 1940), and is influenced by many factors, such as the volumes of gas phase and liquid phase, pH, and buffer composition. When 0.25 µmol of ^{14}C-labelled coelenterazine analogue was luminesced in the presence of $^{17}O_2$ and Renilla luciferase, 40% of the O atom of the product CO_2 was exchanged with the O atom of the solvent water in 4 min (Hart et al., 1978); and when 0.2 µmol of $^{13}CO_2$ was shaken with a buffer containing $H_2{}^{18}O$ for 40 sec, the exchange was 6–8% (Shimomura and Johnson, 1979c).

(3) Contaminating CO_2. The ubiquitous presence of CO_2 causes a great difficulty to the experiment. Normal atmosphere contains 0.03% CO_2, and a 5–6 ml portion of pure water equilibrated with atmosphere contains 0.06 µmol of CO_2. However, the amount of CO_2 in buffers and luciferase solutions is much greater. For example, 5 ml of freshly prepared Tris-HCl buffer, pH 7.8, contained 0.13 µmol of CO_2 after 20 min degassing, and the value increased to 0.23 µmol when the same buffer was tested one week later (Shimomura et al., 1977). It is strongly advised to eliminate all nonessential CO_2 sources from the experimental environment.

(4) Residual molecular oxygen. The process of degassing to remove contaminating CO_2 also removes most of $^{16}O_2$ in the reaction medium, and the amount of residual $^{16}O_2$ after degassing is estimated to be about 10–20 nmol in a solution of 5 ml. This residual $^{16}O_2$ can be removed by luminescence reaction when the solutions of luciferin and luciferase are mixed under evacuated conditions, or can be simply diluted with a large excess of $^{18}O_2$ to reduce its effect to a negligible level.

Considering the effects of the four factors described above, the amount of luciferin required to obtain highly reliable data would be about 1 μmol, and 0.3 μmol seems to be the minimum amount to obtain any conclusive result.

Experimental procedure. The experimental setup we used is illustrated in Fig. A.5. The reaction vessel shown is for a luminescence reaction that requires the mixing of two solutions. One solution (typically a luciferase solution) is placed at the bottom and the other solution (typically a luciferin solution) in the side arm. The reaction vessel and traps can be fabricated from ready-made components such as a flask, stopcocks, tubing, etc. using the basic skills of glass blowing. The trap immersed in a dry ice acetone bath removes all volatile substances produced by the bioluminescence reaction (as well as the solvent vapor) except CO_2. Thus, only the molecules of CO_2 are collected in the trap immersed in liquid nitrogen. If there is any possibility of CO_2 contamination from the vacuum line and vacuum gauge, another trap immersed in liquid nitrogen can be added at the right side of the existing liquid nitrogen trap (Shimomura, 1982).

The system is continuously evacuated to a pressure of about 1 μm Hg (0.133 Pa) without coolant of the traps for at least 2 hr before

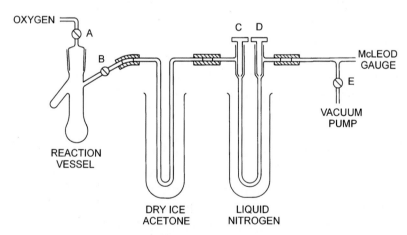

Fig. A.5. An apparatus for the ^{18}O-labelling of the CO_2 produced in bioluminescence reactions. The stopcocks A–E and the ground joint of reaction vessel are lubricated with high-vacuum grease, and the connections between the glass tubes are made with short pieces of thick-walled silicon rubber tubing. The stopcocks C and D are high-vacuum type, such as Ace Glass, Cat. No. 8197-04 and 8195-236, or Corning, Cat. No. 7473-3.

experiment (stopcock A closed; all others open; $^{18}O_2$ container not connected to A). To begin the experiment, close stopcock B, and then open A to introduce air to the reaction vessel. Remove the top part of the reaction vessel from the ground joint. Place a solution of luciferase (about 4–5 ml) at the bottom of the vessel, and a solution of luciferin (0.3–0.5 ml) in the side arm. Re-attach the top part of the vessel, and attach $^{18}O_2$ container to stopcock A. Open stopcocks A, C, D and E (the stopcock of $^{18}O_2$ container, not shown, is closed). Then, with great care, slowly open stopcock B partially to evacuate the reaction vessel and to induce bubbling of the solutions. Bubbling ceases within 2–3 min. The vessel is further degassed with intermittent evacuation (by manipulation of stopcock B), stirring inside the vessel with a swivel motion for 20 min.

At this point, two options are available if required. (1) If more complete degassing is desired, freeze the liquid contents by soaking the appropriate part of reaction vessel in dry ice/acetone bath, evacuate to 10 μm Hg, then thaw the contents, followed by an additional 10 min degassing. (2) If the removal of residual $^{16}O_2$ is desired, mix the two solutions by tilting the vessel to consume the $^{16}O_2$ by luciferin-luciferase luminescence reaction. The resulting reaction will cease in 2–3 min. Then, degas the content for 1–2 min to remove the CO_2 produced in the luminescence reaction.

Now adjust the temperature of the contents to 20–25°C. With stopcock B closed, introduce $^{18}O_2$ to the reaction vessel (through stopcock A), and then close the stopcock A. Agitate the vessel briefly to dissolve $^{18}O_2$ in the solution, and then mix the two solutions vigorously to start luminescence reaction (if not previously mixed). When light emission ceases (typically in 30–60 s), freeze the solution in a dry ice/acetone bath for 15 min. Immerse the two traps in the respective coolants. Close stopcock D, and then open stopcock B. When stopcock D is slightly opened, the CO_2 produced will be collected in the liquid nitrogen trap. Close stopcock C at the pressure of 100 μm Hg to minimize contamination with H_2O vapor, and then close stopcock D at 10 μm Hg. Remove the coolants immersing the traps, and disconnect the liquid nitrogen trap containing CO_2 at the silicon rubber connections. The CO_2 sample is analyzed by mass spectrometry. See Shimomura *et al.* (1977) and Shimomura (1982) for further details.

Calculations. The atoms of incorporated ^{18}O is calculated from mass spectral data. When the ratio of the peak heights at mass-to-charge ratio (m/e) 44, 46 and 48 for CO_2 is $X:Y:Z$, assuming that the height of each peak is strictly proportional to the number of CO_2 molecules, the atom fraction of ^{18}O in CO_2, C, is given by:

$$C = \frac{Y + 2Z}{2(X + Y + Z)}$$

The luminescence reaction does not produce $C^{18}O_2$, thus Z is negligible, therefore:

$$C = \frac{Y}{2(X + Y)}$$

When Y/X is defined equal to R,

$$C = \frac{R}{2(R + 1)}$$

The atoms of incorporated oxygen per mole of CO_2, N, is

$$N = \frac{2(100C - 0.20)}{E - 0.20}$$

where E is the atom % of ^{18}O in $^{18}O_2$ or $H_2^{18}O$ used in experiment, and 0.20 is the natural abundance (percent) of ^{18}O. It should be noted that the equation used in some previous reports, $C = R/(2 + R)$ (DeLuca and Dempsey, 1973; Tsuji *et al.*, 1977), will result in an error that increases as the value of R increases.

C7. Glass Blowing

According to the author's experience, few recent investigators have knowledge or experience of glass blowing. This is probably due to the availability of various, convenient ready-made products for a wide range of applications. However, the ranges of the ready-made products are limited, and it is often difficult to fulfill the requirements of innovative experiments. A special order of glassware is not only cumbersome but also takes a long time, at least 1–2 months. In the chemical study of bioluminescence, some simple techniques of glass blowing, such as cutting, pulling, bending, and connecting glass tubing, are often greatly helpful.

The author recalls that, a half century ago, the glass blowing technique was an absolute essential for experimental chemists, and had been described in various publications in that period. For example, *Experiments in Organic Chemistry* by Louis F. Fieser (1955) contained a section on glass blowing, which explained concisely the basic techniques. However, in its revised version *Organic Experiments* by the same author (1968), the description was reduced to a minimum. Detailed instructions on scientific glass blowing can be presently found on the World Wide Web sites of the Glass Blowing Service of East Carolina University and of Michigan State University. In glass blowing, usually borosilicate glass is used, and a gas-oxygen burner is preferable to a gas-compressed air burner.

D.   Advice to Students Who are Interested in Studying the Chemistry of Bioluminescence

In a sense, bioluminescence is a treasure box of interesting and unusual chemistry. Some novel, epoch-making findings have been made in this field, giving us important knowledge as well as immeasurable benefits through the applications of the findings. For example, the assay method of ATP now widely used is an application of the firefly luminescence, and the unique imidazopyrazinone chemistry has been developed from the studies of *Cypridina* luciferin and coelenterazine. In the case of the luminous jellyfish *Aequorea*, the study resulted in the finding of two important proteins: the extraordinary energy-precharged photoprotein aequorin, highly useful as a calcium probe in biological systems, and the green fluorescent protein that is now an indispensable genetic marker. There are many potentially important bioluminescence systems to be explored.

Since the chemical mechanisms involved in bioluminescence are very diverse, there is no established protocol or methodology for isolating and studying new bioluminescent systems. Therefore, the most important factor in the study of the chemistry of bioluminescence is that the investigator has a creative and innovative mind, although effort would always be essential. The author believes that amateurism is not a handicap in this field; it could be a valuable asset, provided the investigator has the necessary basic knowledge.

Knowledge. The phenomenon of bioluminescence is not a simple chemical reaction, and involves various fields of science. To do creative work in the chemistry of bioluminescence, an investigator must have a solid understanding of a broad range of basic sciences, including physics, biology, and mathematics in addition to chemistry, at least at the undergraduate level. The author once had postdoctoral fellows in biochemistry who could not calculate the volume of a chromatographic column from its dimensions, and who did not know what to expect when $BaCl_2$ is dissolved in a saline solution containing sulfate. Though these are probably extreme examples, it is obvious that creativity gives no merit in the absence of adequate basic knowledge. Accurate and undistorted knowledge is most important. It will be disastrous if one's mind is preoccupied with incorrect or distorted preconceptions. The fundamental laws of science and chemistry are absolute and fully trustable. However, various theories, rules and hypotheses made on experiential bases are not laws; these should be trusted only with reservations especially when studying an unusual phenomenon like bioluminescence.

References. Before beginning a study, previously reported data should be critically examined, as to their reliability and correctness. Then, use only those data and information that are absolutely correct. It is a sad reality that errors at the fundamental level are surprisingly common in the field of bioluminescence. If you carry out your research on a faulty foundation, the resulting work cannot be correct and, therefore, is likely to be worthless. Reported experimental data are usually correct and trustable under the conditions described, but the interpretation of the data and the conclusions might be erroneous. Sometimes, the method used to obtain data might be faulty, making the reported data worthless. Although spotting errors, shortfalls or inconsistencies in reported papers might not be easy for many students, the effort will be worthwhile to avoid regret later.

Raw material of luminous organisms. Obtaining the necessary amount of raw material organisms for a chemical study of a luciferin or a photoprotein is often not an easy task. It requires a good deal of preliminary investigation, physical work and travel. This step might be the most costly part of the project. It is advisable to seek the advice and help of biologists, oceanographers, fishermen, or other experts who are familiar with the organisms. Searching for information on the Internet is a most useful tool that was not available some years ago.

The amount of raw material required for the chemical study of luminescent substances today is much less than that required 20 or 30 years ago, due to the advance in instrumentation. In the case of a luciferin, try to obtain 1–2 mg of purified luciferin. If that is not feasible, try to obtain at least 0.1 mg of purified material, and with some luck, the structure may be obtained.

Extraction and purification. The processes of extraction and purification of luciferins and photoproteins are usually the most critical part of the project, and the work often involves various difficulties and complexities specific to the substance under investigation, such as a highly labile nature and extremely low content. Some discussion on this matter is given in Section C1.

Structure determination of luciferin. Once a luciferin is obtained in a sufficient purity, the determination of luciferin structure should be attempted; most of the important properties of luciferin are usually already obtained during the course of purification as a necessity. The structural study is considerably more straightforward than the extraction and purification, due to the availability of advanced methods, such as high-resolution mass spectrometry and various NMR techniques. If help or collaboration is needed in structure determination, the attractiveness of a luciferin will make it easy to find a good collaborator. However, the purified luciferin is usually an extremely precious material considering the effort spent in preparing it. To avoid accidental loss of the purified material, the chosen collaborator must have solid knowledge and experience in structure determination; a criterion to be considered is that the person has successfully done the structure determination of at least one new natural product.

Confidence. The study of an unknown bioluminescence system is not routine work. During the study, the work will encounter various difficulties and problems that need to be solved in order to make progress. Some of them might easily be solved, but some might take weeks or months to resolve, although all difficulties will eventually be overcome if one persists. If you persist and manage to solve a difficult problem, you will build confidence in yourself, which will empower you to solve further difficulties. Therefore, it is important to solve the first difficult problem that you encounter. If you give up once, you will probably give up the next time.

REFERENCES

Abe, K. (1994). *The Light of Marine Fireflies*, Chikuma Shobo, Tokyo (in Japanese).

AbouKhair, N. K., Ziegler, M. M., and Baldwin, T. O. (1984). The catalytic turnover of bacterial luciferase produces a quasi-stable species of altered conformation. *In* Bray, R. C., *et al.* (eds.), *Flavins Flavoproteins*, Proc. Int. Symp., 8th, pp. 371–374. de Gruyter, Berlin.

AbouKhair, N. K., Ziegler, M. M., and Baldwin, T. O. (1985). Bacterial luciferase: demonstration of a catalytically competent altered conformational state following a single turnover. *Biochemistry* **24**: 3942–3947.

Abu-Soud, H., Mullins, L. S., Baldwin, T. O., and Raushel, F. M. (1992). Stopped-flow kinetic analysis of the bacterial luciferase reaction. *Biochemistry* **31**: 3807–3813.

Airth, R. L., and Foerster, G. E. (1960). Some aspects of fungal bioluminescence. *J. Cell. Comp. Physiol.* **56**: 173–182.

Airth, R. L., and Foerster, G. E. (1962). The isolation of catalytic components required for cell-free fungal bioluminescence. *Arch. Biochem.* **97**: 567–573.

Airth, R. L., and Foerster, G. E. (1964). Enzymes associated with bioluminescence of *Panus stipticus luminescens* and *Panus stipticus nonluminescens*. *J. Bacteriol.* **88**: 1372–1379.

Airth, R. L., and McElroy, W. D. (1959). Light emission from extracts of luminous fungi. *J. Bacteriol.* **77**: 249–250.

Papers with three or more authors are listed in order of publication date, not in alphabetical order of authors. Papers with five or more authors are shown with the name of the first author followed by *et al.* The list contains a considerable number of papers that are not cited in the text.

Airth, R. L., Rhodes, W. C., and McElroy, W. D. (1958). The function of coenzyme A in luminescence. *Biochim. Biophys. Acta* **27**: 519–532.

Airth, R. L., Foerster, G. E., and Behrens, P. Q. (1966). The luminous fungi. *In* Johnson, F. H., and Haneda, Y. (eds.), *Bioluminescence in Progress*, pp. 203–223. Princeton University Press, Princeton, NJ.

Alter, S. C., and DeLuca, M. (1986). The sulfhydryls of firefly luciferase are not essential for activity. *Biochemistry* **25**: 1599–1605.

Anctil, M., and Gruchy, C. G. (1970). Stimulation and photography of bioluminescence in lanternfishes (Mictophidae). *J. Fish. Res. Bd. Can.* **27**: 826–829.

Anctil, M., and Shimomura, O. (1984). Mechanism of photoinactivation and re-activation in the bioluminescence system of the ctenophore. *Mnemiopsis. Biochem. J.* **221**: 269–272.

Anderson, J. M. (1980). Biochemistry of centipede bioluminescence. *Photochem. Photobiol.* **31**: 179–181.

Anderson, J. M., and Cormier, M. J. (1973). Lumisomes, the cellular site of bioluminescence in coelenterates. *J. Biol. Chem.* **248**: 2937–2943.

Anderson, J. M., Charbonneau, H., and Cormier, M. J. (1974). Mechanism of calcium induction of *Renilla* bioluminescence. Involvement of a calcium-triggered luciferin binding protein. *Biochemistry* **13**: 1195–1200.

Anderson, J. M., Faini, G. J., and Wampler, J. E. (1978). Construction of instrumentation for bioluminescence and chemiluminescence assays. *Method. Enzymol.* **57**: 529–540.

Anderson, R. S. (1935). Studies on bioluminescence, II. The partial purification of *Cypridina* luciferin. *J. Gen. Physiol.* **19**: 301–305.

Aoki, M., Hashimoto, K., and Watanabe, H. (1989). The intrinsic origin of bioluminescence in the ascidian. *Clavelina minata. Biol. Bull.* **176**: 57–62.

Arai, M. N., and Brinckman-Voss, A. (1980). Hydromedusae of British Columbia and Puget Sound. *Can. Bull. Fish. Aqua. Sci.* Bulletin **204**: 1–181.

Arnhold, J., Reichl, S., Peikovic, M., and Vocks, A. (2002). Pholasin luminescence of polymorphonuclear leukocytes. *In* Stanley, P. E., and Kricka, L. J. (eds.), *Bioluminescence and Chemiluminescence, Progress and Current Applications*, pp. 233–236. World Scientific, Singapore.

Ashley, C. C. (1970). An estimate of calcium concentration changes during the contraction of single muscle fibres. *J. Physiol.* **210**: 133–134P.

Ashley, C. C., and Campbell, A. K., eds. (1979). *Detection and Measurement of Free Ca²⁺ in Cells*. Elsevier/North-Holland Biomedical Press, Amsterdam.

Azzi, A., and Chance, B. (1969). The "energized state" of mitochondria: lifetime and ATP equivalence. *Biochim. Biophys. Acta* **189**: 141–151.

Bae, Y. M., and Hastings, J. W. (1994). Cloning, sequencing and expression of dinoflagellate luciferase DNA from a marine alga. *Gonyaulax polyedra*. *Biochim. Biophys. Acta* **1219**: 449–456.

Baird, G. S., Zacharias, D. A., and Tsien, R. Y. (2000). Biochemistry, mutagenesis, and oligomerism of DsRed, a red fluorescent protein from coral. *Proc. Natl. Acad. Sci. USA* **97**: 11984–11989.

Baldwin, T. O. (1996). Firefly luciferase: the structure is known, but the mystery remains. *Structure* **4**: 223–228.

Baldwin, T. O., and Ziegler, M. M. (1992). The biochemistry and molecular biology of bacterial bioluminescence. *In* Mueller, F. (ed.), *Chem. Biochem. Flavoenzymes*, pp. 467–530. CRC, Boca Raton, Fla.

Baldwin, T. O., Berends, T., and Treat, M. L. (1984). Cloning of the bacterial luciferase and use of the clone to study the enzyme and reaction *in vivo*. *In* Kricka, L. J. (ed.), *Anal. Appl. Biolumin. Chemilumin.*, Proc. Int. Symp., 3rd, pp. 101–104. Academic, London, UK.

Baldwin, T. O., Holzman, T. F., Holzman, R. B., and Riddle, V. A. (1986). Purification of bacterial luciferase by affinity method. *Method. Enzymol.* **133**: 98–108.

Baldwin, T. O., *et al.* (1987). Applications of the cloned bacterial luciferase genes luxA and luxB to the study of transcriptional promoters and terminators. *In* Schoelmerich, J. (ed.), *Biolumin. Chemilumin.*, Proc. Int. Biolumin. Chemilumin. Symp., 4th, 1986, pp. 373–376. Wiley, Chichester, UK.

Baldwin, T. O., *et al.* (1987). Structural studies of bacterial luciferases: results from recombinant DNA technology. *In* Schoelmerich, J. (ed.), *Biolumin. Chemilumin.*, Proc. Int. Biolumin. Chemilumin. Symp., 4th, 1986, pp. 351–360. Wiley, Chichester, UK.

Baldwin, T. O., *et al.* (1989). Site-directed mutagenesis of bacterial luciferase: analysis of the "essential" thiol. *J. Biolumin. Chemilumin.* **4**: 40–48.

Baldwin, T. O., *et al.* (1989). The complete nucleotide sequence of the lux regulon of *Vibrio fischeri* and the luxABN region of bacterial bioluminescence. *J. Biolumin. Chemilumin.* **4**: 326–341.

Baldwin, T. O., Sinclair, J. F., Clark, A. C., and Ziegler, M. M. (1994). Molecular biology of the folding and assembly of the subunits of bacterial luciferase. *In* Campbell, A. K., *et al.* (eds.), *Biolumin. Chemilumin.*, Proc. Int. Symp., 8th, pp. 501–508. Wiley, Chichester, UK.

Ballou, B., Szent-Gyorgyi, C., and Finley, G. (2000). Properties of a new luciferase from the copepod *Gaussia princeps*. 11th Int. Symp. on Biolumin. & Chemilumin., Abstract p. 34. Asilomar, CA.

Balny, C., and Hastings, J. W. (1975). Fluorescence and bioluminescence of bacterial luciferase intermediates. *Biochemistry* **14**: 4719–4723.

Becker, R. S. (1969). *Theory and Interpretation of Fluorescence and Phosphorescence*, p. 240. Wiley Interscience, New York.

Beijerinck, M. W. (1889). Le Photobacterium luminosum, bacterie lumineu Mer du Nordse de la. *Arch. Neerl. Sci. Exactes Nat.* **23**: 401–415.

Beijerinck, M. W. (1916). Die Leuchtbakterien der Nordsee im August und September. *Folia Microbiol.* **4**: 15–40.

Bellisario, R., and Cormier, M. J. (1971). Peroxide-linked bioluminescence catalyzed by a copper-containing, non-heme luciferase isolated from a bioluminescent earthworm. *Biochem. Biophys. Res. Commun.* **43**: 800–805.

Bellisario, R., Spencer, T. E., and Cormier, M. J. (1972). Isolation and properties of luciferase, a non-heme peroxide from the bioluminescent earthworm, *Diplocardia longa*. *Biochemistry* **11**: 2256–2266.

Bermudes, D., Gerlach, V. L., and Nealson, K. H. (1990). Effects of culture conditions on mycelial growth and luminescence in *Panellus stypticus*. *Mycologia* **82**: 295–305.

Bitler, B., and McElroy, W. D. (1957). The preparation and properties of crystalline firefly luciferin. *Arch. Biochem. Biophys.* **72**: 358–368.

Blinks, J. R. (1985). Detection of Ca^{2+} with photoproteins. *In* Van Dyke, K. (ed.), *Bioluminescence and Chemiluminescence; Instrumental Applications*, Vol. 2, pp. 185–226. CRC, Boca Raton, FL.

Blinks, J. R. (1989). Use of calcium-regulated photoproteins as intracellular Ca^{2+} indicators. *Method. Enzymol.* **172**: 164–203.

Blinks, J. R. (1990). Use of photoproteins as intracellular calcium indicators. *Environ. Health Perspect* **84**: 75–81.

Blinks, J. R., and Harrer, G. C. (1975). Multiple forms of the calcium-sensitive bioluminescent protein aequorin. *Fed. Proc.* **34**: 474.

Blinks, J. R., and Moore, E. D. W. (1986). Practical aspects of the use of photoproteins as biological calcium indicators. *Soc. Gen. Physiol. Ser.* **40**: 229–238.

Blinks, J. R., Prendergast, F. G., and Allen, D. G. (1976). Photoproteins as biological calcium indicators. *Pharmacol. Rev.* **28**: 1–93.

Blinks, J. R., *et al.* (1978). Practical aspects of the use of aequorin as a calcium indicator: Assay, preparation, microinjection, and interpretation of signals. *Method. Enzymol.* **57**: 292–328.

Blinks, J. R., Wier, W. G., Hess, P., and Prendergast, F. G. (1982). Measurement of Ca^{2+} concentrations in living cells. *Prog. Biophys. Mol. Biol.* **40**: 1–114.

Bode, V. C., and Hastings, J. W. (1963). The purification and properties of the bioluminescent system in *Gonyaulax polyedra*. *Arch. Biochem. Biophys.* **103**: 488–499.

Boden, B. P., and Kampa, E. M. (1959). Spectral composition of the luminescence of the euphausiid *Thysanoessa raschii*. *Nature* **184**: 1321–1322.

Boden, B. P., and Kampa, E. M. (1964). Planktonic bioluminescence. *Oceanogr. Mar. Biol.* **2**: 341–371.

Bokman, S. H., and Ward, W. W. (1981). Renaturation of *Aequorea* green-fluorescent protein. *Biochem. Biophys. Res. Commun.* **101**: 1372–1380.

Bondar, V. S., Vysotskii, E. S., Gamalei, I. A., and Kaulin, A. B. (1991). Isolation, properties, and application of calcium activated photoprotein from the hydroid polyp *Obelia longissima*. *Tsitologiya* **33**: 50–59.

Bondar, V. S., Trofimov, K. P., and Vysotskii, E. S. (1992). Physico-chemical properties of the hydroid polyp *Obelia longissima* photoprotein. *Biokhimiya* **57**: 1481–1490.

Bondar, V. S., *et al.* (1995). Cadmium-induced luminescence of recombinant photoprotein obelin. *Biochim. Biophys. Acta* **1231**: 29–32.

Bowden, B. J. (1950). Some observations on a luminescent freshwater limpet from New Zealand. *Biol. Bull.* **99**: 373–380.

Bowie, L. J. (1978). Synthesis of firefly luciferin and analogs. *Method. Enzymol.* **57**: 15–28.

Bowlby, M. R., Widder, E. A., and Case, J. F. (1990). Patterns of stimulated bioluminescence in two pyrosomes (Tunicata: Pyrosomatidae). *Biol. Bull.* **179**: 340–350.

Boylan, M., *et al.* (1989). Lux C, D and E genes of the *Vibrio fischeri* luminescence operon code for the reductase, transferase and synthetase enzymes involved in aldehyde biosynthesis. *Photochem. Photobiol.* **49**: 681–688.

Boyle, R. (1668). Experiments concerning the relation between light and air in shining wood and fish. *Phil. Trans.* **2**: 581–600.

Branchini, B. R. (2000). Chemical synthesis of firefly luciferin analogs and inhibitors. *Method. Enzymol.* **305**: 188–195.

Branchini, B. R., and Rollins, C. B. (1989). High-performance liquid chromatography-based purification of firefly luciferases. *Photochem. Photobiol.* 50: 679–684.

Branchini, B. R., Lusins, J. O., and Zimmer, M. (1997). A molecular mechanics and database analysis of the structural preorganization and activation of the chromophore-containing hexapeptide fragment in green fluorescent protein. *J. Biomol. Struct. Dyn.* 14: 441–448.

Branchini, B. R., *et al.* (1997). Identification of a firefly luciferase active site peptide using a benzophenone-based photooxidation reagent. *J. Biol. Chem.* 272: 19359–19364.

Branchini, B. R., *et al.* (2002). Yellow-green and red firefly bioluminescence from 5,5-dimethoxyloxyluciferin. *J. Am. Chem. Soc.* 124: 2112–2113.

Branchini, B. R., *et al.* (2003). A mutagenesis study of the putative luciferin binding site residues of firefly luciferase. *Biochemistry* 42: 10429–10436.

Branchini, B. R., *et al.* (2004). An alternative mechanism of bioluminescence color determination in firefly luciferase. *Biochemistry* 43: 7255–7262.

Branchini, B. R., *et al.* (2005). Mutagenesis evidence that the partial reactions of firefly bioluminescence are catalyzed by different conformations of the luciferase C-terminal domain. *Biochemistry* 44: 1385–1393.

Brand, L., and Witholt, B. (1967). Fluorescence measurements. *Method. Enzymol.* 11: 776–856.

Brehm, P., and Morin, J. G. (1977). Localization and characterization of luminescent cells in *Ophiopsila californica* and *Amphipholis squamata* (Echinodermata: Ophiuroidea). *Biol. Bull.* 152: 12–25.

Brejc, K., *et al.* (1997). Structural basis for dual excitation and photoisomerization of the *Aequorea victoria* green fluorescent protein. *Proc. Natl. Acad. Sci. USA* 94: 2306–2311.

Brovko, L. Y., Gandelman, O. A., and Savich, W. I. (1994). Fluorescent and quantum-chemical evaluation of emitter structure in firefly bioluminescence. *In* Campbell, A. K., *et al.* (eds.), *Biolumin. Chemilumin.*, Proc. Int. Symp., 8th, pp. 525–527. Wiley, Chichester, UK.

Brovko, L. Y., Dementieva, E. I., and Ugarova, N. N. (1997). Structure of the catalytic site of firefly luciferase and bioluminescence color. *In* Hastings, J. W., *et al.* (eds.), *Biolumin. Chemilumin.*, Proc. Int. Symp., 9th 1996, pp. 206–211. Wiley, Chichester, UK.

Bublitz, G., King, B. A., Nurse, P., and Boxer, S. G. (1998). Electronic structure of the chromphore in green fluorescent protein (GFP). *J. Am. Chem. Soc.* 120: 9370–9371.

Buck, J. B. (1978). Functions and evolutions of bioluminescence. *In* Herring, P. J. (ed.), *Biolumnescence in Action*, pp. 419–460. Academic Press, London.

Buck, J., and Buck, E. (1976). Synchronous fireflies. *Scient. Am.* **234**: No. 5, 74–85.

Buchner, P. (1914). Sind die Leuchtorgane Pilzorgane? *Zool. Anz.* **45**: 17–21.

Campbell, A. K. (1974). Extraction, partial purification and properties of obelin, the calcium-activated luminescent protein from the hydroid *Obelia geniculata. Biochem. J.* **143**: 411–418.

Campbell, A. K., and Herring, P. J. (1987). A novel red fluorescent protein from the deep sea luminous fish *Malacosteus niger. Comp. Biochem. Physiol.* **86B**: 411–417.

Campbell, A. K., and Herring, P. J. (1990). Imidazopyrazine bioluminescence in copepods and other marine organisms. *Mar. Biol.* **104**: 219–225.

Campbell, A. K., et al. (1981). Application of the photoprotein obelin to the measurement of free Ca^{2+} in cells. *In* DeLuca, M. A., and McElroy, W. D. (eds.), Bioluminescence and Chemiluminescence, Basic Chemistry and Analytical Application, pp. 601–607. Academic Press, New York.

Campbell, A. K., Patel, A. K., Razavi, Z. S., and McCapra, F. (1988). Formation of the calcium-activated photoprotein obelin from apo-obelin and mRNA inside human neutrophils. *Biochem. J.* **252**: 143–149.

Campbell, A. K., et al. (1989). Photoproteins as indicators of intracellular free Ca^{2+}. *J. Biolumin. Chemilumin.* **4**: 463–474.

Campbell, A. K., et al. (1993). Targeting aequorin and firefly luciferase to defined compartments of plant and animal cells. *In* Szalay, A. A., et al. (eds.), *Biolumin. Chemilumin.*, Proc. Int. Symp., 7th, pp. 55–59. Wiley, Chichester, UK.

Cannell, J. P. (1998). *Natural Products Isolation*. Humana Press, Totowa, NJ.

Cao, J. G., and Meighen, E. A. (1989). Purification and structural identification of an autoinducer for the luminescence system of *Vibrio harveyi. J. Biol. Chem.* **264**: 21670–21676.

Chalfie, M. (1995). Green fluorescent protein. *Photochem. Photobiol.* **62**: 651–656.

Chalfie, M., and Kain, S. (1998). *Green Fluorescent Protein: Properties, Applications, and Protocols*. Wiley-Liss, New York, NY.

Chalfie, M., et al. (1994). Green fluorescent protein as a marker for gene expression. Science 263: 802–805.

Charbonneau, H., and Cormier, M. J. (1979). Ca^{2+}-induced bioluminescence in Renilla reniformis. Purification and characterization of a calcium-triggered luciferin-binding protein. J. Biol. Chem. 254: 769–780.

Charbonneau, H., et al. (1985). Amino acid sequence of the calcium-dependent photoprotein aequorin. Biochemistry 24: 6762–6771.

Chase, A. M. (1960). The measurement of luciferin and luciferase. In Glick, D. (ed.), Method of Biochemical Analysis, Vol. VIII, pp. 61–117. Interscience Publishers, New York.

Chase, A. M., and Brigham, E. H. (1951). The ultraviolet and visible absorption spectra of Cypridina luciferin solutions. J. Biol. Chem. 190: 529–536.

Chase, A. M., and Langridge, R. (1960). The sedimentation constant and molecular weight of Cypridina luciferase. Arch. Biochem. Biophys. 88: 294–297.

Chen, F.-Q., et al. (1994). Synthesis and preliminary chemi- and bioluminescence studies of a novel photolabile coelenterazine analogue with a trifluoromethyl diazirine group. Chem. Commun. 2405–2406.

Chen, L. H., and Baldwin, T. O. (1989). Random and site-directed mutagenesis of bacterial luciferase: investigation of the aldehyde binding site. Biochemistry 28: 2684–2689.

Chiba, K., Hoshi, M., Isobe, M., and Hirose, E. (1998). Bioluminescence in the tunic of the colonial ascidian Clavelina miniata: identification of luminous cells in vitro. J. Exp. Zool. 281: 546–553.

Chittock, R. S., et al. (1993). The quantum yield of luciferase is dependent on ATP and enzyme concentrations. Mol. Cryst. Liq. Cryst. Sci. Technol., Sect. A 236: 59–64.

Cho, K. W., Lee, H. J., and Shim, S. C. (1986). Bioluminescence of acetaldehyde catalyzed by bacterial luciferase. Photochem. Photobiol. 43: 477–480.

Cho, K. W., Lee, H. J., and Shim, S. C. (1986). Regeneration of the monooxygenating intermediate by hydrogen peroxide in bacterial bioluminescent reaction of Vibrio fischeri. Han'guk Saenghwa Hakhoechi 19: 151–154.

Cho, K. W., Lee, H. J., and Shim, S. C. (1987). Acetaldehyde-induced bacterial bioluminescence of Photobacterium fischeri luciferase and stimulation by long chain alkyl compounds. *Han'guk Saenghwa Hakhoechi* **20**: 45–50.

Cho, K. W., Colepicolo, P., and Hastings, J. W. (1989). Autoinduction and aldehyde chain-length effects on the bioluminescent emission from the yellow protein associated with luciferase in *Vibrio fischeri* strain Y-1b. *Photochem. Photobiol.* **50**: 671–677.

Choi, H., Tang, C.-K., and Tu, S. C. (1995). Catalytically active forms of the individual sub-units of *Vibrio harveyi* luciferase and their kinetic and binding properties. *J. Biol. Chem.* **270**: 16813–16819.

Cobbold, P. H. (1980). Cytoplasmic free calcium and amoeboid movement. *Nature* **285**: 441–446.

Cobbold, P. H., and Bourne, P. K. (1984). Aequorin measurement of free calcium in single heart cells. *Nature* **312**: 444–446.

Cobbold, P. H., and Rink, T. J. (1987). Fluorescence and biolumines-cence measurement of cytoplasmic free calcium. *Biochem. J.* **248**: 313–328.

Cody, C. W., *et al.* (1993). Chemical structure of the hexapeptide chro-mophore of the *Aequorea* green-fluorescent protein. *Biochemistry* **32**: 1212–1218.

Cohn, D. H., *et al.* (1985). Nucleotide sequence of the luxA gene of *Vibrio harveyi* and the complete amino acid sequence of the α subunit of bacterial luciferase. *J. Biol. Chem.* **260**: 6139–6146.

Cohn, M., and Urey, H. C. (1938). Oxygen exchange reactions of organic compounds and water. *J. Am. Chem. Soc.* **60**: 679–687.

Colepicolo, P., *et al.* (1990). A sensitive and specific assay for superoxide anion released by neutrophils or macrophages based on bioluminescence of polynoidin. *Anal. Biochem.* **184**: 369–374.

Colepicolo, P., Camarero, V. C. P. C., and Hastings, J. W. (1992). A circadian rhythm in the activity of superoxide dismutase in the photosynthetic alga *Gonyaulax polyedra*. *Chronobiol. Int.* **9**: 266–268.

Colepicolo, P., Camarero, V. C. P. C., Eckstein, J., and Hastings, J. W. (1992). Induction of bacterial luciferase by pure oxygen. *J. Gen. Micro-biol.* **138**: 831–836.

Colepicolo-Neto, P., Costa, C., and Bechara, E. J. H. (1986). Brazilian species of luminescent Elateridae. Luciferin identification and bioluminescence spectra. *Insect Biochem.* **16**: 803–810.

Colepicolo-Neto, P., Viviani, V. R., Barros, M. P., and Costa, C. (1999). Colors and biological functions of beetle bioluminescence. *An. Acad, Bras. Cienc.* **71**: 169–174.

Coles, S. L., DeFelice, R. C., Eldredge, L. G., and Carlton, J. T. (1997). Biodiversity of marine communities in Pearl Harbor, Oahu, Hawaii with observations on introduced exotic species. Bishop Museum Technical Report No. 10, Honolulu, Hawaii.

Conti, E., Franks, N. P., and Brick, P. (1996). Crystal structure of firefly luciferase throws light on a superfamily of adenylate-forming enzymes. *Structure* 4: 287–298.

Cormack, B. P., Valdivia, R. H., and Falkow, S. (1997). Mutants of green fluorescent protein (GFP) with enhanced fluorescence characteristics. *In* Hastings, J. W., *et al.* (eds.), *Biolumin. Chemilumin.*, Proc. Int. Symp., 9th, 1996, pp. 387–390. Wiley, Chichester, UK.

Cormier, M. J. (1961). Biochemistry of *Renilla reniformis* luminescence. *In* McElroy, W. D., and Glass, B. (eds.), *Light and Life*, pp. 274–293. Johns Hopkins Press, Baltimore.

Cormier, M. J. (1962). Studies on the bioluminescence of *Renilla reniformis*. II. Requirement for 3′,5′-diphosphoadenosine in the luminescent reaction. *J. Biol. Chem.* 237: 2032–2037.

Cormier, M. J. (1978). Comparative biochemistry of animal systems. *In* Herring, P. J. (ed.), *Bioluminescence in Action*, pp. 75–108. Academic Press, London.

Cormier, M. J., and Charbonneau, H. (1977). Isolation, properties and function of a calcium-triggered luciferin binding protein. *In* Wasserman, R. H., *et al.* (eds.), *Calcium Binding Proteins and Calcium Function*, pp. 481–489. Elsevier North-Holland.

Cormier, M. J., and Dure, L. S. (1963). Studies on the bioluminescence of *Balanoglossus biminiensis* extracts. I. Requirement for hydrogen peroxide and characteristics of the system. *J. Biol. Chem.* 238: 785–789.

Cormier, M. J., and Hori, K. (1963). Studies on the bioluminescence of *Renilla reniformis*. IV. Non-enzymatic activation of *Renilla* luciferin. *Biochim. Biophys. Acta* 88: 99–104.

Cormier, M. J., and Strehler, B. L. (1953). The identification of KCF: requirement of long-chain aldehydes for bacterial extract luminescence. *J. Am. Chem. Soc.* 75: 4864–4865.

Cormier, M. J., and Totter, J. R. (1957). Quantum efficiency determinations on components of the bacterial luminescence system. *Biochim. Biophys. Acta* 25: 229–237.

Cormier, M. J., Kreiss, P., and Prichard, P. M. (1966). Bioluminescence systems of the peroxidase type. *In* Johnson, F. H., and Haneda, Y. (eds.), *Bioluminescence in Progress*, pp. 363–384. Princeton University Press, Princeton, NJ.

Cormier, M. J., Crane, J. M., and Nakano, Y. (1967). Evidence for the identity of the luminescence systems of *Porichthys prosissimus* (fish) and *Cypridina hilgendorfii* (crustacean). *Biochem. Biophys. Res. Commun.* **29**: 747–752.

Cormier, M. J., Hori, K., and Karkhanis, Y. D. (1970). Studies on the bioluminescence of *Renilla reniformis*. VII. Conversion of luciferin into luciferyl sulfate by luciferin sulfokinase. *Biochemistry* **9**: 1184–1189.

Cormier, M. J., *et al.* (1973). Evidence for similar biochemical requirements for bioluminescence among the coelenterates. *J. Cell. Physiol.* **81**: 291–298.

Cormier, M. J., Hori, K., and Anderson, J. M. (1974). Bioluminescence in coelenterates. *Biochim. Biophys. Acta* **346**: 137–164.

Cormier, M. J., Prasher, D. C., Longinaru, M., and McCann, R. O. (1989). The enzymology and molecular biology of the calcium-activated photoprotein, aequorin. *Photochem. Photobiol.* **49**: 509–512.

Cotton, B., Allshire, A., Cobbold, P. H., Muller, T., and Campbell, A. K. (1989). Pholasin: a novel bioluminescent probe for monitoring oxidative stress in single cardiomyocytes. *Biochem. Soc. Trans.* **17**: 705–706.

Cousineau, J., and Meighen, E. A. (1976). Chemical modification of bacterial luciferase with ethoxyformic anhydride: evidence for an essential histidyl residue. *Biochemistry* **15**: 4992–5000.

Cousineau, J., and Meighen, E. A. (1977). Sequential chemical modification of a histidyl and a cysteinyl residue in bacterial luciferase. *Can. J. Biochem.* **55**: 433–438.

Crane, J. M. (1968). Bioluminescence in the batfish *Dibranchus atlanticus*. *Copeia*, 410–411.

Cutler, M. W. (1995). Characterization and energy transfer mechanism of green-fluorescent protein from *Aequorea victoria*, Ph.D. Dissertation, Rutgers University, New Brunswick, NJ.

Cutler, M. W., and Ward, W. W. (1993). Protein-protein interaction in *Aequorea* bioluminescence. Bioluminescence Symposium, Maui, Hawaii, Nov. 3–10, 1993.

Cutler, M. W., and Ward, W. W. (1997). Spectral analysis and proposed model for GFP dimerization. *In* Hasting, J.W., *et al.* (eds.), *Bioluminescence and Chemiluminescence; Molecular Reporting with Photons*, pp. 403–406. Wiley, New York.

Danilov, V. S., and Malkov, Y. A. (1984). Bacterial luciferase from *Beneckea harveyi* — an iron-containing enzyme. *Dokl. Akad. Nauk SSSR* **275**: 206–209.

Danilov, V. S., and Shibilkina, O. K. (1985). Induction of bacterial luciferase synthesis by phenobarbital. *Mikrobiologiya* **54**: 750–754.

Daub, M. E., Leisman, G. B., Clark, R. A., and Bowden, E. F. (1992). Reductive detoxification as a mechanism of fungal resistance to singlet oxygen-generating photosensitizers. *Proc. Natl. Acad. Sci. USA* **89**: 9588–9592.

Daubner, S. C., and Baldwin, T. O. (1989). Interaction between luciferase from various species of bioluminescent bacteria and the Yellow Fluorescent Protein of *Vibrio fischeri* strain Y-1. *Biochem. Biophys. Res. Commun.* **161**: 1191–1198.

Daubner, S. C., Astorga, A. M., Leisman, G. B., and Balwin, T. O. (1987). Yellow light emission of *Vibrio harveyi* strain Y-1: purification and characterization of the energy-accepting yellow fluorescent protein. *Proc. Natl. Acad. Sci. USA* **84**: 8912–8916.

Davenport, D., and Nicol, J. A. C. (1955). Luminescence of hydromedusae. *Proc. Roy. Soc., B* **144**: 399–411.

Davenport, D., Wootton, T. M., and Cushing, J. E. (1952). The biology of the Sierra luminous millipede, *Luminodesmus sequoiae* Loomis and Davenport. *Biol. Bull.* **102**: 100–110.

De Wet, J. R., Wood, K. V., Helinski, D. R., and DeLuca, M. (1985). Cloning of firefly luciferase cDNA and the expression of active luciferase in *Escherichia coli. Biochemistry* **82**: 7870–7873.

De Wet, J. R., Wood, K. B., Helinski, D. R., and DeLuca, M. (1986). Cloning firefly luciferase. *Method. Enzymol.* **133**: 3–14.

De Wet, J. R., *et al.* (1987). Cloning and expression of the firefly luciferase gene in mammalian cells. *In* Schoelmerich, J. (ed.), *Biolumin. Chemilumin.*, Proc. Int. Biolumin. Chemilumin. Symp., 4th, 1986, pp. 369–372. Wiley, Chichester, UK.

DeFeo, T. T., and Morgan, K. G. (1986). A comparison of two different indicators: quin 2 and aequorin in isolated single cells and intact strips of ferret portal vein. *Pfluegers Arch.* **406**: 427–429.

DeLuca, M., and Dempsey, M. E. (1970). Mechanism of oxidation in firefly luminescence. *Biochem. Biophys. Res. Commun.* **40**: 117–122.

DeLuca, M., and Dempsey, M. E. (1973). Mechanism of bioluminescence and chemiluminescence elucidated by use of oxygen-18. *In* Cormier, M. J., *et al.* (eds.), *Chemiluminescence and Bioluminescence*, pp. 345–355. Plenum Press, New York.

DeLuca, M., and McElroy, W. D. (1978). Purification and properties of firefly luciferase. *Method. Enzymol.* **57**: 3–15.

Deluca, M., and McElroy, W. D. (1984). Two kinetically distinguishable ATP sites in firefly luciferase. *Biochem. Biophys. Res. Commun.* **123**: 764–770.

DeLuca, M., Wirtz, G. W., and McElroy, W. D. (1964). Role of sulfhydryl groups in firefly luciferase. *Biochemistry* **3**: 935–939.

DeLuca, M., *et al.* (1971). Mechanism of oxidative carbon dioxide production during *Renilla reniformis* bioluminescence. *Proc. Natl. Acad. Sci. USA* **68**: 1658–1660.

DeLuca, M., Dempsey, M. E., Hori, K., and Cormier, M. J. (1976). Source of oxygen in CO_2 produced during chemiluminescence of firefly luciferyl-adenylate and Renilla luciferin. *Biochem. Biophys. Res. Commun.* **69**: 262–267.

Dement'eva, E. I., *et al.* (1986). The pH-dependence of the bioluminescence spectra and kinetic constant for luciferase of the firefly *Luciola mingrelica. Biokhimiya* **51**: 130–139.

Deng, L., *et al.* (2001). Structural basis for the emission of violet bioluminescence from a W92F obelin mutant. *FEBS Lett.* **506**: 281–285.

Deng, L., *et al.* (2002a). High resolution (1.17 A) structure of obelin with coelenterazine h. *Luminescence* **17**: 87.

Deng, L., *et al.* (2002b). The crystal structure of the calcium-regulated photoprotein obelin from *Obelia geniculata. Luminescence* **17**: 87–88.

Deng, L., *et al.* (2004a). Preparation and X-ray crystallographic analysis of the Ca^{2+}-discharged photoprotein obelin. *Acta Crystallogr. Ser. D Biol. Crystallogr.* **60**: 512–514.

Deng, L., *et al.* (2004b). Crystal structure of a Ca^{2+}-discharged photoprotein: Implications for mechanisms of the calcium trigger and bioluminescence. *J. Biol. Chem.* **279**: 33647–33652.

Denton, E. J., Herring, P. J., Widder, E. A., and Latz, M. F. (1985). The roles of filters in the photophores of oceanic animals and their relation to vision in the oceanic environment. *Proc. R. Soc. Lond.* B **225**: 63–97.

DeSa, R., Hastings, J. W., and Vatter, A. E. (1963). Luminescent "crystalline" particles: an organized subcellular bioluminescent system. *Science* **141**: 1269–1270.

DeSa, R., and Hastings, J. W. (1968). The characterization of scintillons. Bioluminescent particles from the marine dinoflagellate, *Gonyaulax polyedra. J. Gen. Physiol.* **51**: 105–122.

Deschamps, J. R., Miller, C. E., and Ward, K. B. (1995). Rapid purification of recombinant green fluorescent protein using the hydrophobic properties of an HPLC size-exclusion column. *Protein Expression and Purification* **6**: 555–558.

Deutscher, M. P., ed. (1990). Protein purification. *Method. Enzymol.*, Vol. 182.

Devine, J. H., *et al.* (1993). Luciferase from the east European firefly *Luciola mingrelica*: cloning and nucleotide sequence of the cDNA over-expression in *Escherichia coli* and purification of enzyme. *Biochim. Biophys. Acta* **1173**: 121–132.

Dey, A. C., Rahal, S. R., Rimsay, R. L., and Senciall, I. R. (1981). A simple procedure for the solubilization of NADH-cytochrome b$_5$ reductase. *Anal. Biochem.* **110**: 373–379.

Duane, W., and Hastings, J. W. (1975). Flavin mononucleotide reductase of luminous bacteria. *Mol. Cell. Biochem.* **6**: 53–64.

Dubois, R. (1885). Note sur la physiologie des Pyrophores. *Compt. Rend. Soc. Biol.* **37**: 559–562.

Dubois, R. (1887). Fonction photogenique chez le *Pholas dactylus*. *Compt. Rend. Soc. Biol.* **39**: 564–566.

Dukhovich, A. F., Ugarova, N. N., Berezin, I. V., and Filippova, N. Y. (1986). Role of magnesium ions in the regulation of catalytic activity of the firefly luciferase. *Dokl. Akad. Nauk SSSR* **288**: 1500–1504.

Dukhovich, A. F., *et al.* (1988). Choline-containing phospholipids as specific activators and stabilizers of firefly luciferase. *Dokl. Akad. Nauk SSSR* **298**: 1257–1260.

Dunlap, J. C. (1979). Circadian organization of bioluminescence in *Gonyaulax polyedra*. Ph.D. Dissertation, Harvard University, Cambridge, MA.

Dunlap, J. C., and Hastings, J. W. (1981). The biological clock in *Gonyaulax* controls luciferase activity by regulating turnover. *J. Biol. Chem.* **256**: 10509–10518.

Dunlap, J. C., Hastings, J. W., and Shimomura, O. (1980). Crossreactivity between the light-emitting systems of distantly related organisms: Novel type of light-emitting compound. *Proc. Natl. Acad. Sci. USA* **77**: 1394–1397.

Dunlap, J. C., Hastings, J. W., and Shimomura, O. (1981). Dinoflagellate luciferin is structurally related to chlorophyll. *FEBS Lett.* **135**: 273–276.

Dunlap, P. V. (1991). Organization and regulation of bacterial luminescence genes. *Photochem. Photobiol.* **54**: 1157–1170.

Dunlap, P. V., and Kita-Tsukamoto, K. (2001). Luminous bacteria. *In* Dworkin, M. (ed.), *The Prokaryotes*, electronic edn., Chapter 329. Springer-Verlag, New York.

Dunn, D. K., Michaliszyn, G. A., Bogacki, I. G., and Meighen, E. A. (1973). Conversion of aldehyde to acid in the bacterial bioluminescent reaction. *Biochemistry* **12**: 4911–4918.

Dunstan, S. L., *et al.* (2000). Cloning and expression of the bioluminescent photoprotein pholasin from the bivalve mollusc *Pholas dactylus*. *J. Biol. Chem.* **275**: 9403–9409.

Dure, L. S., and Cormier, M. J. (1961). Requirements for luminescence in extracts of a balanoglossid species. *J. Biol. Chem.* **236**: PC48–50.

Dure, L. S., and Cormier, M. J. (1963). Studies on the bioluminescence of *Balanoglossus biminiensis* extracts. II. Evidence for the peroxidase nature of *Balanoglossus* luciferase. *J. Biol. Chem.* **238**: 790–793.

Eberhard, A. (1972). Inhibition and activation of bacterial luciferase synthesis. *J. Bacteriol.* **109**: 1101–1105.

Eberhard, A., and Hastings, J. W. (1972). A postulated mechanism for the bioluminescent oxidation of reduced flavin mononucleotide. *Biochem. Biophys. Res. Commun.* **47**: 348–353.

Eberhard, A., *et al.* (1981). Structural identification of autoinducer of *Photobacterium fischeri* luciferase. *Biochemistry* **20**: 2444–2449.

Ebrahimzadeh, M. H., and Haddadchie, G. R. (1993). Intrinsic fluorescent compounds in Armillaria mellea: hyphae and rhizomorph. *J. Sci. Islamic Repub. Iran* **4**: 241–246.

Eckstein, J. W., and Ghisla, S. (1991). On the mechanism of bacterial luciferase. 4a,5-Dihydroflavins as model compounds for reaction intermediates. *In Flavins Flavoproteins*, Proc. Int. Symp., 10th, 1990, 269–272.

Eckstein, J. W., *et al.* (1990). A time-dependent bacterial bioluminescence emission spectrum in an *in vitro* single turnover system: energy transfer alone cannot account for the yellow emission of *Vibrio fischeri* Y-1. *Proc. Natl. Acad. Sci. USA* **87**: 1466–1470.

Eckstein, J. W., Hastings, J. W., and Ghisla, S. (1993). Mechanism of bacterial bioluminescence. 4a,5-Dihydroflavin analogs as models for luciferase hydroperoxide intermediates and the effect of substituents at the 8-position of flavin on luciferase kinetics. *Biochemistry* **32**: 404–411.

Eley, M., *et al.* (1970). Bacterial bioluminescence. Comparisons of bioluminescence emission spectra, the fluorescence of luciferase reaction mixtures, and the fluorescence of flavin cations. *Biochemistry* **9**: 2902–2908.

Elowitz, M. B., *et al.* (1997). Photoactivation turns green fluorescent protein red. *Curr. Biol.* 7: 809–812.

Endo, M., Kajiwara, M., and Nakanishi, K. (1970). Fluorescent constituents and cultivation of *Lampteromyces japonicus*. *Chem. Commun.*, pp. 309–310.

Escher, A., O'Kane, D. J., Lee, J., and Szalay, A. A. (1989). Bacterial luciferase $\alpha\beta$ fusion protein is fully active as a monomer and highly sensitive *in vivo* to elevated temperature. *Proc. Natl. Acad. Sci. USA* 86: 6528–6532.

Fabiato, A. (1988). Computer programs for calculating total from specified free or free from specified total ionic concentrations in aqueous solutions containing multiple metals and ligands. *Method. Enzymol.* 157: 378–417.

Fabiato, A., and Fabiato, F. (1979). Calculator program for computing the composition of the solutions containing multiple metals and ligands used for experiments in skinned muscle cells. *J. Physiol., Paris* 75: 463–505.

Fagan, T. F., *et al.* (1993). Cloning, expression and sequence analysis of cDNA for the Ca^{2+}-binding photoprotein, mitrocomin. *FEBS Lett.* 333: 301–305.

Feigl, F. (1960). *Spot Tests in Organic Analysis*. Elsevier.

Ferri, S. R., and Meighen, E. A. (1991). A Lux-specific myristoyl transferase in luminescent bacteria related to eukaryotic serine esterases. *J. Biol. Chem.* 266: 12852–12857.

Ferri, S. R., Soly, R. R., Szittner, R. B., and Meighen, E. A. (1991). Structure and properties of luciferase from *Photobacterium phosphoreum*. *Biochem. Biophys. Res. Commun.* 176: 541–548.

Fieser, L. F. (1955). *Experiments in Organic Chemistry*. D.C. Heath & Company, Lexington, MA.

Fieser, L. F. (1968). *Organic Experiments*. D.C. Heath & Company, Lexington, MA.

Fisher, A. J., Raushel, F. M., Baldwin, T. O., and Rayment, I. (1995). Three-dimensional structure of bacterial luciferase from *Vibrio harveyi* at 2.4 Å resolution. *Biochemistry* 34: 6581–6586.

Fisher, A. J., *et al.* (1996). The 1.5-Å resolution crystal structure of bacterial luciferase in low salt conditions. *J. Biol. Chem.* 271: 21956–21968.

Flood, P. R., Bassot, J.-M., and Herring, P. J. (1996). The microscopical structure of the bioluminescence system in the medusa *Periphylla*

periphylla. In Hastings, J. W., *et al.* (eds.), Bioluminescence and Chemi-luminescence: Molecular Reporting with Photons, pp. 149–153. John Wiley. Chichester, UK.

Flynn, G. C., Beckers, C. J. M., Baase, W. A., and Dahlquist, F. W. (1993). Individual subunits of bacterial luciferase are molten globules and interact with molecular chaperones. *Proc. Natl. Acad. Sci. USA* **90**: 10826–10830.

Fogel, M., and Hastings, J. W. (1971). A substrate binding molecule in the *Gonyaulax* bioluminescence reaction. *Arch. Biochem. Biophys.* **142**: 310–321.

Fogel, M., Schmitter, R. E., and Hastings, J. W. (1972). On the physical identity of scintillons: bioluminescent particles in *Gonyaulax polyedra. J. Cell. Sci.* **11**: 305–317.

Fontes, R., Dukhovich, A., Sillero, A., and Günther Sillero, M. A. (1997). Synthesis of dehydroluciferin by firefly luciferase, effect of dehydrolu-ciferin, coenzyme A and nucleoside triphosphates on the luminescence reaction. *Biochem. Biophys. Res. Commun.* **237**: 445–450.

Fontes, R., *et al.* (1998). Dehydroluciferin-AMP is the main intermediate in the luciferin dependent synthesis of Ap4A catalyzed by firefly luciferase. *FEBS Lett.* **438**: 190–194.

Foran, D. R., and Brown, W. M. (1988). Nucleotide sequence of the LuxA and LuxB genes of the bioluminescent marine bacterium *Vibrio fischeri. Nucleic Acids Res.* **16**: 777.

Ford, S. R., Hall, M. S., and Leach, F. R. (1992). Enhancement of fire-fly luciferase activity by cytidine nucleotides. *Anal. Biochem.* **204**: 283–291.

Ford, S. R., Buck, L. M., and Leach, F. R. (1995). Does the sulfhydryl or the adenine moiety of CoA enhance firefly luciferase activity? *Biochim. Biophys. Acta* **1252**: 180–184.

Forskal, P. (1775). Descriptiones animalium avium, amphibiorum, piscium, insectorum, vermium; quae in itinere orientali observavit Petrus Forskal, Post mortem auctoris edidit Carsten niebuhr ed. Hauniae.

Fossa, J. H. (1992). Mass occurrence of *Periphylla periphylla* in a Norwegian fjord. *Sarsia* **77**: 237–251.

Fracheboud, M. G., Shimomura, O., Hill, R. K., and Johnson, F. H. (1969). Synthesis of *Latia* luciferin. *Tetrahedron Lett.*, pp. 3951–3952.

Frackman, S., Anhalt, M., and Nealson, K. H. (1990). Cloning, organiza-tion, and expression of the bioluminescence genes of *Xenorhabdus lumi-nescens. J. Bacteriol.* **172**: 5767–5773.

Fradkov, A. F. (2000). Novel fluorescent protein from *Discosoma* coral and its mutants possesses a unique far-red fluorescence. *FEBS Lett.* **479**: 127–130.

Francisco, W. A., *et al.* (1996). Interaction of bacterial luciferase with 8-substituted flavin mononucleotide derivatives. *J. Biol. Chem.* **271**: 104–110.

Freeman, G., and Ridgway, E. B. (1987). Endogenous photoproteins, calcium channels and calcium transients during metamorphosis in hydrozoans. *Roux's Arch. Dev. Biol.* **196**: 30–50.

Fridovich, I. (1975). Oxygen: boon and bane. *American Scientist* **63**: 54–59.

Fridovich, I. (1986). Biological effects of the superoxide radical. *Arch. Biochem.* **247**: 1–11.

Fried, A., and Tu, S. C. (1984). Affinity labeling of the aldehyde site of bacterial luciferase. *J. Biol. Chem.* **259**: 10754–10759.

Friedlander, J., and Hastings, J. W. (1967). The reversibility of the denaturation of bacterial luciferase. *Biochemistry* **6**: 2893–2900.

Fujii, T., *et al.* (2002). A novel photoprotein from oceanic squid (*Symplectoteuthis oualaniensis*) with sequence similarity to mammalian carbon-nitrogen hydrolase domains. *Biochem. Biophys. Res. Commun.* **293**: 874–879.

Fukasawa, S., Dunlap, P. V., Baba, M., and Osumi, M. (1987). Identification of an agar-digesting, luminous bacterium. *Agric. Biol. Chem.* **51**: 265–268.

Fukasawa, S., Suda, T., and Kubota, S. (1988). Identification of luminous bacteria isolated from the light organ of the fish. *Aeropoma japonicum*. *Agric. Biol. Chem.* **52**: 285–286.

Fuqua, W. C., Winans, S. C., and Greenberg, E. P. (1994). Quorum sensing in bacteria: the LuxR-LuxI family of cell density-responsive transcriptional regulators. *J. Bacteriol.* **176**: 269–275.

G'andelman, O. A., Brovko, L. U., Ugarova, N. N., and Shchegolev, A. A. (1990). The bioluminescence system of firefly. A fluorescence spectroscopy study of the interaction of the reaction product, oxyluciferin, and its analogs with luciferase. *Biokhimiya* **55**: 1052–1058.

Gast, R., and Lee, J. (1978). Isolation of the *in vivo* emitter in bacterial bioluminescence. *Proc. Natl. Acad. Sci. USA* **75**: 833–837.

Gates, B. J., and DeLuca, M. (1975). The production of oxyluciferin during the firefly luciferase light reaction. *Arch. Biochem. Biophys.* **169**: 616–621.

Ghisla, S., Entsch, H., Massey, V., and Husein, M. (1977). On the structure of flavin-oxygen intermediates involved in enzymic reactions. *Eur. J. Biochem.* **76**: 139–148.

Gilchrist, J. D. F. (1919). Luminosity and its origin in a South African earthworm (*Chitota* sp.). *Trans. Roy. Soc. S. Afr.* **7**: 203–212.

Gilkey, J. C., Jaffe, L. F., Ridgway, E. B., and Reynolds, G. T. (1978). A free calcium wave traverses the activating egg of the medaka, *Oryzias latipes*. *J. Cell Biol.* **76**: 448–466.

Gilroy, S., Hughes, W. A., and Trewavas, A. J. (1989). A comparison between Quin-2 and aequorin as indicators of cytoplasmic calcium levels in higher plant cell protoplasts. *Plant Physiol.* **90**: 482–491.

Girsch, S. J., and Hastings, J. W. (1978). The properties of mnemiopsin, a bioluminescent and light sensitive protein purified by hollow fiber techniques. *Mol. Cell. Biochem.* **19**: 113–124.

Girsch, S., Herring, P. J., and McCapra, F. (1976). Structure and preliminary biochemical characterization of the bioluminescent system of *Ommastrephes pteropus* (Steenstrup) (Mollusca: Cephalopoda). *J. Mar. Biol. Ass., U.K.* **56**: 707–722.

Gitelzon, G. I., Tugai, V. A., and Zakharchenko, A. N. (1990). Production of obelin, a calcium-activated photoprotein, from *Obelia longissima* and its application for registration of the calcium efflux from the fragmented sarcoplasmic reticulum of skeletal muscles. *Ukr. Biokhim. Zh.* **62**: 69–76.

Godeheu de Riville (1760). Memoire sur la mer lumineuse. *Mem. de Math. et Phys. Acad. Roy. Sci., Paris* **3**: 269–276.

Gonzalez, D. S., Sawyer, A., and Ward, W. W. (1997). Spectral perturbations of mutants of recombinant *Aequorea victoria* green-fluorescent protein (GFP). *Photochem. Photobiol.* **65**: 21S.

Gorokhovatsky, A. Y., et al. (2004). Fusion of *Aequorea victoria* GFP and aequorin provides their Ca^{2+}-induced interaction that results in red shift of GFP absorption and efficient bioluminescence energy transfer. *Biochem. Biophys. Res. Commun.* **320**: 703–711.

Goto, T. (1968). Chemistry of bioluminescence. *Pure Appl. Chem.* **17**: 421–441.

Goto, T., and Fukatsu, H. (1969). *Cypridina* bioluminescence VII. Chemiluminescence in micelle solutions — A model system for *Cypridina* bioluminescence. *Tetrahedron Lett.*, pp. 4299–4302.

Goto, T., Kubota, I., Suzuki, N., and Kishi, Y. (1973). Aspects of the mechanism of bioluminescence. *In* Cormier, M. J., et al. (eds.), *Chemiluminescence and Bioluminescence*, pp. 325–335. Plenum Press, New York.

Goto, T., Iio, H., Inoue, S., and Kakoi, H. (1974). Squid luminescence I. Structure of Watasenia oxyluciferin, a possible light-emitter in the bioluminescence of *Watasenia scintillans*. *Tetrahedron Lett.*, pp. 2321–2324.

Green, A. A., and McElroy, W. D. (1956). Crystalline firefly luciferase. *Biochim. Biophys. Acta* **20**: 170–176.

Gross, L. A. (2000). The structure of the chromophore within DsRed a red fluorescent protein from coral. *Proc. Natl. Acad. Sci. USA* **22**: 11990–11995.

Gruber, M. G., Kutuzova, G. D., and Wood, K. V. (1997). Cloning and expression of a *Phengodes* luciferase. *In* Hastings, J. W., *et al.* (eds.), *Biolumin. Chemilumin.*, Proc. Int. Symp., 9th, 1996, pp. 244–247. Wiley, Chichester, UK.

Guillard, R. R. L. (1974). *In* Smith, W. L., and Chanley, M. H. (eds.), *Culture of Marine Invertebrate Animals*, pp. 29–60. Plenum Press, New York.

Guillard, R. R. L., and Ryther, J. H. (1962). Studies of marine planktonic diatoms 1. *Cyclotella nana* Hustedt and *Detonula confervacea* (cleve). *Canadian J. Microbiol.* **8**: 229–239.

Gunsalus-Miguel, A., *et al.* (1972). Purification and properties of bacterial luciferases. *J. Biol. Chem.* **247**: 398–404.

Haddock, S. H. D., and Case, J. F. (1994). A bioluminescent chaetognath. *Nature* **367**: 225–226.

Haddock, S. H. D., and Case, J. F. (1995). Not all ctenophores are bioluminescent: *Pleurobrachia. Biol. Bull.* **189**: 356–362.

Haddock, S. H. D., Rivers, T. J., and Robison, B. H. (2001). Can coelenterates make coelenterazine? Dietary requirement for luciferin in cnidarian bioluminescence. *Proc. Natl. Acad. Sci. USA* **98**: 11148–11151.

Hall, M. S., and Leach, F. R. (1988). Stability of firefly luciferase in Tricine buffer and in a commercial enzyme stabilizer. *J. Biolumin. Chemilumin.* **2**: 41–44.

Hamman, J. P., and Seliger, H. H. (1972). The mechanical triggering of bioluminescence in marine dinoflagellates: chemical basis. *J. Cell. Physiol.* **80**: 397–408.

Haneda, Y. (1939). Luminosity of *Rocellaria grandis* Deshayes (Lamellibranchia). *Kagaku Nanyo* **2**: 36–39 (in Japanese).

Haneda, Y. (1955). Luminous organisms of Japan and the Far East. *In* Johnson, F. H. (ed.), *The Luminescence of Biological Systems*, pp. 335–385. American Association for the Advancement of Science, Washington, D.C.

Haneda, Y. (1958). Studies on luminescence in marine snails. *Pacific Science* **12**: 152–156.

Haneda, Y. (1970). Luminous fishes (in Japanese). *In* Kawamoto, N. (ed.), *Fish Physiology*, pp. 515–539. Kosei-sha, Tokyo.

Haneda, Y. (1985). *Luminous Organisms* (Hakko Seibutsu) (in Japanese). Koseisha-koseikaku, Tokyo.

Haneda, Y., and Johnson, F. H. (1958). The luciferin-luciferase reaction in a fish, *Parapriacanthus beryciformis*, of newly discovered luminescence. *Proc. Natl. Acad. Sci. USA* **44**: 127–129.

Haneda, Y., and Johnson, F. H. (1962). The comparative anatomy of the indirect type of photogenic system of luminescent fishes, with special reference to *Parapriacanthus beryciformes*. *Science Report of the Yokosuka City Museum*, No. 7, 1–11.

Haneda, Y., and Johnson, F. H. (1962a). The photogenic organs of *Parapriacanthus beryciformes* Franz and other fish with the indirect type of luminescent system. *J. Morph.* **110**: 187–198.

Haneda, Y., and Tsuji, F. I. (1971). Light production in the luminous fishes *Photoblepheron* and *Anomalops* from the Band Islands. *Science* **173**: 143–145.

Haneda, Y., Johnson, F. H., and Sie, E. H.-C. (1958). Luciferin and luciferase extracts of a fish, *Apogon marginatus*, and their luminescent cross-reactions with those of a crustacean, *Cypridina hilgendorfii*. *Biol. Bull.* **115**: 336.

Haneda, Y., *et al.* (1961). Crystalline luciferin from live *Cypridina*. *J. Cell. Comp. Physiol.* **57**: 55–62.

Hannick, L. I., *et al.* (1993). Preparation and initial characterization of crystals of the photoprotein aequorin from *Aequorea victoria*. *Proteins: Struct., Funct., Genet.* **15**: 103–107.

Hart, R. C., Matthews, J. C., Hori, K., and Cormier, M. J. (1979). *Renilla reniformis* bioluminescence: Luciferase-catalyzed production of nonradiating excited states from luciferin analogues and elucidation of the excited state species involved in energy transfer to *Renilla* green fluorescent protein. *Biochemistry* **18**: 2204–2210.

Hart, R. C., Stempel, K. E., Boyer, P. D., and Cormier, M. J. (1978). Mechanism of the enzyme-catalyzed bioluminescent oxidation of coelenterate-type luciferin. *Biochem. Biophys. Res. Commun.* **81**: 980–986.

Hartman, P. E., Hartman, Z., and Ault, K. T. (1990). Scavenging of singlet molecular oxygen by imidazole compounds: high and sustained activities of carboxy terminal histidine dipeptides and exceptional activity of imidazole-4-acetic acid. *Photochem. Photobiol.* **51**: 59–66.

Harvey, E. N. (1916). The mechanism of light production in animals. *Science* **44**: 208–209.

Harvey, E. N. (1917). What substance is the source of the light in the fireflies? *Science* **46**: 241–243.

Harvey, E. N. (1917). The chemistry of light production in luminous organisms. *Publ. Carneg. Instn.* **251**: 171–234.

Harvey, E. N. (1920). *The Nature of Animal Light.* L. B. Lippincott, Philadelphia.

Harvey, E. N. (1921). Studies on bioluminescence. XIII. Luminescence in the coelenterates. *Biol. Bull.* **41**: 280–287.

Harvey, E. N. (1922). Studies on bioluminescence. XIV. The specificity of luciferin and luciferase. *J. Gen. Physiol.* **4**: 285–295.

Harvey, E. N. (1926a). Additional data on the specificity of luciferin and luciferase, together with a general survey of this reaction. *Am. J. Physiol.* **77**: 548–554.

Harvey, E. N. (1926b). Oxygen and luminescence, with a description of methods for removing oxygen from cells and fluids. *Biol. Bull.* **51**: 89–97.

Harvey, E. N. (1931). Chemical aspects of the luminescence of deep-sea shrimp. *Zoologica, N.Y.* **12**: 71–74.

Harvey, E. N. (1940). *Living Light.* Hafner, New York.

Harvey, E. N. (1952). *Bioluminescence.* Academic Press, New York.

Harvey, E. N. (1957). *A History of Luminescence.* The American Philosophical Society, Philadelphia.

Hastings, J. W. (1968). Bioluminescence. *Ann. Rev. Biochem.* **37**: 597–630.

Hastings, J. W. (1978). Bacterial and dinoflagellate luminescent systems. *In* Herring, P. J. (ed.), *Bioluminescence in Action*, pp. 129–170. Academic Press, London.

Hastings, J. W. (1986). Bioluminescence in bacteria and dinoflagellates. *In* Govindjee, *et al.* (eds.), *Light Emission by Plants and Bacteria*, pp. 363–398. Academic Press, Orlando.

Hastings, J. W. (1987). Dinoflagellate bioluminescence: biochemistry, cell biology, and circadian control. *In* Schoelmerich, J. (ed.), *Biolumin. Chemilumin.*, Proc. Int. Biolumin. Chemilumin. Symp., 4th, 1986, pp. 343–350. Wiley, Chichester, UK.

Hastings, J. W. (1989). Chemistry, clones, and circadian control of the dinoflagellate bioluminescent system. The Marlene DeLuca Memorial Lecture. *J. Biolumin. Chemilumin.* **4**: 12–19.

Hastings, J. W. (1996). Chemistries and colors of bioluminescent reactions: a review. *Gene* **173** (1, Fluorescent Proteins and Applications): 5–11.

Hastings, J. W., and Bode, V. C. (1961). Ionic effects upon bioluminescence in *Gonyaulax* extracts. *In* McElroy, W. D., and Glass, B. (eds.), *Light and Life*, pp. 294–306. Johns Hopkins Press, Baltimore.

Hastings, J. W., and Davenport, D. (1957). The luminescence of the millipede, *Luminodesmus sequoiae*. *Biol. Bull.* **113**: 120–128.

Hastings, J. W., and Dunlap, J. C. (1986). Cell-free components in dinoflagellate bioluminescence. The particulate activity: scintillons; the soluble components: luciferase, luciferin, and luciferin-binding protein. *Method. Enzymol.* **133**: 307–327.

Hastings, J. W., and Gibson, Q. H. (1963). Intermediates in the bioluminescent oxidation of reduced flavin mononucleotide. *J. Biol. Chem.* **238**: 2537–2554.

Hastings, J. W., and Morin, J. G. (1968). Calcium activated bioluminescent protein from ctenophores (*Mnemiopsis*) and colonial hydroids (*Obelia*). *Biol. Bull.* **135**: 422.

Hastings, J. W., and Morin, J. C. (1969). Calcium-triggered light emission in *Renilla*. A unitary biochemical scheme for coelenterate bioluminescence. *Biochem. Biophys. Res. Commun.* **37**: 493–498.

Hastings, J. W., and Morin, J. G. (1969a). Comparative biochemistry of calcium-activated photoproteins from the ctenophore, *Mnemiopsis* and the coelenterates *Aequorea*, *Obelia*, *Pelagia* and *Renilla*. *Biol. Bull.* **137**: 402.

Hastings, J. W., and Nealson, K. H. (1977). Bacterial bioluminescence. *Ann. Rev. Microbiol.* **31**: 549–595.

Hastings, J. W., and Reynolds, G. T. (1966). The preparation and standardization by different methods of liquid light sources. *In* Johnson, F. H., and Haneda, Y. (eds.), *Bioluminescence in Progress*, pp. 45–50. Princeton University Press, Princeton, NJ.

Hastings, J. W., and Sweeney, B. M. (1957). The luminescence reaction in extracts of the marine dinoflagellate, *Gonyaulax polyedra*. *J. Cell. Comp. Physiol.* **49**: 209–226.

Hastings, J. W., and Weber, G. (1963). Total quantum flux of isotopic sources. *J. Opt. Soc. Am.* **53**: 1410–1415.

Hastings, J. W., Vergin, M., and DeSa, R. (1966). Scintillons: the biochemistry of dinoflagellate luminescence. *In* Johnson, F. H., and Haneda, Y. (eds.), *Bioluminescence in Progress*, pp. 301–329. Princeton University Press, Princeton, NJ.

Hastings, J. W., *et al.* (1969). Response of aequorin bioluminescence to rapid changes in calcium concentration. *Nature* **222**: 1047–1050.

Hastings, J. W., *et al.* (1969a). Structurally distinct bacterial luciferase. *Biochemistry* **8**: 4681–4689.

Hastings, J. W., Balny, C., Le Peuch, C., and Douzou, P. (1973). Spectral properties of an oxygenated luciferase-flavin intermediate isolated by low-temperature chromatography. *Proc. Natl. Acad. Sci. USA* **70**: 3468–3472.

Hastings, J. W., Baldwin, T. O., and Nicoli, M. Z. (1978). Bacterial luciferase: assay, purification and properties. *Method. Enzymol.* **57**: 135–152.

Hastings, J. W., *et al.* (1985). Biochemistry and physiology of bioluminescent bacteria. *Adv. Microb. Physiol.* **26**: 235–291.

Haygood, M. G., and Cohn, D. H. (1986). Luciferase genes cloned from the unculturable luminous bacteroid symbiont of the Caribbean flashlight fish, *Kryptophanaron alfredi*. *Gene* **45**: 203–209.

Haygood, M. G., and Nealson, K. H. (1985). Mechanisms of iron regulation of luminescence in *Vibrio fischeri*. *J. Bacteriol.* **162**: 209–216.

Head, J. F., Inouye, S., Teranishi, K., and Shimomura, O. (2000). The crystal structure of the photoprotein aequorin at 2.3A resolution. *Nature* **405**: 372–376.

Heim, R., and Tsien, R. Y. (1996). Engineering green fluorescent protein for improved brightness, longer wavelengths and fluorescence resonance energy transfer. *Curr. Biol.* **6**: 178–182.

Henry, J. P., and Michelson, A. M. (1970). Studies in bioluminescence. IV. Properties of luciferin from *Pholas dactylus*. *Biochim. Biophys. Acta* **205**: 451–458.

Henry, J. P., and Monny, C. (1977). Protein-protein interaction in the *Pholas dactylus* system of bioluminescence. *Biochemistry* **16**: 2517–2525.

Henry, J. P., Isambert, M. F., and Michelson, A. M. (1970). Studies in bioluminescence. III. The *Pholas dactylus* system. *Biochim. Biophys. Acta* **205**: 437–450.

Henry, J. P., Isambert, M. F., and Michelson, A. M. (1973). Studies in bioluminescence. IX. Mechanism of the *Pholas dactylus* system. *Biochimie* **55**: 83–93.

Henry, J. P., Monny, C., and Michelson, A. M. (1975). Characterization and properties of *Pholas* luciferase as metalloglycoprotein. *Biochemistry* **14**: 3458–3466.

Herrera, A. A., Hastings, J. W., and Morin, J. G. (1974). Bioluminescence in cell free extracts of the scale worm *Harmothoe* (Annelida, Polynoidae). *Biol. Bull.* **147**: 480–481.

Herring, P. J. (1974). New observations on the bioluminescence of echinoderms. *J. Zool., Lond.* **172**: 401–418.

Herring, P. J., ed. (1978). *Bioluminescence in Action*. Academic Press, London.

Herring, P. J. (1978a). Bioluminescence of invertebrates other than insects. *In* Herring, P. J. (ed.), *Bioluminescence in Action*, pp. 199–240. Academic Press, London.

Herring, P. J. (1978b). A classification of luminous organisms. *In* Herring, P. J. (ed.), *Bioluminescence in Action*, pp. 461–476. Academic Press, London.

Herring, P. J. (1981). Studies on bioluminescent marine amphipods. *J. Mar. Biol. Ass., UK* **61**: 161–176.

Herring, P. J. (1982). Aspects of bioluminescence of fishes. *Oceanogr. Mar. Biol. Ann. Rev.* **20**: 415–470.

Herring, P. J. (1983). The spectral characteristics of luminous marine organisms. *Proc. R. Soc. Lond.* **B 220**: 183–217.

Herring, P. J. (1985). Bioluminescence in the crustacea. *J. Crustacean Biol.* **5**: 557–573.

Herring, P. J. (1990). Bioluminescence response of the deep-sea scyphozoan *Atolla wyvillei*. *Mar. Biol.* **196**: 413–417.

Herring, P. J., and Locket, N. A. (1978). The luminescence and photophores of euphausiid crustaceans. *J. Zool., Lond.* **186**: 431–462.

Herring, P. J., and Morin, J. G. (1978). Bioluminescence in fishes. *In* Herring, P. J. (ed.), *Bioluminescence in Action*, pp. 273–329. Academic Press, London.

Herring, P. J., Bassot, J.-M., and Flood, P. R. (1997). Bioluminescent responses of the scyphozoan *Periphylla periphylla* from a Norwegian fjord. *In* Hastings, J. W., *et al.* (eds.), *Bioluminescence and Chemiluminescence; Molecular Reporting with Photons*, pp. 154–157. John Wiley and Sons, Chichester.

Hill, F., Wood, K., and DeLuca, M. (1987). Firefly luciferase. *In* Schoelmerich, J. (ed.), *Biolumin. Chemilumin.*, Proc. Int. Biolumin. Chemilumin. Symp., 4th, 1986, pp. 397–400. Wiley, Chichester, UK.

Hiller-Adams, P., Widder, E. A., and Case, J. F. (1988). The visual pigments of four deep-sea crustacean species. *J. Comp. Physiol.* **163A**: 63–72.

Hirano, T., *et al.* (1993). Structure elucidation of the light emitter in aequorin bioluminescence. *Tennen Yuki Kagobutsu Toronkai Koen Yoshishu* **35**: 551–558.

Hirano, T., *et al.* (1994). Revision of the structure of the light-emitter in aequorin bioluminescence. *Chem. Commun.*, pp. 165–167.

Hirano, T., *et al.* (1998). Bioluminescent properties of fluorinated semi-synthetic aequorins. *Tetrahedron Lett.* **39**: 5541–5544.

Hirata, Y., Shimomura, O., and Eguchi, S. (1959). The structure of *Cypridina* luciferin. *Tetrahedron Lett.*, pp. 4–9.

Hirose, E., Aoki, M., and Chiba, K. (1996). Fine structures of tunic cells and distribution of bacteria in the tunic of the luminescent ascidian *Clavelina miniate*. *Zool. Sci.* **13**: 519–523.

Holzman, T. F., and Baldwin, T. O. (1982). Isolation of bacterial luciferases by affinity chromatography on 2,2-diphenylpropylamine-Sepharose: phosphate-mediated binding to immobilized substrate analogue. *Biochemistry* **21**: 6194–6201.

Hopkins, T. A., Seliger, H. H., White, E. H., and Cass, M. W. (1967). The chemiluminescence of firefly luciferin. A model for the bioluminescence reaction and identification of the product excited state. *J. Am. Chem. Soc.* **89**: 7148–7150.

Hori, K., and Cormier, M. J. (1965). Studies on the bioluminescence of Renilla reniformis. V. Absorption and fluorescence characteristics of chromatographically pure luciferin. *Biochim. Biophys. Acta* **102**: 386–396.

Hori, K., and Cormier, M. J. (1966). Studies on the bioluminescence of *Renilla reniformis*. VI. Some chemical properties and the tentative partial structure of luciferin. *Biochim. Biophys. Acta* **130**: 420–425.

Hori, K., and Cormier, M. J. (1972). Structure and chemical synthesis of a biologically active form of *Renilla* (sea pansy) luciferin. *Proc. Natl. Acad. Sci. USA* **70**: 120–123.

Hori, K., Nakano, Y., and Cormier, M. J. (1972). Studies on the bioluminescence of *Renilla reniformis*. XI. Location of the sulfate group in luciferyl sulfate. *Biochim. Biophys. Acta* **256**: 638–644.

Hori, K., Wampler, J. E., Matthews, J. C., and Cormier, M. J. (1973). Identification of the product excited states during the chemiluminescent and bioluminescent oxidation of *Renilla* (sea pansy) luciferin and certain of its analogs. *Biochemistry* **12**: 4463–4468.

Hori, K., Wampler, J. E., and Cormier, M. J. (1973a). Chemiluminescence of *Renilla* (sea pansy) luciferin and its analogues. *Chem. Commun.*, pp. 492–493.

Hori, K., Anderson, J. M., Ward, W. W., and Cormier, M. J. (1975). *Renilla* luciferin as the substrate for calcium induced photoprotein luminescence. Assignment of luciferin tautomers in aequorin and mnemiopsin. *Biochemistry* **14**: 2371–2376.

Hori, K., Charbonneau, H., Hart, R. C., and Cormier, M. J. (1977). Structure of native *Renilla reniformis* luciferin. *Proc. Natl. Acad. Sci. USA* **74**: 4285–4287.

Illarionov, B. A., *et al.* (1992). Cloning and expression of cDNA coding for the calcium-activated photoprotein obelin from the hydroid polyp *Obelia longissima. Dokl. Akad. Nauk SSSR* **326**: 911–913.

Illarionov, B. A., Bondar, V. S., Illarionova, V. A., and Vysotski, E. S. (1995). Sequence of the cDNA encoding the Ca^{2+}-activated photoprotein obelin from the hydroid polyp *Obelia longissima. Gene* **153**: 273–274.

Illarionova, V. A., *et al.* (1997). Removal of essential ligand in N-terminal calcium-binding domain of obelin does not inactivate the photoprotein or reduce its calcium sensitivity, but dramatically alters the kinetics of the luminescent reaction. *In* Hastings, J. W., *et al.* (eds.), *Biolumin. Chemilumin.*, Proc. Int. Symp., 9th, 1996, pp. 431–434. Wiley, Chichester, UK.

Illarionov, B. A., *et al.* (2000). Recombinant obelin: cloning and expression of cDNA, purification, and characterization as a calcium indicator. *Method. Enzymol.* **305**: 223–249.

Imai, Y., *et al.* (2001). Fluorescence properties of phenolate anions of coelenteramide analogues: the light-emitter structure in aequorin bioluminescence. *J. Photochem. Photobiol., A: Chemistry* **146**: 95–107.

Inoue, S., and Kakoi, H. (1976). *Oplophorus* luciferin, bioluminescent substance of the decapod shrimp *Oplophorus spinosus* and *Heterocarpus laevigatus. Chem. Commun.*, pp. 1056–1057.

Inoue, S., Sugiura, S., Kakoi, H., and Goto, T. (1969). *Cypridina* bioluminescence. VI A new route for the synthesis of Cypridina luciferin and its analogs. *Tetrahedron Lett.*, pp. 1609–1610.

Inoue, S., Sugiura, S., Kakoi, H., and Hashizume, K. (1975). Squid bioluminescence. II. Isolation from *Watasenia scintillans* and synthesis of 2-(p-hydroxybenzyl)-6-(p-hydroxyphenyl)-3,7-dihydroimidazo[1,2-a]pyrazin-3-one. *Chem. Lett.*, pp. 141–144.

Inoue, S., Kakoi, H., and Goto, T. (1976). Squid bioluminescence. III. Isolation and structure of *Watasenia* luciferin. *Tetrahedron Lett.*, pp. 2971–2974.

Inoue, S., *et al.* (1977a). Complete structure of *Renilla* luciferin and luciferyl sulfate. *Tetrahedron Lett.*, pp. 2685–2688.

Inoue, S., *et al.* (1977b). Squid bioluminescence IV. Isolation and structural elucidation of *Watasenia* dehydropreluciferin. *Chem. Lett.*, pp. 259–262.

Inoue, S., Okada, K., Kakoi, H., and Goto, T. (1977c). Fish bioluminescence. I. Isolation of a luminescent substance from a mictophid fish, *Neoscopelus microchir*, and identification of it as *Oplophorus* luciferin. *Chem. Lett.*, pp. 257–258.

Inoue, S., *et al.* (1983). Trace characterization of the Watasenia luciferin in eye and skin photophores and in liver of *Watasenia scintillans. Agric. Biol. Chem.* **47**: 635–636.

Inoue, S., Okada, K., Tanino, H., and Kakoi, H. (1987). Chemical studies of myctophina fish bioluminescence. *Chem. Lett.*, pp. 417–418.

Inoue, S., *et al.* (2002). Fluorescence polarization of green fluorescence protein. *Proc. Natl. Acad. Sci. USA* **99**: 4272–4277.

Inouye, S. (2004). Blue fluorescent protein from the calcium-sensitive photoprotein aequorin is a heat resistant enzyme, catalyzing the oxidation of coelenterazine. *FEBS Lett.* **577**: 105–110.

Inouye, S., and Shimomura, O. (1997). The use of *Renilla* luciferase, *Oplophorus* luciferase, and apoaequorin as bioluminescent reporter protein in the presence of coelenterazine analogues as substrate. *Biochem. Biophys. Res. Commun.* **233**: 349–353.

Inouye, S., and Tsuji, F. I. (1993). Cloning and sequence analysis of cDNA for the Ca^{2+}-activated photoprotein, clytin. *FEBS Lett.* **315**: 343–346.

Inouye, S., and Tsuji, F. I. (1994). Evidence for redox forms of the *Aequorea* green fluorescent protein. *FEBS Lett.* **351**: 211–214.

Inouye, S., and Tsuji, F. I. (1994a). *Aequorea* green fluorescence protein. Expression of the gene and fluorescence characteristics of the recombinant protein. *FEBS Lett.* **341**: 277–280.

Inouye, S., *et al.* (1985). Cloning and sequence analysis of cDNA for the luminescent protein aequorin. *Proc. Natl. Acad. Sci. USA* **82**: 3154–3158.

Inouye, S., Sakaki, Y., Goto, T., and Tsuji, F. I. (1986). Expression of apoaequorin complementary DNA in *Escherichia coli. Biochemistry* **25**: 8425–8429.

Inouye, S., *et al.* (1989). Overexpression and purification of the recombinant calcium-binding protein, apoaequorin. *J. Biochem.* **105**: 473–477.

Inouye, S., Zenno, S., Sakaki, Y., and Tsuji, F. I. (1991). High-level expression and purification of apoaequorin. *Protein Expression and Purification* **2**: 122–126.

Inouye, S., Watanabe, K., Nakamura, H., and Shimomura, O. (2000). Secretional luciferase of the luminous shrimp *Oplophorus gracilirostris*: cDNA cloning of a novel imidazopyrazinone luciferase. *FEBS Lett.* **481**: 19–25.

Isobe, M., Uyakul, D., and Goto, T. (1987). *Lampteromyces* bioluminescence. I. Identification of riboflavin as the light emitter in mushroom *Lampteromyces japonicus. J. Biolumin. Chemilumin.* **1**: 181–188.

Isobe, M., Uyakul, D., and Goto, T. (1988). *Lampteromyces* bioluminescence. 2. Lampteroflavin, a light emitter in the luminous mushroom *Lampteromyces japonicus. Tetrahedron Lett.* **44**: 1169–1172.

Isobe, M., *et al.* (1991). Fluorescence substance in the luminous land snail, *Dyakia striata. Agric. Biol. Chem.* **55**: 1947–1951.

Isobe, M., *et al.* (1994). Bioluminescence mechanism on new systems. *Pure Appl. Chem.* **66**: 765–772.

Isobe, M., *et al.* (2002). ^{19}F-Dehydrocoelenterazine as probe to investigate the active site of symplectin. *Tetrahedron* **58**: 2117–2126.

Iwai, T., and Asano, H. (1958). On the luminous cardinal fish, *Apogon ellioti* Day. *Sci. Rep. Yokosuka City museum* **3**: 5–12.

Izutsu, K. T., *et al.* (1972). Aequorin: its ionic specificity. *Biochem. Biophys. Res. Commun.* **49**: 1034–1039.

Jockers, R., Ziegler, T., and Schmid, R. D. (1995). Interaction between aldehyde derivatives and the aldehyde binding site of bacterial luciferase. *J. Biolumin. Chemilumin.* **10**: 21–27.

Johnsen, S., Balser, E. J., Fisher, E. C., and Widder, E. A. (1999). Bioluminescence in the deep-sea cirrate octapod *Strauroteuthis syrtensis* Verrill (Mollusca: Cephalopoda). *Biol. Bull.* **197**: 26–39.

Johnson, F. H. (1967). Edmund Newton Harvey, 1988–1959. *Biographical Memoirs* **39**: 193–266.

Johnson, F. H. (1970). The bioluminescence protein "Aequorin". *Naval Research Reviews* **February**, pp. 16–23.

Johnson, F. H., and Haneda, Y. (1958). The luciferin-luciferase reaction in a fish, *Parapriacanthus beryciformis*, of newly discovered luminescence. *Sci. Rep. Yokosuka City Museum* **3**: 25–30.

Johnson, F. H., and Shimomura, O. (1972). Preparation and use of aequorin for rapid microdetermination of Ca^{2+} in biological systems. *Nature* **237**: 287–288.

Johnson, F. H., and Shimomura, O. (1978). Introduction to the bioluminescence of medusae, with special reference to the photoprotein aequorin. *Method. Enzymol.* **57**: 271–291.

Johnson, F. H., *et al.* (1961a). Crystalline luciferin from a luminescent fish, *Parapriacanthus beryciformes*. *Proc. Natl. Acad. Sci. USA* **47**: 486–489.

Johnson, F. H., Shimomura, O., and Saiga, Y. (1961b). Luminescence potency of the *Cypridina* system. *Science* **134**: 1755–1756.

Johnson, F. H., *et al.* (1962). Quantum efficiency of *Cypridina* luminescence, with a note on that of *Aequorea*. *J. Cell. Comp. Physiol.* **60**: 85–104.

Johnson, F. H., Shimomura, O., and Saiga, Y. (1962a). Action of cyanide on *Cypridina* luciferin. *J. Cell. Comp. Physiol.* **59**: 265–272.

Johnson, F. H., Stachel, H. D., Shimomura, O., and Haneda, Y. (1966). Partial purification of the luminescence system of a deep-sea shrimp *Hoplophorus gracilorostris*. *In* Johnson, F. H., and Haneda, Y. (eds.), *Bioluminescence in Progress*, pp. 523–532. Princeton University Press, Princeton, NJ.

Johnson, M. E., and Snook, H. J. (1927). *Seashore Animals of the Pacific Coast*. Macmillan, New York.

Johnston, T. C., Thompson, R. B., and Baldwin, T. O. (1986). Nucleotide sequence of the luxB gene of *Vibrio harveyi* and the complete amino acid sequence of the β subunit of bacterial luciferase. *J. Biol. Chem.* **261**: 4805–4811.

Johnston, T. C., *et al.* (1990). The nucleotide sequence of the luxA and luxB genes of *Xenorhabdus luminescence* HM and a comparison of the amino acid sequences of luciferases from four species of bioluminescent bacteria. *Biochem. Biophys. Res. Commun.* **170**: 407–415.

Kaaret, T. W., and Bruice, T. C. (1990). Electrochemical luminescence with N(5)-ethyl-4a-hydroxy-3-methyl-4a,5-dihydrolumiflavin. The mechanism of bacterial luciferase. *Photochem. Photobiol.* **51**: 629–633.

Kampa, E. M., and Boden, B. P. (1956). Light generation in a sonic scattering layer. *Deep-Sea Res.* **4**: 73–92.

Kamzolkina, O. V., Bekker, Z. E., and Egorov, N. S. (1984). Extraction of the luciferin-luciferase system from the fungus *Armillariella mellea*. *Biologicheskie Nauki (Moscow)* **1984**: 73–77.

Kamzolkina, O. V., Danilov, V. S., and Egorov, N. S. (1983). Nature of luciferase from the bioluminescent fungus *Armillariella mellea*. *Dokl. Akad. Nauk SSSR* **271**: 750–752.

Kanakubo, A., and Isobe, M. (2005). Isolation of brominated quinones showing chemiluminescence activity from luminous acorn worm, *Ptychodera flava*. *Bioorg. Med. Chem.* **13**: 2741–2747.

Kanakubo, A., Koga, K., Isobe, M., and Yoza, K. (2005). Tetrabromohydroquinone and riboflavin are possibly responsible for green luminescence in the luminous acorn worm, *Ptychodera flava*. *Luminescence* 20: 397–400.

Kanashiro, M., Matsubara, T., Goto, T., and Sakamoto, N. (1993). *Cypridina* luciferin analog reduces the incidence of ischemia/reperfusion-induced ventricular fibrillation. *Jpn. J. Pharmacol.* 63: 47–52.

Kanda, S. (1935). *Hotaru (The Fireflies)*. Nihon Hakko Seibutsu Kenkuy Kai (Maruzen), Tokyo.

Kaplan, H. B., and Greenberg, E. P. (1985). Diffusion of autoinducer is involved in regulation of the *Vibrio fischeri* luminescence system. *J. Bacteriol.* 163: 1210–1214.

Kaplan, H. B., and Greenberg, E. P. (1987). Overproduction and purification of the luxR gene product: transcriptional activator of the *Vibrio fischeri* luminescence system. *Proc. Natl. Acad. Sci. USA* 84: 6639–6643.

Karatani, H., and Hastings, J. W. (1993). Two active forms of the accessory yellow fluorescence protein of the luminous bacterium *Vibrio fischeri* strain Y1. *J. Photochem. Photobiol., B* 18: 227–232.

Karkhanis, Y. D., and Cormier, M. J. (1971). Isolation and properties of *Renilla reniformis* luciferase, a low molecular weight energy conversion enzyme. *Biochemistry* 10: 317–326.

Karpetsky, T. P., and White, E. H. (1971). An unambiguous synthesis of *Cypridina* etioluciferamine. An application of titanium tetrachloride to the synthesis of pyrazine N-oxide. *J. Am. Chem. Soc.* 93: 2333–2335.

Kato, S., Oba, Y., Ojika, M., and Inouye, S. (2004). Identification of the biosynthetic units of *Cypridina* luciferin in *Cypridina (Vargula) hilgendorfii* by LC/ESI-TOF-MS. *Tetrahedron* 60: 11427–11434.

Kemple, M. D., et al. (1990). Manganese(II)-EPR measurements of cation binding by aequorin. *Eur. J. Biochem.* 187: 131–135.

Kendall, J. M., Dormer, R. L., and Campbell, A. K. (1992). Targeting aequorin to the endoplasmic reticulum of living cells. *Biochem. Biophys. Res. Commun.* 189: 1008–1016.

Kendall, J. M., et al. (1992). Engineering the Ca^{2+}-activated photoprotein aequorin with reduced affinity for calcium. *Biochem. Biophys. Res. Commun.* 187: 1091–1097.

Kendall, J. M., et al. (1992). Engineering aequorin to measure calcium in defined compartments of living cells. *Biochem. Soc. Trans.* 20: 144S.

Kendall, J. M., *et al.* (1996). Recombinant apoaequorin acting as a pseudo-luciferase reports micromolar changes in the endoplasmic reticulum free Ca^{2+} of intact cells. *Biochem. J.* **318**: 383–387.

Keynan, A., and Hastings, J. W. (1961). The isolation and characterization of dark mutants of luminous bacteria. *Biol. Bull.* **121**: 375.

Kihara, Y., and Morgan, J. P. (1989). A comparative study of three methods for intracellular loading of the calcium indicator aequorin in ferret papillary muscles. *Biochem. Biophys. Res. Commun.* **162**: 402–407.

Kishi, Y., *et al.* (1966a). *Cypridina* bioluminescence I: structure of *Cypridina* luciferin. *Tetrahedron Lett.*, pp. 3427–3436.

Kishi, Y., *et al.* (1966b). The structure of *Cypridina* luciferin. *In* Johnson, F. M., and Haneda, Y. (eds.), *Bioluminescence in Progress*, pp. 89–113. Princeton University Press, Princeton, NJ.

Kishi, Y., *et al.* (1966c). *Cypridina* Bioluminescence. III. Total synthesis of *Cypridina* luciferin. *Tetrahedron Lett.*, pp. 3445–3450.

Kishi, Y., *et al.* (1968). Luciferin and luciopterin isolated from the Japanese firefly, *Luciola cruciata*. *Tetrahedron Lett.*, pp. 2847–2850.

Kishi, Y., Tanino, H., and Goto, T. (1972). The structure confirmation of the light-emitting moiety of bioluminescent jellyfish *Aequorea*. *Tetrahedron Lett.* **27**: 2747–2748.

Knight, H., *et al.* (1994). Luminescent signals in plants. *In* Campbell, A. K., *et al.* (eds.), *Biolumin. Chemilumin.*, Proc. Int. Symp., 8th, pp. 97–100. Wiley, Chichester, UK.

Knight, M. R., Campbell, A. K., Smith, S. M., and Trewavas, A. J. (1991). Transgenic plant aequorin reports the effects of touch and cold-shock and elicitors on cytoplasmic calcium. *Nature* **352**: 524–526.

Knight, M. R., Read, N. D., Campbell, A. K., and Trewavas, A. J. (1993). Imaging calcium dynamics in living plants using semi-synthetic recombinant aequorins. *J. Cell Biol.* **121**: 83–90.

Kobayashi, K., *et al.* (2000). Purification and properties of the luciferase from the marine ostracod *Vargula hilgendorfii*. *In* Case, J. F., *et al.* (eds.), *Biolumin. Chemilumin.*, Proc. Int. Symp., 11th, pp. 87–90. World Scientific, Singapore.

Koda, P., and Lee, J. (1979). Separation and structure of the prosthetic group of the blue fluorescence protein from the bioluminescent bacterium *Photobacterium phosphoreum*. *Proc. Natl. Acad. Sci. USA* **76**: 3068–3072.

Kohama, Y., Shimomura, O., and Johnson, F. H. (1971). Molecular weight of the photoprotein aequorin. *Biochemistry* **10**: 4149–4152.

Kojima, S., *et al.* (1997). Mechanism of the redox reaction of the *Aequorea* green fluorescent protein (GFP). *Tetrahedron Lett.* **38**: 2875–2878.

Kojima, S., *et al.* (1998). Fluorescent properties of model chromophores of tyrosine-66 substitute mutants of Aequorea green fluorescent protein (GFP). *Tetrahedron Lett.* **39**: 5239–5242.

Kojima, S., *et al.* (2000a). Purification and characterization of the luciferase from the freshwater snail *Latia*. Abstract, 11th Int. Symp. on Biolumin. Chemilumin., Asilomar, CA, p. 57.

Kojima, S., *et al.* (2000b). Bioluminescence activity of *Latia* luciferin analogs. *Tetrahedron Lett.* **41**: 4409–4413.

Kojima, S., *et al.* (2000c). Molecular basis on *Latia* bioluminescence. *Tennen Yuki Kagobutsu Toronkai Koen Yoshishu* **42**: 553–558 (Japanese).

Koncz, C., *et al.* (1987). Expression and assembly of functional bacterial luciferase in plants. *Proc. Natl. Acad. Sci. USA* **84**: 131–135.

Koo, J.-Y., Schmidt, S. P., and Schuster, G. B. (1978). Bioluminescence of the firefly: key steps in the formation of the electronically excited state for model systems. *Proc. Natl. Acad. Sci. USA* **75**: 30–33.

Kornicker, L. S., and Baker, J. H. (1977). *Vargula tsujii*, A new species of luminescent Ostracoda from lower and southern California (Myodocopa: Cyprininae). *Proc. Biol. Soc. Washington* **90**: 218–231.

Kramp, P. L. (1959). *The Hydromedusae of the Atlantic Ocean and Adjacent Waters*, Dana-Report No. 46. Carlsberg Foundation, Copenhagen.

Kramp, P. L. (1965). *The Hydromedusae of the Pacific and Indian Oceans*, Dana Report No. 63. Carlsberg Foundation, Copenhagen.

Kramp, P. L. (1968). *The Hydromedusae of the Pacific and Indian Oceans*, sections II and III, Dana-Report No. 72. Carlsberg Foundation, Copenhagen.

Kreiss, P., and Cormier, M. J. (1967). Inhibition of *Renilla reniformis* bioluminescence by light: effects on luciferase and its substrates. *Biochim. Biophys. Acta* **141**: 181–183.

Krieger, N., and Hastings, J. W. (1968). Bioluminescence: pH activity profiles of related luciferase fractions. *Science* **161**: 586–589.

Krieger, N., Njus, D., and Hastings, J. W. (1974). An active proteolytic fragment of *Gonyaulax polyedra*. *Biochemistry* **13**: 2871–2877.

Kulinski, T., Visser, A. J. W. G., O'Kane, D. J., and Lee, J. (1987). Spectroscopic investigations of the single tryptophan residue and of riboflavin and 7-oxolumazine bound to lumazine apoprotein from *Photobacterium leiognathi*. *Biochemistry* **26**: 540–549.

Kumar, S., *et al.* (1990). Amino acid sequence of the calcium-triggered luciferin binding protein of *Renilla reniformis*. *FEBS Lett.* **268**: 287–290.

Kurfuerst, M., Hastings, J. W., Ghisla, S., and Macheroux, P. (1984). Identification of the luciferase-bound flavin-4a-hydroxide as the primary emitter in the bacterial bioluminescence reaction. *In* Bray, R. C., *et al.* (eds.), *Flavins Flavoproteins*, Proc. Int. Symp. 8th, pp. 657–667. de Gruyter, Berlin.

Kurfuerst, M., Ghisla, S., and Hastings, J. W. (1986). Bacterial luciferase intermediates: the neutral flavin semiquinone, its reaction with superoxide, and the flavin 4a-hydroxide. *Method. Enzymol.* **133**: 140–149.

Kurfuerst, M., Macheroux, P., Ghisla, S., and Hastings, J. W. (1987). Isolation and characterization of the transient, luciferase-bound flavin-4a-hydroxide in the bacterial luciferase reaction. *Biochim. Biophys. Acta* **924**: 104–110.

Kurfuerst, M., Macheroux, P., Ghisla, S., and Hastings, J. W. (1989). Bioluminescence emission of bacterial luciferase with 1-deaza-FMN. Evidence for the noninvolvement of N(1)-protonated flavin species as emitters. *Eur. J. Biochem.* **181**: 453–457.

Kurian, E., Fisher, P. J., Ward, W. W., and Prendergast, F. G. (1994). Characterization of secondary and tertiary structure of the green fluorescent protein from *A. victoria*. *J. Biolumin. Chemilumin.* **9**: 333.

Kurose, K., Inouye, S., Sakaki, Y., and Tsuji, F. I. (1989). Bioluminescence of the calcium-binding photoprotein aequorin after cysteine modification. *Proc. Natl. Acad. Sci. USA* **86**: 80–84.

Kuse, M., *et al.* (2001). 7,8–Dihydropterin-6-carboxylic acid as light emitter of luminous millipede, *Luminodesmus sequoiae*. *Bioorg. Med. Chem. Lett.* **11**: 1037–1040.

Kutuzova, G. D., Hannah, R. R., and Wood, K. V. (1997). Bioluminescence color variation and kinetic behavior relationships among beetle luciferases. *In* Hastings, J. W., *et al.* (eds.), *Biolumin. Chemilumin.*, Proc. Int. Symp., 9th, 1996, pp. 248–252. Wiley. Chichester, UK.

Kuwabara, S., and Wassink, E. C. (1966). Purification and properties of the active substance of fungal luminescence. *In* Johnson, F. H., and Haneda, Y. (eds.), *Bioluminescence in Progress*, pp. 233–245. Princeton University Press, Princeton, NJ.

La, S. Y., and Shimomura, O. (1982). Fluorescence polarization study of the Ca^{2+}-sensitive photoprotein aequorin. *FEBS Lett.* **143**: 49–51.

Lamola, A. A. (1969). Electronic energy transfer in solutions: theory and application. *In* Leermakers, P. A., and Weissberger, A. (eds.), *Energy Transfer and Organic Photochemistry, Technique of Organic Chemistry* 14: 17–132. Interscience Publishers, New York.

Lang, D., Erdmann, H., and Schmid, R. D. (1992). Bacterial luciferase of *Vibrio harveyi* MAV: purification, characterization and crystallization. *Enzyme Microb. Technol.* 14: 479–485.

Latz, M. I., Frank, T., Case, J. F., Swift, E., and Bidigare, R. R. (1987). Bioluminescence of colonial Radiolaria in the Western Sargasso Sea. *J. Exp. Mar. Biol. Ecol.* 109: 25–38.

Lavelle, M. M. F., Durosay, P., and Michelson, A. M. (1972). Bioluminescence. Luminescence des champignons lumineux. *C. R. Acad. Sci. Paris* 275: Serie D 1227–1230.

Leach, F. R. (1981). ATP determination with firefly luciferase. *J. Appl. Biochem.* 3: 473–517.

Leach, F. R., Ye, L., Schaeffer, H. J., and Buck, L. M. (1997). Cloning and sequencing of a firefly luciferase from *Photuris pennsylvanica*. *In* Hastings, J. W., *et al.* (eds.), *Biolumin. Chemilumin.*, Proc. Int. Symp., 9th, 1996, pp. 240–243. Wiley, Chichester, UK.

Lecuyer, B., and Arrio, B. (1975). Some spectral characteristics of the light emitting system of the polynoid worms. *Photochem. Photobiol.* 22: 213–215.

Lee, D. H., *et al.* (1993). Molecular cloning and genomic organization of a gene for luciferin-binding protein from the dinoflagellate *Gonyaulax polyedra*. *J. Biol. Chem.* 268: 8842–8850.

Lee, J. (1972). Bacterial bioluminescence. Quantum yields and stoichiometry of the reactants reduced flavin mononucleotide, dodecanal and oxygen, and of a product hydrogen peroxide. *Biochemistry* 11: 3350–3359.

Lee, J. (1976). Bioluminescence of the Australian glow-worm, *Arachnocampa richardsae* Harrison. *Photochem. Photobiol.* 24: 279–285.

Lee, J. (1990). Lumazine protein and the bioluminescence of Photobacterium. *In* Curtius, H.-C., *et al.* (eds.), *Chem. Biol. Pteridines*, Proc. Int. Symp. Pteridines Folic Acid Deriv., 9th, 1989, pp. 445–456. de Gruyter, Berlin.

Lee, J. (1993). Lumazine protein and the excitation mechanism in bacterial bioluminescence. *Biophys. Chem.* 48: 149–158.

Lee, J., and Seliger, H. H. (1965). Absolute spectral sensitivity of phototubes and the application to the measurement of the absolute quantum yields of chemiluminescence and bioluminescence. *Photochem. Photobiol.* 4: 1015–1048.

Lee, J., and Seliger, H. H. (1972). Quantum yields of the luminol chemiluminescence reaction in aqueous and aprotic solvents. *Photochem. Photobiol.* **15**: 227–237.

Lee, J., Wesley, A. S., Ferguson, J. F., and Seliger, H. H. (1966). The use of luminol as a standard of photon emission. *In* Johnson, F. H., and Haneda, Y. (eds.), *Bioluminescence in Progress*, pp. 35–43. Princeton University Press, Princeton, NJ.

Lee, J., O'Kane, D. J., and Visser, A. J. W. G. (1985). Spectral properties and function of two lumazine proteins from *Photobacterium*. *Biochemistry* **24**: 1476–1483.

Lee, J., O'Kane, D. J., and Gibson, D. G. (1988). Dynamic fluorescence properties of bacterial luciferase intermediates. *Biochemistry* **27**: 4862–4870.

Lee, J., O'Kane, D. J., and Gibson, B. G. (1989). Bioluminescence spectral and fluorescence dynamics study of the interaction of lumazine protein with the intermediates of bacterial luciferase bioluminescence. *Biochemistry* **28**: 4263–4271.

Lee, J., O'Kane, D. J., and Gibson, B. G. (1989). Dynamic fluorescence study of the interaction of lumazine protein with bacterial luciferases. *Biophys. Chem.* **33**: 99–111.

Lee, J., Wang, Y., and Gibson, B. G. (1990). Recovery of components of fluorescence spectra of mixtures by intensity- and anisotropy decay-associated analysis: the bacterial luciferase intermediates. *Anal. Biochem.* **185**: 220–229.

Lee, J., Wang, Y., and Gibson, B. G. (1991a). Electronic excitation transfer in the complex of lumazine protein with bacterial bioluminescence intermediates. *Biochemistry* **30**: 6825–6835.

Lee, J., *et al.* (1991). The mechanism of bacterial bioluminescence. *In* Muller, F. (ed.), *Chemistry and Biochemistry of Flavoenzymes*, pp. 109–151. CRC Press, Orlando.

Lee, R., and McElroy, W. D. (1969). Role and reactivity of sulfhydryl groups in firefly luciferase. *Biochemistry* **8**: 130–136.

Legocki, R. P., Legocki, M., Baldwin, T. O., and Szalay, A. A. (1986). Bioluminescence in soybean root nodules: demonstration of a general approach to assay gene expression *in vivo* by using bacterial luciferase. *Proc. Natl. Acad. Sci. USA* **83**: 9080–9084.

Lei, B. F., and Becvar, J. E. (1991). A new reducing agent of flavins and its application to the assay of bacterial luciferase. *Photochem. Photobiol.* **54**: 473–476.

Lei, B., Ding, Q., and Tu, S.-C. (2004). Identity of the emitter in the bacterial luciferase luminescence reaction: binding and fluorescence quantum yield studies of 5-decyl-4a-hydroxy-4a,5-dihydroriboflavin-5'-phosphate as a model. *Biochemistry* **43**: 15975–15982.

Leisman, G., Cohn, D. H., and Nealson, K. H. (1980). Bacterial origin of luminescence in marine animals. *Science* **208**: 1271–1273.

Lembert, N. (1996). Firefly luciferase can use L-luciferin to produce light. *Biochem. J.* **317**: 273–277.

Levine, L. D., and Ward, W. W. (1982). Isolation and characterization of a photoprotein, "phialidin," and a spectrally unique green-fluorescent protein from the bioluminescent jellyfish *Phialidium gregarium. Comp. Biochem. Physiol.* **72B**: 77–85.

Letunov, V. N., and Vysotski, E. S. (1988). Bioluminescence activity and content of calcium-dependent photoprotein in *Obelia longissima* (Pallas) hydroid colonies. *Zh. Obshch. Biol.* **49**: 381–387.

Li, L. (2000). *Gonyaulax* luciferase: gene structure, protein expression, and purification from recombinant sources. *Method. Enzymol.* **305**: 249–258.

Li, L., and Hastings, J. W. (1998). The structure and organization of the luciferase gene in the photosynthetic dinoflagellate *Gonyaulax polyedra. Plant Mol. Biol.* **36**: 275–284.

Li, Z., and Meighen, E. A. (1994). The turnover of bacterial luciferase is limited by a slow decomposition of the ternary enzyme-product complex of luciferase, FMN, and fatty acid. *J. Biol. Chem.* **269**: 6640–6644.

Li, L., Hong, R., and Hastings, J. W. (1997). The single polypeptide chain of *Gonyaulax* luciferase has three enzymically active repeat units. *In* Hastings, J. W., *et al.* (eds.), *Biolumin. Chemilumin.*, Proc. Int. Symp., 9th, 1996, pp. 74–77. Wiley, Chichester, UK.

Li, L., Hong, R., and Hastings, J. W. (1997). Three functional luciferase domains in a single polypeptide chain. *Biochemistry* **94**: 8954–8958.

Limaye, N. M., and Santhanam, K. S. V. (1986). Effect of calcium on electrobioluminescence of *Lampito mauritii. Bioelectrochem. Bioenerg.* **15**: 341–351.

Lingle, W. L. (1994). Bioluminescence of *Panellus* during dikaryotization. *In* Campell, A. K., *et al.* (eds.), *Biolumin. Chemilumin.*, Proc. Int. Symp., 8th, 1994, pp. 544–547. Wiley, Chichester, UK.

Liu, J., O'Kane, D. J., and Escher, A. (1997). Secretion of functional *Renilla reniformis* luciferase by mammalian cells. *Gene* **203**: 141–148.

Liu, Z.-J., Vysotski, E. S., Chen, C.-J., Rose, J. B., Lee, J., and Wang, B.-C. (2000). Structure of the Ca^{2+}-regulated photoprotein obelin at 1.7 Å resolution determined directly from its sulfur substructure. *Protein Science* 9: 2085–2093.

Liu, Z.-J., *et al.* (2003). Atomic resolution structure of obelin: soaking with calcium enhances electron density of the second oxygen atom substituted at the C2-position of coelenterazine. *Biochem. Biophys. Res. Commun.* 311: 433–439.

Llinas, R., Sugimori, M., and Silver, R. B. (1992). Microdomains of high calcium concentration in a presynaptic terminal. *Science* 256: 677–679.

Loomis, H. F., and Davenport, D. (1951). A luminescent new xystodesmid milliped from California. *J. Wash. Acad. Sci.* 41: 270–272.

Lorenz, W. W., McCann, R. O., Longiaru, M., and Cormier, M. J. (1991). Isolation and expression of a cDNA encoding *Renilla reniformis* luciferase. *Proc. Natl. Acad. Sci. USA* 88: 4438–4442.

Lorenz, W. W., *et al.* (1993). Overexpression and purification of recombinant *Renilla reniformis* luciferase. *In* Szalay, A. A., *et al.* (eds.), *Biolumin. Chemilumin.*, Proc. Int. Symp., 7th, pp. 191–195. Wiley, Chichester, UK.

Lorenz, W. W., *et al.* (1996). Expression of the *Renilla reniformis* luciferase gene in mammalian cells. *J. Biolumin. Chemilumin.* 11: 31–37.

Loschen, G., and Chance, B. (1971). Rapid kinetic studies of the light emitting protein aequorin. *Nature New Biology* 233: 273–274.

Lucas, M., and Solano, F. (1992). Coelenterazine is a superoxide anion-sensitive chemiluminescent probe: its usefulness in the assay of respiratory burst in neutrophils. *Anal. Biochem.* 206: 273–277.

Macartney, J. (1810). Observations upon luminous animals. *Philos. Trans.* 100: 258–293.

Macheroux, P., *et al.* (1987). Purification of the yellow fluorescent protein from *Vibrio fischeri* and identity of the flavin chromophore. *Biochem. Biophys. Res. Commun.* 146: 101–106.

Macheroux, P., Ghisla, S., and Hastings, J. W. (1993). Spectral detection of an intermediate preceding the excited state in the bacterial luciferase reaction. *Biochemistry* 32: 14183–14186.

Macheroux, P., Ghisla, S., and Hastings, J. W. (1994). Bacterial luciferase: bioluminescence emission using lumazines as substrates. *In* Yagi, K. (ed.), *Flavins Flavoproteins*, Proc. Int. Symp., 11th, 1993, pp. 839–842. de Gruyter, Berlin, Germany.

Mackie, G. O., and Bone, Q. (1978). Luminescence and associated effector activity in *Pyrosoma* (Tunicata:Pyrosomida). *Proc. R. Soc. Lond. (Ser. B)* **202**: 483–495.

Maeda, Y., *et al.* (1997). Engineering of functional chimeric protein G-*Vargula* luciferase. *Anal. Biochem.* **249**: 147–152.

Mager, H. I. X., *et al.* (1990). Electrochemical superoxidation of flavins: generation of active precursors in luminescent model systems. *Photochem. Photobiol.* **52**: 1049–1056.

Makemson, J. C., Hastings, J. W., and Quirke, J. M. E. (1992). Stabilization of luciferase intermediates by fatty amines, amides, and nitriles. *Arch. Biochem.* **294**: 361–366.

Malkov, Y. A., and Danilov, V. S. (1986). Effect of ethylenediaminetetraacetate on bacterial luciferase. *Biokhimiya* **51**: 622–626.

Mallefet, J., and Baguet, F. (1993). Metabolic control of luminescence in isolated photophores of *Porichthys*: effects of glucose on oxygen consumption and luminescence. *J. Exp. Biol.* **181**: 279–293.

Mallefet, J., and Shimomura, O. (1995). Presence of coelenterazine in mesopelagic fishes from the Strait of Messina. *Mar. Biol.* **124**: 381–385.

Mamaev, S. V., Laikhter, A. L., Arslan, T., and Hecht, S. M. (1996). Firefly luciferase: alteration of the color of emitted light resulting from substitutions at position 286. *J. Am. Chem. Soc.* **118**: 7243–7244.

Manuel, D. C., Rangarajan, M., and Cass, A. E. G. (1986). The role of carboxyl groups in the activity and subunit interactions of bacterial luciferase. *Biochem. Soc. Trans.* **14**: 1285–1286.

Markova, S. V., *et al.* (2002). Obelin from the bioluminescent marine hydroid *Obelia geniculata*: cloning, expression, and comparison of some properties with those of other Ca^{2+}-regulated photoproteins. *Biochemistry* **41**: 2227–2236.

Markova, S. V., *et al.* (2004). Cloning and expression of cDNA for a luciferase from the marine copepod *Metridia longa*. *J. Biol. Chem.* **279**: 3212–3217.

Martin, A. R. (1761). Naturlig Phosphorus, eller Ren om Fisk och Kett, som lyser i merkeret. *Kongl. Vetenskaps Academiens Handlinger* **22**: 225–230.

Mashiko, S., *et al.* (1991). Measurement of rate constants for quenching singlet oxygen with a *Cypridina* luciferin analog (2-methyl-6-[p-methoxyphenyl]-3,7-dihydroimidazo[1,2-a]pyrazin-3-one) and sodium azide. *J. Biolumin. Chemilumin.* **6**: 69–72.

Mashiko, S., *et al.* (1991). Chemiluminescence reaction rates of *Cypridina* luciferin analog with superoxide. *In* Stanley, P. E., and Kricka, L. J. (eds.), *Biolumin. Chemilumin.* Proc. Int. Symp., 6th, 1990, pp. 475–478. Wiley, Chichester, UK.

Masuda, H., *et al.* (2003). Chromatography of isoforms of recombinant apoaequorin and method for the preparation of aequorin. *Protein Expression and Purification* **31**: 181–187.

Matheson, I. B. C., Lee, J., and Muller, F. (1981). Bacterial bioluminescence: spectral study of the emitters in the *in vivo* reaction. *Proc. Natl. Acad. Sci. USA* **78**: 948–952.

Matheson, I. B. C., O'Kane, D. J., and Lee, J. (1986). Free radical participation in bacterial bioluminescence. *Free Radical Res. Commun.* **2**: 1–5.

Matsui, S., *et al.* (1988). 4-Hydroxyretinal, a new pigment chromophore found in the bioluminescent squid, *Watasenia scintillans. Biochim. Biophys. Acta* **966**: 370–374.

Matthews, J. C., Hori, K., and Cormier, M. J. (1977a). Purification and properties of *Renilla reniformis* luciferase. *Biochemistry* **16**: 85–92.

Matthews, J. C., Hori, K., and Comier, M. J. (1977b). Substrate and substrate analogue binding properties of *Renilla* luciferase. *Biochemistry* **16**: 5217–5220.

Matveev, S. V., Lewis, J. C., and Daunert, S. (1999). Genetically engineered obelin as a bioluminescent label in an assay for a peptide. *Anal. Biochem.* **270**: 69–74.

Matz, M. V., *et al.* (1999). Fluorescent proteins from nonbioluminescent anthozoa species. *Nature Biotech.* **17**: 969–973.

Mayer, A. G. (1910). *Medusae of the World*, Vol. II (Hydromedusae). Carnegie Inst. Wash. Publ.

Mayerhofer, R., Langridge, W. H. R., Cormier, M. J., and Szalay, A. A. (1995). Expression of recombinant *Renilla* luciferase in transgenic plants results in high levels of light emission. *Plant. J.* **7**: 1031–1038.

McCapra, F. (1977). Alternative mechanism for dioxetane decomposition. *Chem. Commun.*, pp. 946–948.

McCapra, F. (1997). Mechanisms in chemiluminescence and bioluminescence — unfinished business. *In* Hastings, J. W., *et al.* (eds.), *Biolumin. Chemilumin.*, Proc. Int. Symp., 9th, 1996, pp. 7–15. Wiley, Chichester, UK.

McCapra, F., and Chang, Y. C. (1967). The chemiluminescence of a *Cypridina* luciferin analogue. *Chem. Commun.*, pp. 1011–1012.

McCapra, H., and Hart, R. (1980). The origins of marine bioluminescence. *Nature* **286**: 660–661.

McCapra, F., and Hysert, D. W. (1973). Bacterial bioluminescence — identification of fatty acid as product, its quantum yield and a suggested mechanism. *Biochem. Biophys. Res. Commun.* **52**: 298–304.

McCapra, F., and Manning, M. J. (1973). Bioluminescence of coelenterates: chemiluminescent model compounds. *Chem. Commun.*, pp. 467–468.

McCapra, F., Chang, Y. C., and Francois, V. P. (1968). The chemiluminescence of a firefly luciferin analogue. *Chem. Commun.*, pp. 22–23.

McCapra, F., Razavi, Z., and Neary, A. P. (1988). The fluorescence of the chromophore of the green fluorescent protein of *Aequorea* and *Renilla*. *Chem. Commun.*, pp. 790–791.

McCapra, F., *et al.* (1994). The chemical origin of color differences in beetle bioluminescence. *In* Campbell, A. E., *et al.* (eds.), *Biolumin. Chemilumin.*, Proc. Int. Symp., 8th, pp. 387–391. Wiley, Chichester, UK.

McCord, J. M., and Fridovich, I. (1969). Superoxide dismutase: an enzymic function for erythrocuprein (hemocuprein). *J. Biol. Chem.* **244**: 6049–6055.

McElroy, W. D. (1947). The energy source for bioluminescence in an isolated system. *Proc. Natl. Acad. Sci. USA* **33**: 342–345.

McElroy, W. D. (1960). Bioluminescence. *Fed. Proc.* **19**: 941–950.

McElroy, W. D., and Chase, A. M. (1951). Purification of *Cypridina* luciferase. *J. Cell. Comp. Physiol.* **38**: 401–408.

McElroy, W. D., and DeLuca, M. (1978). Chemistry of firefly luminescence. *In* Herring, P. J. (ed.), *Bioluminescence in Action*, pp. 109–127. Academic Press, London.

McElroy, W. D., and Green, A. A. (1955). *Arch. Biochem.* **56**: 240–255.

McElroy, W. D., and Hastings, J. W. (1955). Biochemistry of firefly luminescence. *In* Johnson, F. H. (ed.), *The Bioluminescence of Biological Systems*, pp. 161–198. American association of the advancement of science, Washington, DC.

McElroy, W. D., and Seliger, H. H. (1961). Mechanisms of bioluminescent reactions. *In* McElroy, W. D., and Glass, B. (eds.), *Light and Life*, pp. 219–257. The Johns Hopkins Press, Baltimore.

McElroy, W. D., and Seliger, H. H. (1966). Firefly bioluminescence. *In* Johnson, F. H., and Haneda, Y. (eds.), *Bioluminescence in Progress*, pp. 427–458. Princeton University Press, Princeton, NJ.

McElroy, W. D., and Strehler, B. L. (1949). Factors influencing the response of the bioluminescent reaction to adenosine triphosphate. *Arch. Biochem.* **22**: 420–433.

McElroy, W. D., Hastings, J. W., Coulombre, J., and Sonnenfeld, V. (1953). The mechanism of action of pyrophosphate in firefly luminescence. *Arch. Biochem. Biophys.* **46**: 399–416.

McNeil, P. L., and Taylor, D. L. (1985). Aequorin entrapment in mammalian cells. *Cell Calcium* **6**: 83–93.

Meighen, E. A. (1991). Molecular biology of bacterial bioluminescence. *Microbiol. Rev.* **55**: 123–142.

Meighen, E. A., and Bartlet, I. (1980). Complementation of subunits from different bacterial luciferases. Evidence for the role of the β subunit in the bioluminescent mechanism. *J. Biol. Chem.* **255**: 11181–11187.

Meighen, E. A., and Hastings, J. W. (1971). Binding site determination from kinetic data. *J. Biol. Chem.* **246**: 7666–7674.

Merenyi, G., Lind, J., Mager, H. I. X., and Tu, S. C. (1992). Properties of 4a-hydroxy-4a,5-dihydroflavin radicals in relation to bacterial bioluminescence. *J. Phys. Chem.* **96**: 10528–10533.

Michelson, A. M. (1978). Purification and properties of *Pholas dactylus* luciferin and luciferase. *Method. Enzymol.* **57**: 385–406.

Miller, A. L., *et al.* (1991). Imaging free calcium in cultured *Aplysia* bag cell neurons. *Biol. Bull.* **181**: 325.

Mills, C. E. (2001). Jellyfish blooms: are populations increasing globally in response to changing ocean conditions. *Hydrobiologia* **451**: 55–68.

Mills, G. A., and Urey, H. C. (1940). The kinetics of isotopic exchange between carbon dioxide, bicarbonate ion, carbonate ion and water. *J. Am. Chem. Soc.* **62**: 1019–1026.

Miyamoto, C., Boylan, M., Graham, A., and Meighen, E. (1986). Cloning and expression of the genes from the bioluminescent system of marine bacteria. *Method. Enzymol.* **133**: 70–83.

Miyamoto, C., Byers, D., Graham, A. F., and Meighen, E. A. (1987). Expression of bioluminescence by *Escherichia coli* containing recombinant *Vibrio harveyi* DNA. *J. Bacteriol.* **169**: 247–253.

Miyamoto, C., Boylan, M., Cragg, L., and Meighen, E. (1989). Comparison of the lux systems in *Vibrio harveyi* and *Vibrio fischeri. J. Biolumin. Chemilumin.* **3**: 193–199.

Morgan, J. P., DeFeo, T. T., and Morgan, K. G. (1984). A chemical procedure for loading the calcium indicator aequorin into mammalian working myocardium. *Pfluegers Arch.* **400**: 338–340.

Morin, J. G. (1974). Coelenterate bioluminescence. *In* Muscatine, L., and Lenhoff, H. M. (eds.), *Coelenterate Biology Reviews and Perspectives,* pp. 397–438. Academic Press, New York.

Morin, J. G., and Hastings, J. W. (1971a). Biochemistry of the bioluminescence of colonial hydroids and other coelenterates. *J. Cell. Physiol.* **77**: 305–311.

Morin, J. G., and Hastings, J. W. (1971b). Energy transfer in a bioluminescent system. *J. Cell. Physiol.* **77**: 313–318.

Morin, J. G., and Reynolds, G. T. (1969). Fluorescence and time distribution of photon emission of bioluminescent photocytes in *Obelia geniculata. Biol. Bull.* **137**: 410.

Morin, J. G., and Reynolds, G. T. (1972). Spectral and kinetic characteristics of bioluminescence in *Pelagia noctiluca* and other coelenterates. *Biol. Bull.* **143**: 470–471.

Morin, J. G., *et al.* (1975). Light for all reasons: versatility in the behavioral repertoire of the flashlight fish. *Science* **190**: 74–76.

Morise, H., Shimomura, O., Johnson, F. H., and Winant, J. (1974). Intermolecular energy transfer in the bioluminescent system of *Aequorea. Biochemistry* **13**: 2656-2662.

Morishita, H., *et al.* (2002). Cloning and characterization of an active fragment of luciferase from a luminescent marine alga, *Pyrocystis lunula. Photochem. Photobiol.* **75**: 311–315.

Morse, D., and Mittag, M. (2000). Dinoflagellate luciferin-binding protein. *Method. Enzymol.* **305**: 258–276.

Morse, D., Fritz, L., Pappenheimer, A. M., Jr., and Hastings, J. W. (1989). Properties and cellular localization of luciferin binding protein in the bioluminescence reaction of *Gonyaulax polyedra. J. Biolumin. Chemilumin.* **3**: 79–83.

Morse, D., Pappenheimer, A. M., Jr., and Hastings, J. W. (1989). Role of luciferin-binding protein in the circadian bioluminescent reaction of *Gonyaulax polyedra. J. Biol. Chem.* **264**: 11822–11826.

Morton, R. A., Hopkins, T. A., and Seliger, H. H. (1969). The spectroscopic properties of firefly luciferin and related compounds. An approach to product emission. *Biochemistry* **8**: 1598–1607.

Moss, G. W. J., Franks, N. P., and Lieb, W. R. (1991). Modulation of the general anesthetic sensitivity of a protein: a transition between two forms of firefly luciferase. *Proc. Natl. Acad. Sci. USA* **88**: 134–138.

Mulkerrin, M. G., and Wampler, J. E. (1978). Assaying hydrogen peroxide using the earthworm bioluminescence system. *Method. Enzymol.* **57**: 375–381.

Muller, G. W. (1890). Neue Cypridiniden. *Zoologische Jahrbucher* **5**: 211–252.

Muller, G. W. (1912). *Ostracoda*. Friedlander, Berlin.

Müller, T., and Campbell, A. K. (1990). The chromophore of pholasin: a highly luminescent protein. *J. Biolumin. Chemilumin.* **5**: 25–30.

Murbach, L., and Shearer, C. (1902). Preliminary report on a collection of medusae from the coast of British Columbia and Alaska. *Ann. Mag. Nat. Hist. Ser.* **79**: 71–73.

Musicki, B. G. (1987). *Chemistry of Bioluminescence: Jellyfish and Euphausiid Bioluminescence Systems*, Ph.D. Dissertation, Harvard University, Chemistry Department.

Musicki, B., Kishi, Y., and Shimomura, O. (1986). Structure of the functional part of photoprotein aequorin. *Chem. Commun.* **1986**: 1566–1568.

Nagano, K., and Tsuji, F. I. (1990). Dimeric interaction of calcium-binding photoprotein aequorin. *In* Rivier, J. E., and Marshall, G. R. (eds.), *Pept.: Chem., Struct. Biol.*, Proc. Am. Pept. Symp., 11th, 1989, pp. 508–509. ESCOM Sci. Pub., Leiden, Netherland.

Nakajima, Y., *et al.* (2004). cDNA cloning and characterization of a secreted luciferase from the luminous Japanese ostracod, *Cypridina noctiluca*. *Biosci. Biotechnol. Biochem.* **68**: 565–570.

Nakamura, H., Musicki, B., Kishi, Y., and Shimomura, O. (1988). Structure of the light emitter in krill *Euphausia pacifica*. *J. Am. Chem. Soc.* **110**: 2683-2685.

Nakamura, H., Kishi, Y., and Shimomura, O. (1988a). Panal: a possible precursor of fungal luciferin. *Tetrahedron* **44**: 1597–1602.

Nakamura, H., *et al.* (1989). Structure of dinoflagellate luciferin and its enzymatic and nonenzymatic air-oxidation products. *J. Am. Chem. Soc.* **111**: 7607–7611.

Nakamura, H., Takeuchi, D., and Murai, A. (1995). Synthesys of 5- and 3,5-substituted 2-aminopyrazines by Pd mediated Stille coupling. *Synlett.*, pp. 1227–1228.

Nakamura, H., *et al.* (1997). Efficient bioluminescence of bisdeoxycoelenterazine with the luciferase of a deep-sea shrimp *Oplophorus*. *Tetrahedron Lett.* **38**: 6405–6406.

Nakamura, H., *et al.* (2000). Convergent and short-step synthesis of *dl*-Cypridina luciferin and its analogues based on Pd-mediated cross coupling. *Tetrahedron Lett.* **41**: 2185-2188.

Nakamura, T., and Matsuda, K. (1971). Studies on luciferase from *Photobacterium phosphoreum*. 1. Purification and physical properties. *J. Biochem.* **70**: 35–44.

Nakano, E., and Sugisaki, Y. (1991). Firefly luciferase. Mass production and developments of its application. *Bio. Ind.* 8: 389–394.

Nakano, M. (1990). Assay for superoxide dismutase based on chemiluminescence of luciferin analog. *Method. Enzymol.* 186: 227–232.

Nakano, M. (1990). Determination of superoxide radical and singlet oxygen based on chemiluminescence of luciferin analog. *Method. Enzymol.* 186: 585–591.

Nakano, M. (1998). Detection of active oxygen species in biological systems. *Cell. Mol. Neurobiol.* 18: 565–579.

Nakano, M., Sugioka, K., Ushijima, Y., and Goto, T. (1986). Chemiluminescence probe with *Cypridina* luciferin analogue, 2-methyl-6-phenyl-3,7-dihydroimidazo[1,2-a]pyrazine-3-one, for estimating the ability of human granulocytes to generate superoxide anion. *Anal. Biochem.* 159: 363–369.

Nakatsubo, F., Kishi, Y., and Goto, T. (1970). Synthesis and stereochemistry of Latia luciferin. *Tetrahedron Lett.*, pp. 381–382.

Nealson, K. H. (1978). Isolation, identification and manipulation of luminous bacteria. *Method. Enzymol.* 57: 153–166.

Nealson, K. H., and Hastings, J. W. (1979). Bacterial bioluminescence: its control and ecological significance. *Microbiol. Rev.* **Dec. 1979:** 496–518.

Nealson, K. H., Platt, T., and Hastings, J. W. (1970). The cellular control of the synthesis and activity of the bacterial luminescence system. *J. Bacteriol.* 104: 313–322.

Neering, I. R., and Fryer, M. W. (1986). The effect of alcohols on aequorin luminescence. *Biochim. Biophys. Acta* 882: 39–43.

Neto, P. C., and Bechara, E. J. H. (1984). Chemistry and biology of insect bioluminescence. *Arq. Biol. Tecnol.* 27: 439–464.

Nickerson, W. J., and Strauss, G. (1960). The photochemical cleavage of water by riboflavin. *J. Am. Chem. Soc.* 82: 5007–5008.

Nicol, J. A. C. (1953). Luminescence in polynoid worms. *J. Mar. Biol. Ass., UK* 32: 65–84.

Nicol, J. A. C. (1957). Luminescence in polinoids II. Different modes of response in the elytra. *J. Mar. Biol. Ass., UK* 36: 529–538.

Nicolas, M. T. (1979). Presence de photosomes dans les fractions lumineuses du systeme elytral des Polynoinae (Annelides, Polychetes). *C. R. Acad. Sci. Paris* 289 D: 177–180.

Nicolas, M. T. (1980). Solubilisation du système lumineux des Polynoïniens, comparaison de différents tests d'activité. *Biol. Cellulaire* 39: 5.

Nicolas, M. T., Bassot, J. M., and Shimomura, O. (1982). Polynoidin: a membrane photoprotein isolated from the bioluminescent system of scale-worms. *Photochem. Photobiol.* **35**: 201–207.

Nicolas, M. T., et al. (1991). Luminescence detection of superoxide radicals with the photoprotein polynoidin. *In* Stanley, P. E., and Kricka, L. J. (eds.), *Biolumin. Chemilumin.* Proc. Int. Symp., 6th, 1990, pp. 401–404. Wiley, Chichester, UK.

Nicolas, M.-T., Morse, D., Bassot, J. M., and Hastings, J. W. (1991). Colocalization of luciferin-binding protein and luciferase to the scintillons of *Gonyaulax polyedra* revealed by double immunolabeling after fast-freeze fixation. *Protoplasma* **160**: 159–166.

Nishinaka, Y., et al. (1993). A new sensitive chemiluminescence probe, L-012, for measuring the production of superoxide anion by cells. *Biochem. Biophys. Res. Commun.* **193**: 554–559.

Niwa, H., et al. (1997). *Aequorea* green fluorescent protein: structural elucidation of the chromophore. *In* Hastings, J. W., et al. (eds.), *Biolumin. Chemilumin.*, Proc. Int. Symp., 9th, 1996, pp. 395–398. Wiley, Chichester, UK.

Njus, D. (1975). The control of bioluminescence in *Gonyaulax polyedra*. Ph.D. Thesis, Harvard University, Cambridge, MA.

Noguchi, M., Tsuji, F. I., and Sakaki, Y. (1986). Molecular cloning of apoaequorin cDNA. *Method. Enzymol.* **133**: 298–306.

Nomura, M., Inouye, S., Ohmiya, Y., and Tsuji, F. I. (1991). A C-terminal proline is required for bioluminescence of the calcium-binding photoprotein, aequorin. *FEBS Lett.* **295**: 63–66.

Oba, Y., et al. (2004). Identification of the luciferin-luciferase system and quantification of coelenterazine by mass spectrometry in the deep-see luminous ostracod *Conchoecia pseudodiscophora*. *ChemBioChem* **5**: 1495–1499.

Ohba, N. (1997). Twenty years with fireflies — an outline in research in Japan. *Insectarium* (in Japanese) **34**: 132–146.

Ohba, N. (2004). *Mystery of fireflies* (in Japanese). Yokosuka City Museum, Yokosuka, Japan.

Ohmiya, Y., and Hirano, T. (1996). Shining the light: the mechanism of the bioluminescence reaction of calcium-binding photoproteins. *Chem. Biol.* **3**: 337–347.

Ohmiya, Y., and Tsuji, F. I. (1993). Bioluminescence of the Ca^{2+}-binding photoprotein aequorin, after histidine modification. *FEBS Lett.* **320**: 267–270.

Ohmiya, Y., and Tsuji, F. I. (1997). Mutagenesis of firefly luciferase shows that cysteine residues are not required for bioluminescence activity. *FEBS Lett.* **404**: 115–117.

Ohmiya, Y., Ohashi, M., and Tsuji, F. I. (1992). Two excited states in aequorin bioluminescence induced by tryptophan modification. *FEBS Lett.* **301**: 197–201.

Ohmiya, Y., *et al.* (1993). Mass spectrometric evidence for a disulfide bond in aequorin regeneration. *FEBS Lett.* **332**: 226–228.

Ohmiya, Y., Teranishi, K., Akutagawa, M., and Ohashi, M. (1993a). Bioluminescence activity of coelenterazine analogs after incorporation into recombinant apoaequorin. *Chem. Lett.* **1993**: 2149–2152.

Ohmiya, Y., Ohba, N., Toh, H., and Tsuji, F. I. (1994). Comparative aspects of the bioluminescence reactions of *Hotaria parvula*, *Luciola cruciata*, and *Photinus pyralis*. *In* Campbell, A. K., *et al.* (eds.), *Biolumin. Chemilumin.* Proc. Int. Symp., 8th 1994, pp. 572–575. Wiley, Chichester, UK.

Ohmiya, Y., Ohba, N., Toh, H., and Tsuji, F. I. (1995). Cloning, expression and sequence analysis of cDNA for the luciferases from Japanese fireflies, *Pyrocoelia miyako* and *Hotaria parvula*. *Photochem. Photobiol.* **62**: 309–313.

Ohmiya, Y., Hirano, T., and Ohashi, M. (1996). The structural origin of the color differences in the bioluminescence of firefly luciferase. *FEBS Lett.* **384**: 83–86.

Ohtsuka, H., Rudie, N. G., and Wampler, J. E. (1976). Structural identification and synthesis of luciferin from the bioluminescent earthworm, *Diplocardia longa*. *Biochemistry* **15**: 1001–1004.

O'Kane, D. J. (1994). Determination of coelenterazine with *Renilla* luciferase. *In* Campbell, A. K., *et al.* (eds.), *Biolumin. Chemilumin.*, Proc. Int. Symp., 8th, pp. 139–142. Wiley, Chichester, UK.

O'Kane, D. J., and Lee, J. (1985). Chemical characterization of lumazine protein from *Photobacterium leiognathi*: comparison with lumazine protein from *Photobacterium phosphoreum*. *Biochemistry* **24**: 1467–1475.

O'Kane, D. J., and Lee, J. (1985). Physical characterization of lumazine proteins from *Photobacterium*. *Biochemistry* **24**: 1484–1488.

O'Kane, D. J., and Lee, J. (1986). Purification and properties of lumazine proteins from *Photobacterium* strains. *Method. Enzymol.* **133**: 149–172.

O'Kane, D. J., and Lee, J. (2000). Absolute calibration of luminometers with low-level light standards. *Method. Enzymol.* **305**: 87–96.

O'Kane, D. J., Karle, V. A., and Lee, J. (1985). Purification of lumazine proteins from *Photobacterium leiognathi* and *Photobacterium phosphoreum*: bioluminescence properties. *Biochemistry* **24**: 1461–1467.

O'Kane, D. J., Ahmad, M., Matheson, I. B. C., and Lee, J. (1986). Purification of bacterial luciferase by high-performance liquid chromatography. *Method. Enzymol.* **133**: 109–128.

O'Kane, D. J., Fuhrer, B., and Lingle, W. I. (1994). Spectral studies of fungal bioluminescence. *In* Campbell, A. K., *et al.* (eds.), *Bioluminescence and Chemiluminescence: Fundamentals and Applied Aspects*, pp. 552–555. John Wiley & Sons, Chichester.

Orlova, G., Goddard, J. D., and Brovko, L. U. (2003). Theoretical study of the amazing firefly bioluminescence: the formation and structures of the light emitters. *J. Am. Chem. Soc.* **125**: 6962–6971.

Ormö, M., *et al.* (1996). Crystal structure of the *Aequorea victoria* green fluorescent protein. *Science* **273**: 1392–1395.

Palmer, L. M., and Colwell, R. R. (1991). Detection of luciferase gene sequence in nonluminescent *Vibrio cholerae* by colony hybridization and polymerase chain reaction. *Appl. Environ. Microbiol.* **57**: 1286–1293.

Panceri, P. (1872). The light organ and the light of Pholades. *Quart. J. Micr. Sci.* **12**: 254–259.

Panceri, P. (1878). La luce e gli Organi luminosi di alcuni annelidi. *Atti. Accad. Sci. fis. mat. Napoli* 7 (No. 1): 1–20.

Paquatte, O., Fried, A., and Tu, S. C. (1988). Delineation of bacterial luciferase aldehyde site by bifunctional labeling reagents. *Arch. Biochem.* **264**: 392–399.

Péron, F. (1804). Mémoíre sur le noveau genre *Pyrosoma*. *Ann. Mus. Hist. Nat. Paris* 4: 437–446.

Péron, F., and Lesueur, C. A. (1809). Tableau des caracteres generiques et specifiques de toutes les especes de Meduses connues jusqu'a ce jour. *Ann. Mus. Hist. Nat. Paris* **14**: 325–366.

Perozzo, M. A., Ward, K. B., Thompson, R. B., and Ward, W. W. (1988). X-ray diffraction and time-resolved fluorescence analyses of *Aequorea* green fluorescent protein crystals. *J. Biol. Chem.* **263**: 7713–7716.

Perreault, C. L., Gonzalez-Serratos, H., Litwin, S. E., and Morgan, J. P. (1992). A chemical method for intracellular loading of the calcium indicator aequorin in mammalian skeletal muscle. *Proc. Soc. Exp. Biol. Med.* **199**: 178–182.

Peschke, K., Schmitt, K., and Zinner, K. (1986). Occurrence of electronically excited products during the defensive reaction of bombardier beetles. *Photobiochem. Photobiophys.* **12**: 275–282.

Petushkov, V. N., Gibson, B. G., and Lee, J. (1995). The yellow bioluminescence bacterium, *Vibrio fischeri* Y1, contains a bioluminescence active riboflavin protein in addition to the yellow fluorescence FMN protein. *Biochem. Biophys. Res. Commun.* **211**: 774–779.

Petushkov, V. N., Gibson, G. B., and Lee, J. (1995). Properties of recombinant fluorescent proteins from *Photobacterium leiognathi* and their interaction with luciferase intermediates. *Biochemistry* **34**: 3300–3309.

Petushkov, V. N., Gibson, B. G., Visser, A. J. W. G., and Lee, J. (2000). Purification and ligand exchange protocols for antenna proteins from bioluminescent bacteria. *Method. Enzymol.* **305**: 164–180.

Pierantoni, U. (1921). Organi luminosi batterici nei pesci. *Riv. Biol.* **3**: 342–346.

Plant, P. J., White, E. H., and McElroy, W. D. (1968). The decarboxylation of luciferin in firefly bioluminescence. *Biochem. Biophys. Res. Commun.* **31**: 98–103.

Poulsen, E. M. (1962). *Ostracoda-Myodocopa, Part I, Cyprodiniformes-Cypridinidae*, Dana-Report, Vol. 57. Carlsberg Foundation, Copenhagen.

Prasher, D. C. (1995). Using GFP to see the light. *Trends in Genetics* **11**: 320–323.

Prasher, D., McCann, R. O., and Cormier, M. J. (1985). Cloning and expression of the cDNA coding for aequorin, a bioluminescent calcium-binding protein. *Biochem. Biophys. Res. Commun.* **126**: 1259–1268.

Prasher, D. C., McCann, R. O., and Cormier, M. J. (1986). Isolation and expression of a cDNA coding for aequorin, the calcium-activated photoprotein from *Aequorea victoria*. *Method. Enzymol.* **133**: 288–298.

Prasher, D. C., McCann, R. O., Longiaru, M., and Cormier, M. J. (1987). Molecular biology of aequorin. *In* Schoelmerich, J. (ed.), *Biolumin. Chemilumin.* Symp., 4th, 1986, pp. 365–368. Wiley, Chichester, UK.

Prasher, D. C., McCann, R. O., Longiaru, M., and Cormier, M. J. (1987). Sequence comparisons of complememtary DNAs encoding aequorin isotypes. *Biochemistry* **26**: 1326–1332.

Prasher, D. C., O'Kane, D., Lee, J., and Woodward, B. (1990). The lumazine protein gene in *Photobacterium phosphoreum* is linked to the lux operon. *Nucleic Acids Res.* **18**: 6450.

Prasher, D. C., *et al.* (1992). Primary structure of the *Aequorea victoria* green fluorescent protein. *Gene* **111**: 229–233.

Prendergast, F. G., and Mann, K. G. (1978). Chemical and physical properties of aequorin and the green fluorescent protein isolated from *Aequorea forskalea*. *Biochemistry* **17**: 3448–3453.

Pryor, W. A. (1976). The role of free radical reactions in biological systems. *In* Pryor, W. A. (ed.), *Free Radicals in Biology*, Vol. 1. Academic Press, New York.

Puget, K., and Michelson, A. M. (1972). Studies in bioluminescence. VII Bacterial NADH: flavine mononucleotide oxidoreductase. *Biochimie* **54**: 1197–1204.

Puget, K., and Michelson, A. M. (1974). Iron containing superoxide dismutase from luminous bacteria. *Biochimie* **56**: 1255–1267.

Qi, C. F., *et al.* (1991). Chemi- and bio-luminescence of coelenterazine analogues. Effect of substituents at the C-2 position. *Chem. Commun.*, pp. 1307–1309.

Raushel, F. M., and Baldwin, T. O. (1989). Proposed mechanism for the bacterial bioluminescence reaction involving dioxirane intermediate. *Biochem. Biophys. Res. Commun.* **164**: 1137–1142.

Ray, B. D., *et al.* (1985). Proton NMR of aequorin. Structural changes concomitant with calcium-independent light emission. *Biochemistry* **24**: 4280–4287.

Rees, J. F., and Baguet, F. (1988). Metabolic control of spontaneous glowing in isolated photophores of *Porichthys*. *J. Exp. Biol.* **135**: 289–299.

Rees, J. F., Thompson, E. M., Baguet, F., and Tsuji, F. I. (1990). Detection of coelenterazine and related luciferase activity in the tissues of the luminous fish, *Vinciguerria attenuata*. *Comp. Biochem. Physiol.* **96A**: 425–430.

Rees, J. F., Thompson, E. M., Baguet, F., and Tsuji, F. I. (1992). Evidence for the utilization of coelenterazine as the luminescent substrate in *Argyropelecus* photophores. *Mol. Mar. Biol. Biotechnol.* **1**: 219–225.

Rees, J.-F., *et al.* (1998). The origins of marine bioluminescence: turning oxygen defense mechanism into deep-sea communication tools. *J. Exp. Biol.* **201**: 1211–1221.

Reichl, S. (2000). Reactions of pholasin with peroxidases and hypochlorous acid. *Free Radic. Biol. Med.* **28**: 1555–1563.

Reichl, S., *et al.* (1999). Factors influencing the pholasin chemiluminescence. *In* Roda, A. (ed.), *Biolumin. Chemilumin.* Proc. Int. Symp., 10th, 1998. Wiley, Chichester, UK.

Reid, B. G., and Flynn, G. C. (1997). Chromophore formation in green fluorescent protein. *Biochemistry* **36**: 6786–6791.

Reilly, M. J., and Jenkins, H. C. (1997). Factors affecting the shelf life of freeze-dried firefly luciferase reagents. *In* Hastings, J. W., *et al.* (eds.), *Biolumin. Chemilumin.*, Proc. Int. Symp., 9th, 1996, pp. 257–260. Wiley, Chichester, UK.

Rhodes, W. C., and McElroy, W. D. (1958). The synthesis and function of luciferyl-adenylate and oxyluciferyl-adenylate. *J. Biol. Chem.* **233**: 1528–1537.

Ridgway, E. B., and Ashley, C. C. (1967). Calcium transients in single muscle fibers. *Biochem. Biophys. Res. Commun.* **29**: 229–234.

Ridgway, E. B., and Snow, A. E. (1983). Effects of EGTA on aequorin luminescence. *Biophys. J.* **41**: 244a.

Riendeau, D., and Meighen, E. A. (1980). Co-induction of fatty acid reductase and luciferase during development of bacterial bioluminescence. *J. Biol. Chem.* **255**: 12060–12065.

Rizzuto, R., Simpson, A. W. M., Brini, M., and Pozzan, T. (1992). Rapid changes of mitochondrial Ca^{2+} revealed by specifically targeted recombinant aequorin. *Nature* **358**: 325–327.

Rizzuto, R., *et al.* (1995). Photoprotein-mediated measurement of calcium ion concentration in mitochondria of living cells. *Method. Enzymol.* **260**: 417–428.

Rizzuto, R., Marisa, B., Pizzo, P., Murgia, M., and Pozzan, T. (1995). Chimeric green fluorescent protein as a tool for visualizing subcellular organelles in living cells. *Curr. Biol.* **5**: 635–642.

Roberts, P. A., Knight, J., and Campbell, A. K. (1987). Pholasin — a bioluminescent indicator for detecting activation of single neutrophils. *Anal. Biochem.* **160**: 139–148.

Rodriguez, J. F., *et al.* (1988). Expression of the firefly luciferase gene in vaccinia virus: a highly sensitive gene marker to follow virus dissemination in tissues of infected animals. *Proc. Natl. Acad. Sci. USA* **85**: 1667–1671.

Robinson, B. H., Reisenbichler, J. C., Hunt, J. C., and Haddock, S. H. D. (2003). Light production by the arm tips of the deep-sea cephalopod *Vampyroteuthis infernalis*. *Biol. Bull.* **205**: 102–109.

Rogers, P., and McElroy, W. D. (1955). Biochemical characteristics of aldehyde and luciferase mutants of luminous bacteria. *Proc. Natl. Acad. Sci. USA* **41**: 67–70.

Roosen-Runge, E. C. (1970). Life cycle of the hydromedusa *Phialidium gregarium* (A. Agassiz, 1862) in the laboratory. *Biol. Bull.* **139**: 203–221.

Rosenow, M. A., Huffman, H. A., Phail, M. E., and Wachter, R. M. (2004). The crystal structure of the Y66L variant of green fluorescent protein supports a cyclization-oxidation-dehydration mechanism for chromophore maturation. *Biochemistry* **43**: 4464–4472.

Rost, M., Karge, E., and Klinger, W. (1998). What do we measure with luminol-, lucigenin- and penicillin-amplified chemiluminescence? 1. Investigations with hydrogen peroxide and sodium hypochlorite. *J. Biolumin. Chemilumin.* **13**: 355–363.

Roth, A. (1985). *Purification and Protease Susceptibility of the Green-fluorescent Protein of Aequorea aequorea with a note on Halistaura*, M.S. Thesis, Rutgers University, New Brunswick, NJ.

Roth, A. F., and Ward, W. W. (1983). Conformational stability after protease treatment in *Aequorea* GFP. *Photochem. Photobiol.* **37S**: S71.

Roth, J. A., and Kaeberle, M. L. (1980). Chemiluminescence by *Listeria monocytogenes*. *J. Bacteriol.* **144**: 752–757.

Ruby, E. G., and Nealson, K. H. (1977). A luminous bacterium which emits yellow light. *Science* **196**: 432–434.

Ruby, E. G., and Nealson, K. H. (1978). Seasonal changes in the species composition of luminous bacteria in nearshore seawater. *Limnolo. Oceanogr.* **23**: 530–533.

Rudie, N. G. (1977). Studies of the physiology and chemistry of the bioluminescent earthworm. *Diplocardia longa*. Ph.D. Dissertation, University of Georgia, Athens, GA.

Rudie, N. G., Ohtsuka, H., and Wampler, J. E. (1976). Purification and properties of luciferin from the bioluminescent earthworm, *Diplocardia longa*. *Photochem. Photobiol.* **23**: 71–73.

Rudie, N. G., Mulkerrin, M. G., and Wampler, J. E. (1981). Earthworm bioluminescence: characterization of high specific activity *Diplocardia longa* luciferase and the reaction it catalyzes. *Biochemistry* **20**: 344–350.

Russell, F. S. (1953). *The Medusae of the British Isles*, Vol. I: Anthomedusae, Leptomedusae, Limnomedusae, Trchymedusae and Narcomedusae. Cambridge University Press, London.

Russell, F. S. (1970). *The Medusae of the British Isles*, Vol. II: Pelagic Scypho-zoa with a supplement to the first volume on hydromedusae. Cambridge University Press, London.

Sacchetti, A., *et al.* (1999). Red GFP and endogenous porphyrins. *Curr. Biol.* 9: R391–R393.

Saito, R., Hirano, T., Niwa, H., and Ohashi, M. (1997). Solvent and substituent effects on the fluorescent properties of coelenteramide analogues. *J. Chem. Soc., Perkin Trans. 2*, pp. 1711–1716.

Sakaki, Y., *et al.* (1988). Structure and function of the calcium-binding photoprotein aequorin: studies by recombinant DNA technology. *In* Yagi, Y., and Miyazaki, T. (eds.), *Calcium Signal Cell Response*, pp. 151–156. Jpn. Sci. Soc. Press: Tokyo, Japan.

Sakharov, G. N., Ismailov, A. D., and Danilov, V. S. (1988). Temperature dependencies of the reaction of bacterial luciferases from *Beneckea harveyi* and *Photobacterium fischeri*. *Biokhimiya* 53: 891–898.

Sala-Newby, G. B., Thomson, C. M., and Campbell, A. K. (1994). Cloning and characterization of the luciferase from the glow-worm. *In* Campbell, A. K., *et al.* (eds.), *Biolumin. Chemilumin.*, Proc. Int. Symp., 8th, pp. 588–591. Wiley, Chichester, UK.

Samuel, D. (1962). Methodology of oxygen isotopes. *In* Hayaishi, O. (ed.), *Oxygenases*, pp. 31–86. Academic Press, New York.

Santhanum, K. S. V., and Limaye, N. M. (1989). A proposed scheme for the activated luminescence of *Lampito mauritii* by ferrous ion injection. *Bioelectrochem. Bioenerg.* 22: 231–240.

Sawin, K. E., and Nurse, P. (1997). Photoactivation of green fluorescent protein. *Curr. Biol.* 7: R606–R607.

Schaefer, A. L., Hanzelka, B. L., Parsek, M. R., and Greenberg, E. P. (2000). Detection, purification, and structural elucidation of the acylhomoserine lactone inducer of *Vibrio fischeri* luminescence and other related molecules. *Method. Enzymol.* 305: 288–301.

Schmidt, T. M., Kopecky, K., and Nealson, K. H. (1989). Bioluminescence of the insect pathogen *Xenorhabdus luminescens*. *Appl. Environ. Microbiol.* 55: 2607–2612.

Schmitter, R. E., *et al.* (1976). Dinoflagellate luminescence: a comparative study of *in vitro* components. *J. Cell. Physiol.* 87: 123–134.

Schultz, L. W., Lie, L., Cegielski, M., and Hastings, J. W. (2005). Crystal structure of a pH-regulated luciferase catalyzing the bioluminescent oxidation of an open tetrapyrrole. *Proc. Natl. Acad. Sci. USA* 102: 1378–1383.

Schuster, G. B. (1979). Chemiluminescence of organic peroxides. Conversion of ground-state reactants to excited-state products by chemically initiated electron-exchange luminescence mechanism. *Acc. Chem. Res.* 12: 366–373.

Scopes, R. K. (1993). *Protein Purification, Principles and Practice*, 3rd ed. Springer-Verlag, New York.

Sedbrook, J. C., *et al.* (1996). Transgenic aequorin reveals organ-specific cytosolic Ca^{2+} responses to anoxia in Arabidopsis thaliana seedlings. *Plant Physiol.* 111: 243–257.

Seliger, H. H. (1987). The evolution of bioluminescence in bacteria. *Photochem. Photobiol.* 45: 291–297.

Seliger, H. H., and McElroy, W. D. (1959). Quantum yield in the oxidation of firefly luciferin. *Biochem. Biophys. Res. Commun.* 1: 21–24.

Seliger, H. H., and McElroy, W. D. (1960). Spectral emission and quantum yield of firefly bioluminescence. *Arch. Biochem. Biophys.* 88: 136–141.

Seliger, H. H., and McElroy, W. D. (1962). Chemiluminescence of firefly luciferin without enzyme. *Science* 138: 683–685.

Seliger, H. H., and McElroy, W. D. (1964). The colors of firefly bioluminescence: enzyme configuration and species specificity. *Proc. Natl. Acad. Sci. USA* 52: 75–81.

Seliger, H. H., and Morton, R. A. (1968). A physical approach to bioluminescence. *Photophysiology* 4: 253.

Seliger, H. H., Biggley, W. H., and Swift, E. (1969). Absolute values of photon emission from the marine dinoflagellates *Pyrodinium bahamense, Gonyaulax polyedra* and *Pyrocystis lunula*. *Photochem. Photobiol.* 10: 227–232.

Sheu, Y. A., Kricka, L. J., and Pritchett, D. B. (1993). Measurement of intracellular calcium using bioluminescent aequorin expressed in human cells. *Anal. Biochem.* 209: 343–347.

Shimomura, O. (1960). Structure of *Cypridina* luciferin, II. *J. Chem. Soc. Japan, Pure Chem. Section* 81: 179–182.

Shimomura, O. (1979). Structure of the chromophore of *Aequorea* green fluorescent protein. *FEBS Lett.* 104: 220–222.

Shimomura, O. (1980). Chlorophyll-derived bile pigment in bioluminescent euphausiids. *FEBS Lett.* 116: 203–206.

Shimomura, O. (1981). A new type of ATP-activated bioluminescent system in the millipede *Luminodesmus sequoiae*. *FEBS Lett.* 128: 242–244.

Shimomura, O. (1982). Mechanism of bioluminescence. *In* Adam, W., and Cilento, G. (eds.), *Chemical and Biological Generation of Excited States*, pp. 249–276. Academic Press, New York.

Shimomura, O. (1984). Porphyrin chromophore in *Luminodesmus* photoprotein. *Comp. Biochem. Physiol.* **79B**: 565–567.

Shimomura, O. (1985). Bioluminescence in the sea: photoprotein systems. *Symp. Soc. Exp. Biol.* **39**: 351–372.

Shimomura, O. (1986). Rechargeable energy storage in nature: coelenterate photoproteins. *In* Imura, H., *et al.* (eds.), *Natural Products and Biological Activities*, pp. 33–44. University of Tokyo Press, Tokyo.

Shimomura, O. (1986a). Isolation and properties of various molecular forms of aequorin. *Biochem. J.* **234**: 271–277.

Shimomura, O. (1986b). Bioluminescence of the brittle star *Ophiopsila californica*. *Photochem. Photobiol.* **44**: 671–674.

Shimomura, O. (1987). Presence of coelenterazine in non-bioluminescent marine organisms. *Comp. Biochem. Physiol.* **86B**: 361–363.

Shimomura, O. (1989). Chemiluminescence of panal (a sesquiterpene) isolated from the luminous fungus *Panellus stipticus*. *Photochem. Photobiol.* **49**: 355–360.

Shimomura, O. (1991a). Preparation and handling of aequorin solutions for the measurement of cellular Ca^{2+}. *Cell Calcium* **12**: 635–643.

Shimomura, O. (1991b). Superoxide-triggered chemiluminescence of the extract of luminous mushroom *Panellus stipticus* after treatment with methylamine. *J. Exp. Botany* **42**: 555–560.

Shimomura, O. (1992). The role of superoxide dismutase in regulating the light emission of luminescent fungi. *J. Exp. Botany* **43**: 1519–1525.

Shimomura, O. (1993). The role of superoxide anion in bioluminescence. *In* Shima, A., *et al.* (eds.), *Frontiers of Photobiology*, pp. 249–254. Elsevier Science Publishers, Amsterdam.

Shimomura, O. (1995a). The roles of the two highly unstable components F and P involved in the bioluminescence of euphausiid shrimps. *J. Biolumin. Chemilumin.* **10**: 91–101.

Shimomura, O. (1995b). A short story of aequorin. *Biol. Bull.* **189**: 1–5.

Shimomura, O. (1995c). Luminescence of aequorin is triggered by the binding of two calcium ions. *Biochem. Biophys. Res. Commun.* **211**: 359–363.

Shimomura, O. (1995d). Cause of spectral variation in the luminescence of semisynthetic aequorins. *Biochem. J.* **306**: 537–543.

Shimomura, O. (1997). Membrane permeability of coelenterazine analogues measured with fish eggs. *Biochem. J.* **326**: 297–298.

Shimomura, O. (1998). The discovery of green fluorescent protein. *In* Chalfie, M., and Kain, S. (eds.), *Green Fluorescent Protein*, pp. 3–15. Wiley-Liss, New York.

Shimomura, O. (2005a). The discovery of aequorin and green fluorescent protein. *J. Microscopy* **217**: 3–15.

Shimomura, O. (2005b). Aequorin and GFP: an historical account. *In* Tsuji, A., *et al.* (eds.), *Bioluminescence and Chemiluminescence, Progress and Perspectives*, pp. 27–34. World Scientific, Singapore.

Shimomura, O. (2006). Discovery of green fluorescent protein. *In* Chalfie, M., and Kain, S. (eds.), *Green Fluorescent Protein*, pp. 1–13. Wiley, Hoboken, NJ.

Shimomura, O., and Eguchi, S. (1960). Studies on 5-imidazolone. I-II. *Nippon Kagaku Zasshi* **81**: 1434–1439.

Shimomura, O., and Flood, P. R. (1998). Luciferase of the scyphozoan medusa *Periphylla periphylla*. *Biol. Bull.* **194**: 244–252.

Shimomura, O., and Inouye, S. (1996). Titration of recombinant aequorin with calcium chloride. *Biochem. Biophys. Res. Commun.* **221**: 77–81.

Shimomura, O., and Inouye, S. (1999). The *in situ* regeneration and extraction of recombinant aequorin from *Escherichia coli* cells and the purification of extracted aequorin. *Protein Expression and Purification* **16**: 91–95.

Shimomura, O., and Johnson, F. H. (1966). Partial purification and properties of the *Chaetopterus* luminescence system. *In* Johnson, F. H., and Haneda, Y. (eds.), *Bioluminescence in Progress*, pp. 495–521. Princeton University Press, Princeton, NJ.

Shimomura, O., and Johnson, F. H. (1967). Extraction, purification and properties of the bioluminescence system of the euphausiid shrimp *Meganyctiphanes norvegica*. *Biochemistry* **6**: 2293–2306.

Shimomura, O., and Johnson, F. H. (1968a). Light emitting molecules in a new photoprotein type of luminescence system from the euphausiid shrimp *Meganyctiphanes norvegica*. *Proc. Natl. Acad. Sci. USA* **59**: 475–477.

Shimomura, O., and Johnson, F. H. (1968b). The structure of *Latia* luciferin. *Biochemistry* **7**: 1734–1738.

Shimomura, O., and Johnson, F. H. (1968c). Purification and properties of the luciferase and of a protein cofactor in the bioluminescence system of *Latia neritoides*. *Biochemistry* **7**: 2574–2580.

Shimomura, O., and Johnson, F. H. (1968d). *Chaetopterus* photoprotein: crystallization and cofactor requirements for bioluminescence. *Science* 159: 1239–1240.

Shimomura, O., and Johnson, F. H. (1969). Properties of the bioluminescent protein aequorin. *Biochemistry* 8: 3991–3997.

Shimomura, O., and Johnson, F. H. (1970a). Mechanisms in the quantum yield of *Cypridina* bioluminescence. *Photochem. Photobiol.* 12: 291–295.

Shimomura, O., and Johnson, F. H. (1970b). Calcium binding, quantum yield, and emitting molecule in aequorin bioluminescence. *Nature* 227: 1356–1357.

Shimomura, O., and Johnson, F. H. (1971). Mechanism of the luminescent oxidation of *Cypridina* luciferin. *Biochem. Biophys. Res. Commun.* 44: 340–346.

Shimomura, O., and Johnson, F. H. (1972). Structure of the light-emitting moiety of aequorin. *Biochemistry* 11: 1602–1608.

Shimomura, O., and Johnson, F. H. (1973a). Exchange of oxygen between solvent H_2O and the CO_2 produced in *Cypridina* bioluminescence. *Biochem. Biophys. Res. Commun.* 51: 558–563.

Shimomura, O., and Johnson, F. H. (1973b). Mechanism of the luminescent oxidation of *Cypridina* luciferin. *In* Cormier, M. J., *et al.* (eds.), *Chemiluminescence and Bioluminescence*, pp. 337–343. Plenum Press, New York.

Shimomura, O., and Johnson, F. H. (1973c). Chemical nature of light emitter in bioluminescence of aequorin. *Tetrahedron Lett.*, pp. 2963–2966.

Shimomura, O., and Johnson, F. H. (1973d). Further data on the specificity of aequorin luminescence to calcium. *Biochem. Biophys. Res. Commun.* 53: 490–494.

Shimomura, O., and Johnson, F. H. (1975a). Influence of buffer system and pH on the amount of oxygen exchanged between solvent H_2O and the CO_2 produced in the aerobic oxidation of *Cypridina* luciferin catalyzed by *Cypridina* luciferase. *Anal. Biochem.* 64: 601–605.

Shimomura, O., and Johnson, F. H. (1975b). Chemical Nature of bioluminescence systems in coelenterates. *Proc. Natl. Acad. Sci. USA* 72: 1546–1549.

Shimomura, O., and Johnson, F. H. (1975c). Regeneration of the photoprotein aequorin. *Nature* 256: 236–238.

Shimomura, O., and Johnson, F. H. (1975d). Specificity of aequorin bioluminesce to calcium. *In* Chappelle, E. W., and Picciolo, G. L. (eds.),

Analytical Application of Bioluminescence and Chemiluminescence, NASA SP-388 ed., pp. 89–94. National Aeronautics and Space Administration, Washington, DC.

Shimomura, O., and Johnson, F. H. (1976). Calcium-triggered luminescence of the photoprotein aequorin. *Symp. Soc. Exp. Biol.* **30**: 41–54.

Shimomura, O., and Johnson, F. H. (1978). Peroxidized coelenterazine, the active group in the photoprotein aequorin. *Proc. Natl. Acad. Sci. USA* **75**: 2611–2615.

Shimomura, O., and Johnson, F. H. (1979a). Chemistry of the calcium-sensitive photoprotein aequorin. *In* Ashley, C. C., and Campbell, A. K. (eds.), *Detection and Measurement of Free Calcium Ions in Cells*, pp. 73–83. Elsevier/North-Holland, Amsterdam.

Shimomura, O., and Johnson, F. H. (1979b). Comparison of the amounts of key components in the bioluminescence system of various coelenterates. *Comp. Biochem. Physiol.* **64B**: 105–107.

Shimomura, O., and Johnson, F. H. (1979c). Elimination of the effect of contaminating CO_2 in the ^{18}O-labeling of the CO_2 produced in bioluminescent reactions. *Photochem. Photobiol.* **30**: 89–91.

Shimomura, O., and Shimomura, A. (1981). Resistivity to denaturation of the apoprotein of aequorin and reconstitution of the luminescent photoprotein from the partially denatured apoprotein. *Biochem. J.* **199**: 825–828.

Shimomura, O., and Shimomura, A. (1982). EDTA-binding and acylation of the Ca^{2+}-sensitive photoprotein aequorin. *FEBS Lett.* **138**: 201–204.

Shimomura, O., and Shimomura, A. (1984). Effect of calcium chelators on the Ca^{2+}-dependent luminescence of aequorin. *Biochem. J.* **221**: 907–910.

Shimomura, O., and Shimomura, A. (1985). Halistaurin, phialidin and modified forms of aequorin as Ca^{2+} indicator in biological systems. *Biochem. J.* **228**: 745–749.

Shimomura, O., and Teranishi, K. (2000). Light-emitters involved in the luminescence of coelenterazine. *Luminescence* **15**: 51–58.

Shimomura, O., Goto, T., and Hirata, Y. (1957). Crystalline *Cypridina* luciferin. *Bull. Chem. Soc. Japan* **30**: 929–933.

Shimomura, O., Johnson, F. H., and Saiga, Y. (1961). Purification and properties of *Cypridina* luciferase. *J. Cell. Comp. Physiol.* **58**: 113–124.

Shimomura, O., Johnson, F. H., and Saiga, Y. (1962). Extraction, purification and properties of aequorin, a bioluminescent protein from the luminous hydromedusan, *Aequorea. J. Cell. Comp. Physiol.* **59**: 223–239.

Shimomura, O., Johnson, F. H., and Saiga, Y. (1963a). Extraction and properties of halistaurin, a bioluminescent protein from the hydromedusan *Halistaura. J. Cell. Comp. Physiol.* **62**: 9–16.

Shimomura, O., Johnson, F. H., and Saiga, Y. (1963b). Further data on the bioluminescent protein, aequorin. *J. Cell. Comp. Physiol.* **62**: 1–8.

Shimomura, O., Johnson, F. H., and Saiga, Y. (1963c). Microdetermination of calcium by aequorin luminescence. *Science* **140**: 1339–1340.

Shimomura, O., Johnson, F. H., and Saiga, Y. (1963d). Partial purification and properties of the *Odontosyllis* luminescence system. *J. Cell. Comp. Physiol.* **61**: 275–292.

Shimomura, O., Beers, J. R., and Johnson, F. H. (1964). The cyanide activation of *Odontosyllis* luminescence. *J. Cell. Comp. Physiol.* **64**: 15–22.

Shimomura, O., Johnson, F. H., and Haneda, Y. (1966a). Observations on the biochemistry of luminescence in the New Zealand glowworm, *Arachnocampa luminosa. In* Johnson, F. H., and Haneda, Y. (eds.), *Bioluminescence in Progress*, pp. 487–494. Princeton University Press, Princeton, NJ.

Shimomura, O., Johnson, F. H., and Haneda, Y. (1966b). Isolation of the luciferin of the New Zealand fresh-water limpet, *Latia neritoides* Gray. *In* Johnson, F. H., and Haneda, Y. (eds.), *Bioluminescence in Progress*, pp. 391–404. Princeton University Press, Princeton, NJ.

Shimomura, O., Johnson, F. H., and Masugi, T. (1969). *Cypridina* bioluminescence: light-emitting oxyluciferin-luciferase complex. *Science* **164**: 1299–1300.

Shimomura, O., Johnson, F. H., and Kohama, Y. (1972). Reactions involved in bioluminescence systems of limpet (*Latia neritoides*) and luminous bacteria. *Proc. Natl. Acad. Sci. USA* **69**: 2086–2089.

Shimomura, O., Johnson, F. H., and Morise, H. (1974a). The aldehyde content of luminous bacteria and of an "aldehydeless" dark mutant. *Proc. Natl. Acad. Sci. USA* **71**: 4666–4669.

Shimomura, O., Johnson, F. H., and Morise, H. (1974b). Mechanism of the luminescent intramolecular reaction of aequorin. *Biochemistry* **13**: 3278–3286.

Shimomura, O., Inoue, S., and Goto, T. (1975). The light-emitter in bioluminescence of the sea cactus *Cavernularia obesa. Chem. Lett.*, 247–248.

Shimomura, O., Goto, T., and Johnson, F. H. (1977). Source of oxygen in the CO_2 produced in the bioluminescent oxidation of firefly luciferin. *Proc. Natl. Acad. Sci. USA* **74**: 2799–2802.

Shimomura, O., Masugi, T., Johnson, F. H., and Haneda, Y. (1978). Properties and reaction mechanism of the bioluminescence system of the deep-sea shrimp *Oplophorus gracilorostris. Biochemistry* **17**: 994–998.

Shimomura, O., Inoue, S., Johnson, F. H., and Haneda, Y. (1980). Widespread occurrence of coelenterazine in marine bioluminescence. *Comp. Biochem. Physiol.* **65B**: 435–437.

Shimomura, O., Musicki, B., and Kishi, Y. (1988). Semi-synthetic aequorin: an improved tool for the measurement of calcium ion concentration. *Biochem. J.* **251**: 405–410.

Shimomura, O., Musicki, B., and Kishi, Y. (1989). Semi-synthetic aequorins with improved sensitivity to Ca2+ ions. *Biochem. J.* **261**: 913–920.

Shimomura, O., Inouye, S., Musicki, B., and Kishi, Y. (1990). Recombinant aequorin and recombinant semi-synthetic aequorins. *Biochem. J.* **270**: 309–312.

Shimomura, O., Kishi, Y., and Inouye, S. (1993). The relative rate of aequorin regeneration from apoaequorin and coelenterazine analogues. *Biochem. J.* **296**: 549–551.

Shimomura, O., Musicki, B., Kishi, Y., and Inouye, S. (1993a). Light-emitting properties of recombinant semi-synthetic aequorins and recombinant fluorescein-conjugated aequorin for measuring cellular calcium. *Cell Calcium* **14**: 373–378.

Shimomura, O., Satoh, S., and Kishi, Y. (1993b). Structure and non-enzymatic light emission of two luciferin precursors isolated from the luminous mushroom *Panellus stipticus. J. Biolumin. Chemilumin.* **8**: 201–205.

Shimomura, O., Wu, C., Murai, A., and Nakamura, H. (1998). Evaluation of five imidazopyrazinone-type chemiluminescent superoxide probes and their application to the measurement of superoxide anion generated by *Listeria monocytogenes. Anal. Biochem.* **258**: 230–235.

Shimomura, O., *et al.* (2001). Isolation and properties of the luciferase stored in the ovary of the scyphozoan medusa *Periphylla periphylla. Biol. Bull.* **201**: 339–347.

Shoji, R. (1919). A physiological study on the luminescence of *Watasenia scintillans* (Berry). *Am. J. Physiol.* **47**: 534–557.

Shushkina, E. A., Nikolaeva, G. G., and Lukasheva, T. A. (1990). Changes in the structure of the Black Sea planktonic community at mass reproduction of sea gooseberries *Mnemiopsis leidyi* (Agassiz). *Oceanology* 51: 54–60.

Sie, E. H.-C., McElroy, W. D., Johnson, F. H., and Haneda, Y. (1961). Spectroscopy of the *Apogon* luminescent system and of its cross reaction with the *Cypridina* system. *Arch. Biochem. Biophys.* 93: 286–291.

Simpson, W. J., and Hammond, J. R. M. (1991). The effect of detergents on firefly luciferase reactions. *J. Biolumin. Chemilumin.* 6: 97–106.

Sinclair, J. F., Waddle, J. J., Waddill, E. F., and Baldwin, T. O. (1993). Purified native subunits of bacterial luciferase are active in the bioluminescence reaction but fail to assemble into the αβ structure. *Biochemistry* 32: 5036–5044.

Sirokman, G., Wilson, T., and Hastings, J. W. (1995). A bacterial luciferase reaction with a negative temperature coefficient attributable to protein-protein interaction. *Biochemistry* 34: 13074–13081.

Skogsberg, T. (1920). *Studies on Marine Ostracoda, Part I (Cypridinids, Halocyprids, and Polycopids)*, Uppsala.

Skowron, S. (1926). On the luminescence of *Microscolex phosphoreus* Dug. *Biol. Bull.* 51: 199–208.

Small, E. D., Koda, P., and Lee, J. (1980). Lumazine protein from the bioluminescent bacterium *Photobacterium phosphoreum*. Purification and characterization. *J. Biol. Chem.* 255: 8804–8810.

Snowdowne, K. W., Ertel, R. J., and Borle, A. B. (1985). Measurement of cytosolic calcium with aequorin in dispersed rat ventricular cells. *J. Mol. Cell. Cardiol.* 17: 233–241.

Spudich, J., and Hastings, J. W. (1963). Inhibition of the bioluminescent oxidation of reduced flavin mononucleotide by 2-decenal. *J. Biol. Chem.* 238: 3106–3108.

Stanley, P. E. (1992). A survey of more than 90 commercially available luminometers and imaging devices for low-light measurements of chemiluminescence and bioluminescence, including instruments for manual, automatic and specialized operation, for HPLC, LC, GLC and microtiter plates. Part I: descriptions. *J. Biolumin. Chemilumin.* 7: 77–108.

Stanley, P. E. (1997). Commercially available luminometers and imaging devices for low-light level measurements and kits and reagents utilizing bioluminescence or chemiluminescence: survey update 5. *J. Biolumin. Chemilumin.* 12: 61–78.

Stanley, P. E. (1999). Commercially available luminometers and imaging devices for low-light level measurements and kits and reagents utilizing bioluminescence or chemiluminescence: survey update 6. *J. Biolumin. Chemilumin.* **14**: 201–213.

Steadman, J., and Syage, J. A. (1991). Time-resolved studies of phenol proton transfer in clusters. 3. solvent structure and ion-pair formation. *J. Phys. Chem.* **95**: 10326–10331.

Stephenson, D. G., and Sutherland, P. J. (1981). Studies on the luminescent response of the Ca^{2+}-activated photoprotein, obelin. *Biochim. Biophys. Acta* **678**: 65–75.

Stiffey, A. V., Blank, D. L., and Loeb, G. I. (1985). An inexpensive solid-state photometer circuit useful in studying bioluminescence. *J. Chem. Educ.* **62**: 360–361.

Stojanovic, M. N. (1995). *Chemistry of Bioluminescence and Chemiluminescence: Dinoflagellate, Fungal and Davis' Oxaziridine Systems*, Ph.D. Dissertation, Department of Chemistry, Harvard University.

Stojanovic, M. N., and Kishi, Y. (1994a). Dinoflagellate bioluminescence: the chromophore of dinoflagellate luciferin. *Tetrahedron Lett.* **35**: 9343–9346.

Stojanovic, M. N., and Kishi, Y. (1994b). Dinoflagellate bioluminescence: the chemical behavior of the chromophore towards oxidants. *Tetrahedron Lett.* **35**: 9347–9350.

Stone, H. (1968). The enzyme catalyzed oxidation of *Cypridina* luciferin. *Biochem. Biophys. Res. Commun.* **31**: 386–391.

Storch, J., and Ferber, E. (1988). Detergent-amplified chemiluminescence of lucigenin for determination of superoxide anion production by NADPH oxidase and xanthine oxidase. *Anal. Biochem.* **169**: 262–267.

Strauss, G., and Nickerson, W. J. (1961). Photochemical cleavage of water by riboflavin. II. Role of activators. *J. Am. Chem. Soc.* **83n**: 3187–3191.

Strehler, B. L. (1953). Luminescence in cell-free extracts of luminous bacteria and its activation by DPN. *J. Am. Chem. Soc.* **75**: 1264.

Strehler, B. L., and Cormier, M. J. (1953). Factors affecting the luminescence of cell-free extracts of the luminous bacterium, *Achromobacter fischeri*. *Arch. Biochem. Biophys.* **47**: 16–33.

Strehler, B. L., and Cormier, M. J. (1954). Isolation, identification, and function of long chain fatty aldehydes affecting the bacterial luciferin-luciferase reaction. *J. Biol. Chem.* **211**: 213–225.

Strum, J. (1969). Photophores of *Porichthys notatus*: ultrastructure of innervation. *Anat. Rec.* **164**: 433–461.

Stults, N. L., *et al.* (1991). Applications of recombinant bioluminescent proteins as probes for proteins and nucleic acids. *In* Stanley, P. E., and Kricka, L. J. (eds.), *Biolumin. Chemilumin.* Proc. Int. Symp., 6th, 1990, pp. 533–536. Wiley, Chichester, UK.

Stults, N. L., *et al.* (1992). Use of recombinant biotinylated aequorin in microtiter and membrane-based assays: Purification of recombinant aequorin from *Escherichia coli. Biochemistry* **31**: 1433–1442.

Stults, N. L., Rivera, H. N., Burke-Payne, J., and Smith, D. F. (1997). Preparation of stable covalent conjugates of recombinant aequorin with proteins and nucleic acids. *In* Hastings, J. W., *et al.* (eds.), *Biolumin. Chemilumin.,* Proc. Int. Symp., 9th, 1996, pp. 423–426. Wiley, Chichester, UK.

Sugioka, K., Nakano, M., Kurashige, S. A., Y, and Goto, T. (1986). A chemiluminescent probe with a *Cypridina* luciferin analog, 2-methyl-6-phenyl-3,7-dihydroimidazo[1,2-a]pyrazin-3-one, specific and sensitive for superoxide anion production in phagocytizing macrophages. *FEBS Lett.* **197**: 27–30.

Suter, H. (1890). Miscellaneous communications on New Zealand land and fresh water molluscs. *Trans. N.Z. Inst.* **23**: 93–96.

Suzuki, N., and Goto, T. (1971). Firefly bioluminescence. II. Identification of 2-(6′-hydroxy benzothiazol-2′-yl)4-hydroxythiazol as a product in the bioluminescence of firefly lanterns and as a product in the chemiluminescence of firefly luciferin in DMSO. *Tetrahedron Lett.* **22**: 2021–2024.

Suzuki, N., Sato, M., Nishikawa, K., and Goto, T. (1969). Synthesis and spectral properties of 2-(6′-hydroxybenzothiazol-2′-yl)-4-hydroxythiazol, a possible emitter species in the firefly bioluminescence. *Tetrahedron Lett.* **55**: 4683–4684.

Suzuki, N., *et al.* (1991). Studies on the chemiluminescent detection of active oxygen species: 9-acridone-2-sulfonic acid, a specific probe for superoxide. *Agric. Biol. Chem.* **55**: 1561–1564.

Suzuki, N., *et al.* (1991). Chemiluminescent detection of active oxygen species, singlet molecular oxygen and superoxide, using *Cypridina* luciferin analogs. *Nippon Suisan Gakkaishi* **57**: 1711–1715.

Suzuki, N., *et al.* (1991). Reaction rates for the chemiluminescence of *Cypridina* luciferin analogs with superoxide: a quenching experiment with superoxide dismutase. *Agric. Biol. Chem.* **55**: 157–160.

Swanson, R., *et al.* (1985). Crystals of luciferase from *Vibrio harveyi. J. Biol. Chem.* **260**: 1287–1289.

Swift, E., and Meunier, V. (1976). Effects of light intensity on division rate, stimulable bioluminescence and cell size of the oceanic dinoflagellates *Dissodinium lunula*, *Pyrocystis fusiformis* and *P. noctiluca*. *J. Phycol.* **12**: 14–22.

Swift, E., Biggley, W. H., and Napora, T. A. (1977). The bioluminescence emission spectra of *Pyrosoma atlanticum*, *P. spinosum* (tunicata), *Euphausia tenera* (Crustacea) and *Gonostoma* sp. (Pisces). *J. Mar. Biol. Assoc. U.K.* **57**: 817–823.

Takahashi, H., and Isobe, M. (1993). *Symplectoteuthis* bioluminescence. (1). Structure and binding form of chromophore in photoprotein of a luminous squid. *Bioorg. Med. Chem. Lett.* **3**: 2647–2652.

Takahashi, H., and Isobe, M. (1994). Photoprotein of luminous squid, *Symplectoteuthis oualaniensis* and reconstruction of the luminous system. *Chem. Lett.* **5**: 843–846.

Takahashi, H., Yasuda, Y., and Isobe, M. (1994). Bioluminescence mechanism on Okinawan squid. *Tennen Yuki Kagobutsu Toronkai Koen Yoshishu* 36th, pp. 144–151.

Tanahashi, H., *et al.* (1990). Photoprotein aequorin: use as a reporter enzyme in studying gene expression in mammalian cells. *Gene* **96**: 249–255.

Tatsumi, H., Masuda, T., and Nakano, E. (1988). Synthesis of enzymically active firefly luciferase in yeast. *Agric. Biol. Chem.* **52**: 1123–1127.

Tatsumi, H., Masuda, T., Kajiyama, N., and Nakano, E. (1989). Luciferase cDNA from Japanese firefly, *Luciola cruciata*: cloning, structure and expression in *Escherichia coli*. *J. Biolumin. Chemilumin.* **3**: 75–78.

Tatsumi, H., Kajiyama, N., and Nakano, E. (1992). Molecular cloning and expression in *Escherichia coli* of a cDNA clone encoding luciferase of a firefly, *Luciola lateralis*. *Biochim. Biophys. Acta* **1131**: 161–165.

Tawa, R., and Sakurai, H. (1997). Determination of four active oxygen species such as H_2O_2, OH, O_2^- and $^1O_2$ by luminol-and CLA-chemiluminescence methods and evaluation of antioxidative effects of hydroxybenzoic acid. *Anal. Lett.* **30**: 2811–2825.

Teranishi, K. (2003). Cyclodextrin-bound 6-(4-methoxyphenyl)imidazo[1,2-α]-pyrazin-3(7H)-one as chemiluminescent probe for superoxide anions. *ITE Letters on Batteries, New Technologies and Medicine* **4**: 201–205.

Teranishi, K., and Goto, T. (1990). Synthesis and chemiluminescence of coelenterazine (*Oplophorus* luciferin) analogues. *Bull. Chem. Soc. Japan* **63**: 3132–3140.

Teranishi, K., Isobe, M., Yamada, T., and Goto, T. (1992). Revision of structure of yellow compound, a reduction product from aequorin, photoprotein in jellyfish, *Aequorea aequorea*. *Tetrahedron Lett.* **33**: 1303–1306.

Teranishi, K., and Shimomura, O. (1997a). Solubilizing coelenterazine in water with hydroxypropyl-β-cyclodextrin. *Biosci. Biotech. Biochem.* **61**: 1219–1220.

Teranishi, K., and Shimomura, O. (1997b). Coelenterazine analogs as chemiluminescent probe for superoxide anion. *Anal. Biochem.* **249**: 37–43.

Terry, B. R., Matthews, E. K., and Haseloff, J. (1995). Molecular characterization of recombinant green fluorescent protein by fluorescent correlation microscopy. *Biochem. Biophys. Res. Commun.* **217**: 21–27.

Thompson, E. M., Nafpaktitis, B. G., and Tsuji, F. I. (1987). Induction of bioluminescence in the marine fish, *Porichthys*, by *Vargula* (crustacean) luciferin. Evidence for *de novo* synthesis or recycling of luciferin. *Photochem. Photobiol.* **45**: 529–533.

Thompson, E. M., Nafpaktitis, B. G., and Tsuji, F. I. (1988). Dietary uptake and blood transport of *Vargula* (crustacean) luciferin in the bioluminescent fish, *Porichthys notatus*. *Comp. Biochem. Physiol., A: comp. Physiol.* **89A**: 203–209.

Thompson, E. M., *et al.* (1988). Induction of bioluminescence capability in the marine fish, *Porichthys notatus*, by *Vargula* (crustacean) [^{14}C]luciferin and unlabeled analogs. *J. Exp. Biol.* **137**: 39–51.

Thompson, E. M., Nagata, S., and Tsuji, F. I. (1989). Cloning and expression of cDNA for the luciferase from the marine ostracod *Vargula hilgendorfii*. *Proc. Natl. Acad. Sci. USA* **86**: 6567–6571.

Thompson, E. M., Nagata, S., and Tsuji, F. I. (1990). *Vargula hilgendorfii* luciferase: a secreted reporter enzyme for monitoring gene expression in mammalian cells. *Gene* **96**: 257–262.

Thompson, J. F., *et al.* (1997). Mutation of a protease-sensitive region in firefly luciferase alters light emission properties. *J. Biol. Chem.* **272**: 18766–18771.

Thomson, C. M., Herring, P. J., and Campbell, A. K. (1995). Evidence for *de novo* biosynthesis of coelenterazine in the bioluminescent midwater shrimp, *Systellaspis debilis*. *J. Mar. Biol. Ass., UK* **75**: 165–171.

Thomson, C. M., Herring, P., and Campbell, A. K. (1995a). Coelenterazine distribution and luciferase characteristics in oceanic decapod crustaceans. *Marine Biology* **124**: 197–207.

Tilbury, R. N., and Quickenden, T. I. (1992). Luminescence from the yeast *Candida utilis* and comparisons across three genera. *J. Biolumin. Chemilumin.* **7**: 245–255.

Toma, S., *et al.* (2004). The crystal structure of semi-synthetic aequorins. *Protein Science* **14**: 409–416.

Tomarev, S. I., *et al.* (1993). Abundant mRNAs in the squid light organ encode proteins with a high similarity to mammalian peroxidases. *Gene* **132**: 219–226.

Totsune, H., Nakano, M., and Inaba, H. (1993). Chemiluminescence from bamboo shoot cut. *Biochem. Biophys. Res. Commun.* **194**: 1025–1029.

Toya, Y. (1992). Chemistry of *Vargula* (formerly *Cypridina*) bioluminescence. *Nippon Nogei Kagaku Kaishi* **66**: 742–747.

Toya, Y., Nakatsuka, S., and Goto, T. (1983). Structure of *Cypridina* luciferinol, "reversibly oxidized *Cypridina* luciferin." *Tetrahedron Lett.*, pp. 5753–5756.

Toya, Y., Nakatsuka, S., and Goto, T. (1985). Structure of *Cypridina* biluciferyl, a dimer of *Cypridina* luciferyl radical having bioluminescent activity. *Tetrahedron Lett.* **26**: 239–242.

Trainor, G. L. (1979). *Studies on the Odontosyllis bioluminescence system*, Ph.D. Dissertation, Harvard University, Cambridge, MA.

Trofimov, K. P., Bondar, V. S., Illarionov, B. A., and Vysotski, E. S. (1994). Light emission of the recombinant obelin from hydroid polyp *Obelia longissima* initiated by Cd^{2+} ions. *In* Campbell, A. K., *et al.* (eds.), *Biolumin. Chemilumin.*, Proc. Int. Symp., 8th, pp. 600–603. Wiley, Chichester, UK.

Tsien, R. Y. (1998). The green fluorescent protein. *Ann. Rev. Biochem.* **67**: 509–544.

Tsuji, F. I. (1955). The absorption spectrum of reduced and oxidized *Cypridina* luciferin, isolated by a new method. *Arch. Biochem. Biophys.* **59**: 452–464.

Tsuji, F. I. (1983). Molluskan bioluminescence. *Mollusca* **2**: 257–279.

Tsuji, F. I. (1985). ATP-dependent bioluminescence in the firefly squid, *Watasenia scintillans*. *Proc. Natl. Acad. Sci. USA* **82**: 4629–4632.

Tsuji, F. I. (2002). Bioluminescence reaction catalyzed by membrane-bound luciferase in the "firefly squid" *Watasenia scintillans*. *Biochim. Biophys. Acta* **1564**: 189–197.

Tsuji, F. I., and Haneda, Y. (1971). Luminescent system in a mictophid fish, *Diaphus elucens* Brauer. *Nature* **233**: 623–624.

Tsuji, F. I., and Haneda, Y. (1971a). Studies on the luminescence reaction of a mictophid fish, *Diaphus elucens* Brauer. *Science Report of the Yokosuka City Museum*, No. 18, pp. 104–109.

Tsuji, F. I., and Leisman, G. B. (1981). K^+/Na^+-triggered bioluminescence in the oceanic squid *Symplectoteuthis oualaniensis*. *Proc. Natl. Acad. Sci. USA* **78**: 6719–6723.

Tsuji, F. I., and Leisman, G. (1982). Membrane-bound bioluminescence in the pelagic squid. *Symplectoteuthis oualaniensis*. *In* Hastings, J. W., and Gitelzon, I. I. (eds.), *Biolyumin. Tikhon Okeane. Mater. Simp. Tikhookean. Nauchn. Kongr.* 14th, 1979, pp. 127–135. Akad. Nauk SSSR, Sib. Otd., Inst. Fiz.: Krasnoyarsk, USSR.

Tsuji, F. I., and Sowinski, R. (1961). Purification and molecular weight of *Cypridina* luciferase. *J. Cell. Comp. Physiol.* 58: 125–130.

Tsuji, F. I., Chase, A. M., and Harvey, E. N. (1955). Recent studies on the chemistry of *Cypridina* luciferin. *In* Johnson F. H. (ed.), *The Luminescence of Biological Systems*, pp. 127–156. American Association for the Advancement of Science, Washington, D.C.

Tsuji, F. I., Barnes, A. T., and Case, J. F. (1972). Bioluminescence in the marine teleost, *Porichthys notatus*, and its induction in a non-luminous form by *Cypridina* luciferin. *Nature* 237: 515–516.

Tsuji, F. I., Lynch, R. V., III, and Stevens, C. L. (1974). Some properties of luciferase from the bioluminescent crustacean, *Cypridina hilgendorfii*. *Biochemistry* 13: 5204–5209.

Tsuji, F. I., et al. (1977). Mechanism of the enzyme-catalyzed oxidation of *Cypridina* and firefly luciferins studied by means of $^{17}O_2$ and $H_2^{18}O$. *Biochem. Biophys. Res. Commun.* 74: 606–613.

Tsuji, F. I., Inouye, S., Goto, T., and Sakaki, Y. (1986). Site-specific mutagenesis of the calcium-binding photoprotein aequorin. *Proc. Natl. Acad. Sci. USA* 83: 8107–8111.

Tsuji, F. I., Inouye, S., Ohmiya, Y., and Ohashi, M. (1993). Bioluminescence of the calcium-binding photoprotein aequorin. *In Frontiers of Photobiology*, Excerpta Medica, Amsterdam, pp. 234–241.

Tsuji, F. I., Ohmiya, Y., Fagan, T. F., Toh, H., and Inouye, S. (1995). Molecular evolution of the Ca^{2+}-binding photoproteins of the hydrozoa. *Photochem. Photobiol.* 62: 657–661.

Tu, S.-C. (1978). Preparation of the subunits of bacterial luciferase. *Method. Enzymol.* 51: 171–174.

Tu, S.-C. (1979). Isolation and properties of bacterial luciferase-oxygenated flavin intermediate complexed with long-chain alcohols. *Biochemistry* 18: 5940–5945.

Tu, S.-C. (1986). Bacterial luciferase 4a-hydroperoxyflavin intermediates: stabilization, isolation, and properties. *Method. Enzymol.* 133: 128–139.

Tu, S.-C. (1991). Oxygenated flavin intermediates of bacterial luciferase and flavoprotein aromatic hydroxylases: enzymology and chemical models. *Adv. Oxygenated Processes* **3**: 115–140.

Tu, S.-C., and Cho, K. W. (1991). On the mechanism of dithionite/hydrogen peroxide-induced bacterial bioluminescence. *In* Curti, B., *et al.* (eds.), *Flavins Flavoproteins*, Proc. Int. Symp., 10th, pp. 281–284. de Gruyter, Berlin.

Tu, S.-C., and Mager, H. I. X. (1991). Recent advances in chemical modeling of bacterial bioluminescence mechanism. *In* Muller, F. (ed.), *Photobiol. Sci. its Appl.*, [Proc. Int. Congr. Photobiol.], 10th, 1988, pp. 319–328. CRC, Boca Raton, FL.

Tu, S.-C., and Mager, H. I. X. (1995). Biochemistry of bacterial bioluminescence. *Photochem. Photobiol.* **62**: 615–624.

Tyul'kova, N. A., and Sandalova, T. P. (1996). A comparative study of the temperature effect on different bacterial luciferase. *Biokhimiya* **61**: 275–287.

Ugarova, N. N., and Dukhovich, A. F. (1987). Function of lipids in firefly luciferase. *In* Schoelmerich, J. (ed.), *Biolumin. Chemilumin.*, Proc. Int. Biolumin. Chemilumin. Symp., 4th, 1986, pp. 409–412. Wiley, Chichester, UK.

Ulitzur, S. (1989). The regulatory control of the bacterial luminescence system — a new view. *J. Biolumin. Chemilumin.* **4**: 317–325.

Ulitzur, S., and Hastings, J. W. (1978). Myristic acid stimulation of bacterial bioluminescence. *Proc. Natl. Acad. Sci. USA* **75**: 266–269.

Ulitzur, S., and Hastings, J. W. (1979). Autoinduction in luminous bacteria: a confirmation of the hypothesis. *Curr. Microbiol.* **2**: 345–348.

Ulitzur, S., and Hastings, J. W. (1979a). Evidence for tetradecanal as the natural aldehyde in bacterial luminescence. *Proc. Natl. Acad. Sci. USA* **76**: 265–267.

Ulitzur, S., and Kuhn, J. (1988). The transcription of bacterial luminescence is regulated by sigma 32. *J. Biolumin. Chemilumin.* **2**: 81–93.

Uri, N. (1961). Mechanism of antioxidation. *In* Lundberg, W. O. (ed.), *Autoxidation and Antioxidants*, pp. 133–169. Interscience/John Wiley, New York.

Usami, K., and Isobe, M. (1995). Two luminescent intermediates of coelenterazine analog, peroxide and dioxetanone, prepared by direct photooxygenation at low temperature. *Tetrahedron Lett.* **36**: 8613–8616.

Usami, K., and Isobe, M. (1996). Chemiluminescent characters of hydroperoxide and dioxetanone of coelenterate luciferin analog prepared by low-temperature photooxygenation. *Chem. Lett.* 3: 215–216.

Uyakul, D., Isobe, M., and Goto, T. (1989). *Lampteromyces* bioluminescence. 3. Structure of lampteroflavin, the light emitter in the luminous mushroom, *Lampteromyces japonicus*. *Bioorganic Chemistry* 17: 454–460.

Van der Burg, A. (1943). *Spektrale onderzoekingen over chemo- en bioluminescentie*, Thesis, University of Utrecht, Utrecht.

Verhaegen, M., and Christopoulos, T. K. (2002). Recombinant *Gaussia* luciferase. Overexpression, purification, and analytical application of a bioluminescent reporter for DNA hybridization. *Anal. Chem.* 74: 4378–4385.

Vervoort, J., *et al.* (1986). Identification of the true carbon-13 nuclear magnetic resonance spectrum of the stable intermediate II in bacterial luciferase. *Biochemistry* 25: 8062–8067.

Vervoort, J., *et al.* (1986). Bacterial luciferase: a carbon-13, nitrogen-15, and phosphorus-31 nuclear magnetic resonance investigation. *Biochemistry* 25: 8067–8075.

Visser, A. J. W. G., and Lee, J. (1980). Lumazine protein from the bioluminescent bacterium *Photobacterium phosphoreum*. A fluorescence study of the protein-ligand equilibrium. *Biochemistry* 19: 4366–4372.

Visser, A. J. W. G., *et al.* (1997). Time-dissolved fluorescence study of the dissociation of FMN from the yellow fluorescence protein from *Vibrio fischeri*. *Photochem. Photobiol.* 65: 570–575.

Viviani, V. R. (2002). The origin, diversity, and structure function relationships of insect luciferases. *Cell. Mol. Life Sci.* 59: 1833–1850.

Viviani, V. R., and Bechara, E. J. H. (1993). Biophysical and biochemical aspects of phengodid (railroad worm) bioluminescence. *Photochem. Photobiol.* 58: 615–622.

Viviani, V. R., and Bechara, E. J. H. (1995). Bioluminescence of Brazilian fireflies (Coleoptera: Lampyridae): Spectral distribution and pH effect on luciferase-elicited colors. Comparison with Elaterid and phengodid luciferases. *Photochem. Photobiol.* 62: 490–495.

Viviani, V. R., and Bechara, E. J. H. (1997). Bioluminescence and biological aspects of Brazilian railroad worms (Coleoptera: Phengodidae). *Ann. Entomol. Soc. Am.* 90: 389–398.

Viviani, V. R., Bechara, E. J. H., and Ohmiya, Y. (1999). Cloning, sequence analysis, and expression of active *Phrixothrix* railroad-worm luciferase: relationship between bioluminescence spectra and primary structures. *Biochemistry* **38**: 8271–8279.

Viviani, V. R., *et al.* (1999a). Cloning and molecular characterization of the cDNA for the Brazilian larval click-beetle *Pyrearinus termitilluminans luciferace. Photochem. Photobiol.* **70**: 254–260.

Viviani, V. R., Uchida, A., Viviani, W., and Ohmiya, Y. (2002). The structural determinants of bioluminesce colors in railroad worm and other pH-insensitive luciferases. *In* Stanley, P. E., and Kricka, L. J. (eds.), *Bioluminescence and Chemiluminescence, Progress and Current Applications*, pp. 19–22. World Scientific, Singapore.

Viviani, V. R., Hastings, J. W., and Wilson, T. (2002a). Two bioluminescent Diptera: the North American *Orfelia fultoni* and the Australian *Arachnocampa flava*. Similar niche, different bioluminescence systems. *Photochem. Photobiol.* **75**: 22–27.

Vysotski, E. S., and Lee, J. (2004). Ca^{2+}-Regulated photoproteins: structural insight into the bioluminescence mechanism. *Accounts of Chemical Research* **37**: 405–415.

Vysotski, E. S., Bondar, V. S., and Letunov, V. N. (1989). Extraction and purification of obelin, a calcium-activated photoprotein from the hydroid polyp *Obelia longissima. Biokhimiya* **54**: 965–973.

Vysotskii, E. S., *et al.* (1990). Extraction, some properties and application of obelin, calcium-activated photoprotein. *In* Jezowska-Trzebiatowska, B. (ed.), *Biol. Lumin.*, Proc. Int. Sch., 1st, 1989, pp. 386–395. World Scientific, Singapore.

Vysotski, E. S., Bondar, V. S., and Gitelzon, I. I. (1991). Isolation and properties of various molecular forms of calcium^{2+}-activated photoprotein obelin. *Dokl. Akad. Nauk SSSR* **321**: 214–217.

Vysotski, E. S., Bondar, V. S., Trofimov, K. P., and Gitelzon, I. I. (1991). Luminescence of the calcium-activated photoprotein obelin under the action of active forms of oxygen. *Dokl. Akad. Nauk SSSR* **321**: 850–854.

Vysotski, E. S., Trofimov, K. P., Bondar, V. S., and Gitelson, J. I. (1993). Luminescence of calcium-activated photoprotein obelin initiated by sodium hypochlorite and manganese chloride. *J. Biolumin. Chemilumin.* **8**: 301–305.

Vysotski, E. S., *et al.* (1995). Mn^{2+}-activated luminescence of the photoprotein obelin. *Arch. Biochem.* **316**: 92–99.

Vysotski, E. S., *et al.* (1999). Preparation and preliminary study of crystals of the recombinant calcium-regulated photoprotein obelin from the bioluminescent hydroid *Obelia longissima*. *Acta Cryst.* **D55**: 1965–1966.

Wada, N. (1989). The active oxygen participating in the primary process of firefly D-(−)-luciferin chemiexcitation in dimethylsulfoxide. *Toyo Daigaku Kogakubu Kenkyu Hokoku* **25**: 27–34.

Wada, N., Mitsuta, K., Kohno, M., and Suzuki, N. (1989). ESR studies of firefly D-(−)-luciferin chemiluminescence in dimethyl sulfoxide. *J. Phys. Soc. Jpn.* **58**: 3501–3504.

Wada, N., *et al.* (1997). A theoretical approach to elucidate a mechanism of O_2 addition to intermediate I in bacterial bioluminescence. *In* Hastings, J. W., *et al.* (eds.), *Biolumin. Chemilumin.*, Proc. Int. Symp., 9th, 1996, pp. 58–61. Wiley, Chichester, UK.

Waddle, J. J., Johnston, T. C., and Baldwin, T. O. (1987). Polypeptide folding and dimerization in bacterial luciferase occur by a concerted mechanism *in vivo*. *Biochemistry* **26**: 4917–4921.

Waddle, J., and Baldwin, T. O. (1991). Individual α and β subunits of bacterial luciferase exhibit bioluminescence activity. *Biochem. Biophys. Res. Commun.* **178**: 1188–1193.

Wampler, J. E. (1978). Measurements and physical characteristics of luminescence. *In* Herring, P. J. (ed.), *Bioluminescence in Action*, pp. 1–48. Academic Press, London.

Wampler, J. E., and Jamieson, B. G. M. (1980). Earthworm bioluminescence: comparative physiology and biochemistry. *Comp. Biochem. Physiol.* **66B**: 43–50.

Wampler, J. E., Hori, K., Lee, J. W., and Cormier, M. J. (1971). Structured bioluminescence. Two emitters during both the *in vitro* and the *in vivo* bioluminescence of the sea pansy, *Renilla*. *Biochemistry* **10**: 2903–2909.

Wampler, J. E., Karkhanis, Y. D., Morin, J. G., and Cormier, M. J. (1973). Similarities in the bioluminescence from the Pennatulacea. *Biochim. Biophys. Acta* **314**: 104–109.

Wannlund, J., DeLuca, M., Stempel, K., and Boyer, P. D. (1978). Use of ^{14}C-carboxyl-luciferin in determining the mechanism of the firefly luciferase catalyzed reactions. *Biochem. Biophys. Res. Commun.* **81**: 987–992.

Ward, W. W. (1979). Energy transfer processes in bioluminescence. *In* Smith, K. C. (ed.), *Photochemical and Photobiological Reviews*, Vol. 4, pp. 1–57. Plenum Press, New York.

Ward, W. W. (1981). Properties of the coelenterate green-fluorescent proteins. *In* DeLuca, M. A., and McElroy, W. D. (eds.), *Bioluminescence*

and Chemiluminescence: Basic Chemistry and Analytical Applications,
pp. 235–242. Academic Press, New York.

Ward, W. W. (1998). Biochemical and physical properties of green fluorescent protein. *In* Chalfie, M., and Kain, S. (eds.), *Green Fluorescent
Protein,* pp. 45–75. Wiley-Liss, New York.

Ward, W. W., and Bokman, S. H. (1982). Reversible denaturation of
Aequorea green-fluorescent protein: physical separation and characterization of the renatured protein. *Biochemistry* 21: 4535–4540.

Ward, W. W., and Cormier, M. J. (1976). *In vitro* energy transfer in *Renilla*
bioluminescence. *J. Phys. Chem.* 80: 2289–2291.

Ward, W. W., and Cormier, M. J. (1978). Energy transfer via protein-
protein interaction in *Renilla* bioluminescence. *Photochem. Photobiol.*
27: 389–396.

Ward, W. W., and Cormier, M. J. (1979). An energy transfer protein in
coelenterate bioluminescence. Characterization of the *Renilla* green fluorescent protein. *J. Biol. Chem.* 254: 781–788.

Ward, W. W., and Seliger, H. H. (1974a). Extraction and purification of
calcium-activated photoproteins from ctenophores. *Biochemistry* 13:
1491–1499.

Ward, W. W., and Seliger, H. H. (1974b). Properties of mnemiopsin
and berovin, calcium-activated photoproteins. *Biochemistry* 13:
1500–1510.

Ward, W. W., and Seliger, H. H. (1976). Action spectrum and quantum
yield for the photoinactvation of mnemiopsin, a bioluminescent photoprotein from the ctenophore *Mnemiopsis* sp. *Photochem. Photobiol.* 23:
351–363.

Ward, W. W., Cody, C. W., Hart, R. C., and Cormier, M. J. (1980). Spec-
trophotometric identity of the energy transfer chromophores in *Renilla*
and *Aequorea* green fluorescent proteins. *Photochem. Photobiol.* 31:
611–615.

Ward, W. W., *et al.* (1982). Spectral perturbations of the *Aequorea* green-
fluorescent protein. *Photochem. Photobiol.* 35: 803–808.

Ward, W. W., Davis, D. F., and Cutler, M. W. (1994). The origin of chromophores in coelenterate bioluminescence. *In* Campbell, A. K., *et al.*
(eds.), *Biolumin. Chemilumin.,* Proc. Int. Symp., 8th, pp. 131–134.
Wiley, Chichester, UK.

Wassink, E. C. (1948). Observations on the luminescence in fungi, 1, including a critical review of the species mentioned as luminescent in literature.
Rev. Trv. Bot. Neerl. 41: 150–212.

Wassink, E. C. (1978). Luminescence in Fungi. *In* Herring, P. J. (ed.), *Bioluminescence in Action*, pp. 171–197. Academic Press, London.

Watanabe, H., and Hastings, J. W. (1987). Enhancement of light emission in the bacterial luciferase reaction by hydrogen peroxide. *J. Biochem.* **101**: 279–282.

Watanabe, H., and Hastings, J. W. (1990). Inhibition of bioluminescence in *Photobacterium phosphoreum* by sulfamethizole and its stimulation by thymine. *Biochim. Biophys. Acta* **1017**: 229–234.

Watanabe, H., *et al.* (1991). Aldehyde-enhanced photon emission from crude extracts of soybean seedlings. *In* Stanley, P. E., and Kricka, L. J. (eds.), *Biolumin. Chemilumin.*, Proc. Int. Symp., 6th 1990, pp. 273–276. Wiley, Chichester, UK.

Watanabe, H., *et al.* (1992). Chemiluminescence in the crude extracts of soybean seedlings. Postulated mechanism on the formation of hydroperoxide intermediates. *Biochim. Biophys. Acta* **1117**: 107–113.

Watanabe, H., Nagoshi, T., and Inaba, H. (1993). Luminescence of a bacterial luciferase intermediate by reaction with H_2O_2: the evolutionary origin of luciferase and source of endogenous light emission. *Biochim. Biophys. Acta* **1141**: 297–302.

Watanabe, T., and Nakamura, T. (1972). Studies on luciferase from *Photobacterium phosphoreum* II. Substrate specificity and stoichitometry of the reaction *in vitro*. *J. Biochem.* **72**: 647–653.

Watanabe, T., and Nakamura, T. (1976). Studies on luciferase from *Photobacterium phosphoreum*. VIII. FMN-H_2O_2 initiated bioluminescence and the thermodynamics of the elementary steps of the luciferase reaction. *J. Biochem.* **79**: 489–495.

Watanabe, T., Mimura, N., Takimoto, A., and Nakamura, T. (1975). Luminescence and respiratory activities of *Photobacterium phosphoreum*. Competition for cellular reducing power. *J. Biochem.* **77**: 1147–1155.

Watase, S. (1905). On the luminous organs of *Abraliopsis* (in Japanese). *Zool. Mag. Tokyo* **17**: 119–122.

Watkins, N. J., and Campbell, A. K. (1993). Requirement of the C-terminal proline residue for stability of the Ca^{2+}-activated photoprotein aequorin. *Biochem. J.* **293**: 181–185.

Watkins, N. J., Herring, P. J., and Campbell, A. K. (1993). An evolutionary pathway for aequorin and coelenterazine bioluminescence. *In* Szalay, A. A., *et al.* (eds.), *Biolumin. Chemilumin.*, Proc. Int. Symp., 7th, pp. 64–68. Wiley, Chichester, UK.

Watkins, N. J., Knight, M. R., Trewavas, A. J., and Campbell, A. K. (1994). Engineering aequorin as an intracellular calcium indicator in *Escherichia coli* and plants. *In* Campbell, A. K., *et al.* (eds.), *Biolumin. Chemilumin.*, Proc. Int. Symp., 8th, pp. 528–531. Wiley, Chichester, UK.

Weitzman, S. H. (1974). Osteology and evolutionary relationships of the Sternoptychidae, with a new classification of stomiatoid fishes. *Bull. Am. Mus. Nat. Hist.* 153: 329–478.

White, E. H., and Roswell, D. F. (1991). Analogs and derivatives of firefly oxyluciferin, the light emitter in firefly bioluminescence. *Photochem. Photobiol.* 53: 131–136.

White, E. H., McCapra, F., Field, G., and McElroy, W. D. (1961). The structure and synthesis of firefly luciferin. *J. Am. Chem. Soc.* 83: 2402–2403.

White, E. H., McCapra, F., and Field, G. F. (1963). The structure and synthesis of firefly luciferin. *J. Am. Chem. Soc.* 85: 337–343.

White, E. H., Rapaport, E., Hopkins, T. A., and Seliger, H. H. (1969). Chemi- and bioluminescence of firefly luciferin. *J. Am. Chem. Soc.* 91: 2178–2180.

White, E. H., Rapaport, E., Seliger, H. H., and Hopkins, T. A. (1971). The chemi- and bioluminescence of firefly luciferin: an efficient chemical production of electoronically excited states. *Bioorg. Chem.* 1: 92–122.

White, E. H., Miano, J. D., and Umbreit, M. (1975). On the mechanism of firefly luciferin luminescence. *J. Am. Chem. Soc.* 97: 198–200.

Widder, E. A., Latz, M. I., and Case, J. F. (1983). Marine bioluminescence spectra measured with an optical multichannel detection system. *Biol. Bull.* 165: 791–810.

Widder, E. A., Herring, P. J., and Case, J. F. (1984). Far red bioluminescence from deep-sea fishes. *Science* 225: 512–514.

Widder, E. A., Latz, M. I., and Herring, P. J. (1986). Temporal shifts in bioluminescence emission spectra from the deep-sea fish, *Searsia koefoedi. Photochem. Photobiol.* 44: 97–101.

Wiedemann, E. (1988). Uber fluorescenz and phosphorescenz. *Ann. der Physik* 34: 446–449.

Witko-Sarsat, V., Nguyen Anh Thu, Knight, J., and Descamps-Latscha, B. (1992). Pholasin: a new chemiluminescent probe for the detection of chloramines derived from human phagocytes. *Free Radic. Biol. Med.* 13: 83–88.

Wood, K. V. (1993). Evolution of bioluminescence in insects. *In* Szalay, A. A., *et al.* (eds.), *Biolumin. Chemilumin.*, Proc. Int. Symp., 7th, pp. 104–108. Wiley, Chichester, UK.

Wood, K. V. (1993a). Luciferin-luciferase reaction in luminous flies. Bioluminescence Symposium abstract, Nov. 5–10, 1993, Kaanapali Beach, Hawaii, p. 64.

Wood, K. V. (1995). The chemical mechanism and evolutionary development of beetle bioluminescence. *Photochem. Photobiol.* **62**: 662–673.

Wood, K. V., De Wet, J. R., Dewji, N., and DeLuca, M. (1984). Synthesis of active firefly luciferase by *in vitro* translation of RNA obtained from adult lanterns. *Biochem. Biophys. Res. Commun.* **124**: 592–596.

Wood, K. V., Lam, Y. A., and McElroy, W. D. (1989). Introduction to beetle luciferase and their applications. *J. Biolumin. Chemilumin.* **4**: 289–301.

Wood, K. V., Lam, Y. A., Seliger, H. H., and McElroy, W. D. (1989). Complementary DNA coding click beetle luciferase can elicit bioluminescence of different colors. *Science* **244**: 700–702.

Woodruff, R. I., Miller, A. L., and Jaffe, L. F. (1991). Difference in free calcium concentration between oocytes and nurse cells revealed by corrected aequorin luminescence. *Biol. Bull.* **181**: 349–350.

Wu, C., Nakamura, H., Murai, A., and Shimomura, O. (2000). Chemi- and bioluminescence of coelenterazine analogues with a conjugated group at the C-8 position. *Tetrahedron Lett.* **42**: 2997–3000.

Wu, C.-K., *et al.* (1997). The three-dimensional structure of green fluorescent protein resembles a lantern. *In* Hastings, J. W., *et al.* (eds.), *Biolumin. Chemilumin.*, Proc. Int. Symp., 9th, 1996, pp. 399–402. Wiley, Chichester, UK.

Xi, L., Cho, K. W., and Tu, S. C. (1991). Cloning and nucleotide sequences of lux genes and characterization of luciferase of *Xenorhabdus luminescens* from a human wound. *J. Bacteriol.* **173**: 1399–1405.

Yamaguchi, A., Suzuki, H., Tanoue, K., and Yamazaki, H. (1986). Simple method of aequorin loading into platelets using dimethyl sulfoxide. *Thromb. Res.* **44**: 165–174.

Yang, F., Moss, L. G., and Phillips, G. N. J. (1996). The molecular structure of green fluorescent protein. *Nature Biotech.* **14**: 1246–1251.

Yang, F., Moss, L. G., and Phillips, G. N. J. (1997). The three-dimensional structure of green fluorescent protein. *In* Hastings, J. W., *et al.* (eds.), *Biolumin. Chemilumin.*, Proc. Int. Symp., 9th, 1996, pp. 387–390. Wiley, Chichester, UK.

Ye, L., Buck, L. M., Schaeffer, H. J., and Leach, F. R. (1997). Cloning and sequencing of a cDNA for firefly luciferase from *Photuris pennsylvanica*. *Biochim. Biophys. Acta* **1339**: 39–52.

Zeng, J., and Jewsbury, R. A. (1993). Enhanced bioluminescence of bacterial luciferase induced by metal ions and their complexes. *In* Szalay, A. A., *et al.* (eds.), *Biolumin. Chemilumin.*, Proc. Int. Symp., 7th, pp. 173–177. Wiley, Chichester, UK.

Ziegler, M. M., and Baldwin, T. O. (1981). Biochemistry of bacterial bioluminescence. *Current Topics in Bioenergetics* **12**: 65–113.

Zinner, K., and Nassi, L. (1985). On the possible bioluminescence of the bombardier beetles. *Rev. Latinoam. Quim.* **16**: 78–81.

Zinner, K., and Vani, Y. S. (1988). Possible correlation of the distribution of certain chemical elements, mucus characteristics and luminescence of *Chaetopterus variopedatus*. *Arq. Biol. Tecnol.* **31**: 281–296.

Zimmer, M., Branchini, B., and Lusins, J. O. (1997). A computational analysis of the preorganization and the activation of the chromophore forming hexapeptide fragment in green fluorescent protein. *In* Hastings, J. W., *et al.* (eds.), *Biolumin. Chemilumin.*, Proc. Int. Symp., 9th, 1996, pp. 407–410. Wiley, Chichester, UK.

INDEX